T0188373

MAKING TELECOMS WORK

SCiENCE MUSEUM APPROVED

MAKING TELECOMS WORK

FROM TECHNICAL INNOVATION TO COMMERCIAL SUCCESS

Geoff Varrall

Director, RTT Programmes Limited, UK

A John Wiley & Sons, Ltd., Publication

Library of Congress Cataloging-in-Publication Data:

Varrall, Geoffrey.
  Making telecoms work : from technical innovation to commercial success / Geoff Varrall.
    p. cm.
  Includes index.
  ISBN 978-1-119-97641-7 (cloth)– ISBN 978-1-119-96714-9 (epub)
  1. Telecommunication systems.   2. Technological innovations.   I. Title.
  TK5101.V328 2012
  621.3845′6–dc23

                          2011038098

A catalogue record for this book is available from the British Library.

Print ISBN: 9781119976417

Set in 10/12pt Times by Aptara Inc., New Delhi, India

Printed and Bound in Great Britain by Antony Rowe Ltd, Chippenham, Wiltshire.

Contents

Foreword

We were delighted when the Science Museum was approached about a new publication that would look at the technical aspects of the telecoms industry, with a nod to the developments and experience of the past.

The ambition of this publication fits with our own development of a major new gallery in information and communications technology. Our planned gallery will trace the dramatic change since the invention of the telegraph, the laying of the first transatlantic cable, the creation of the first digital electronic computer, the development of mobile communications and the advent of the World Wide Web.

This publication, and the gallery, will provide an important record for questioning the speed of change and developing broader narratives on telecommunications history.

Tilly Blyth
Curator of Computing and Information
Science Museum, London
October 2011

List of Acronyms and Abbreviations

3GPP	3G Partnership Project
ACK/NACK	Acknowledgement/Negative Acknowledgement
AF	Autofocus
AM	Amplitude Modulation
AMPD	Average Margin Per Device
AMPU	Average Margin Per User
AMR NB	Adaptive Multirate Narrowband
AMR-WB	Adaptive Multirate Wideband
AP	Access Point
ARPU	Average Revenues Per User
ATC	Ancillary Terrestrial Component
ATSC	Advanced Television Systems Committee
AVC	Adaptive Variable Source Coding
B GAN	Broadband Global Access
CAM	Content Addressable Memory
CCD	Charge Coupled Device
CDMA	Code Division Multiple Access
C/I	Carrier to Interference Ratio
CIF	Common Intermediate Format
CMOS	Complementary Metal-Oxide-Semiconductor
CQI	Channel Quality Indication
CTS	Clear To Send
DAN	Device Access/Device Area Network
DCT	Discrete Cosine Transform
DFT	Discrete Fourier Transform
DPSK	Differential Binary Phase Shift Keying
DRM	Digital Radio Mondiale
DSSS	Direct Sequence Spread Spectrum
DWDM	Dense Wavelength Division Multiplexing
EXIF	Exchangeable Image Format File
FDN	Frequency Dependent Noise
FEXT	Far-End Crosstalk

FHT	Fast Hadamard Transform
GaAs	Gallium Arsenide
GEO	Geostationary
GIF	Graphic Interchange Format
GSO	Geostationary
GSO	Geosynchronous
HC	Hybrid Coordinator
HDMI	High Definition Multi Media Interface
HDSL	High Bit Rate DSL
HLEO	'High' Low-Earth Orbits
HRAN	High Radio Access Network
HWIH	Hear What I Hear
I/O	Input And Output
IDFT	Inverse Discrete Fourier Transform
IF	Intermediate Frequency
IFHT	Inverse Fast Hadamard Transform
ISD	Information Spectral Density
ISM	Industrial Scientific Medical
LEO	Low-Earth Orbit
LLU	Local Loop Unbundling
LNA	Low-Noise Amplifier
LPDDR2	Low-Power Double Data Rate 2
LRAN	Lower Radio Access Network
M2M	Machine To Machine
MAC	Medium Access Control
MEMS	Microelectrical Mechanical System
MEO	Medium Earth
MIDP	Mobile Information Device Profile
MIMO	Multiple Input Multiple Output
MIPI	Mobile Industry Processor Interface
MMS	Multi Media Subsystem
NEXT	Near-End Crosstalk
NRE	Nonrecurring Expenditure
Ofcom	Office Of Communications
OFDM	Orthogonal Frequency Division Multiplexing
OFDMA	Orthogonal Frequency Division Multiple Access
OOK	On/Off Keying
OTA	Over The Air
P25	Project 25
PA	Power Amplifier
PAL	Phase Alternating Line
PAN	Personal Area Network
PAR	Peak To Average Ratio
PC	Point Coordination
PIFA	Planar Inverted F Antenna
PMD	Polarisation Mode Dispersion

PPDR	Public Protection And Disaster Relief
QE	Quantum Efficiency
QoS	Quality Of Service
RF BOM	RF Bill Of Material
ROI	Return On Investment
RSSI	Received Signal Strength Indication
RTS	Request To Send
SC-FDMA	Single-Carrier Frequency Division Multiple Access
SDH	Synchronous Digital Hierarchy
SHDSL	Single Pair HDSL
SiGe	Silicon Germanium
SIM	Subscriber Identity Module
SIMO	Single Input Multiple Output
SIP	Session Initiation Protocol
SONET	Synchronous Optical Network
STiMI	Satellite And Terrestrial Interactive Multi Service Infrastructure
SWIS	See What I See
TDD	Time-Division Duplexed
TDM PON	Time Division Multiplexed Passive Optical Network
TEDS	Tetraenhanced Data Service
TIS	Isotropic Sensitivity
TRP	Total Radiated Power
TSPEC	Traffic Specification
UDP	Unacknowledged Datagram Protocol
USIM	Universal Subscriber Identity Module
VCO	Voltage-Controlled Oscillator
VGA	Video Graphics Array
VOG	Video Object Groupings
VOL	Video Object Layers
VOP	Video Object Planes
VOS	Video Object Sequence
WDM	Wavelength Division Multiplexing
WEP	Wireline Equivalent Privacy
WPA	WiFi Protected Access

Acknowledgements

Having coauthored three previous technical books on telecom related topics, The Mobile Radio Servicing Handbook in 1989, Data Over Radio in 1992 and 3G Handset and Network Design in 2003[1] it occurred to me that it might be useful to write a fourth book documenting the changes that have occurred in the past eight years and the related implications for the industry looking forward.

It then occurred to me that this was a massive amount of work for a minimal return and the idea was shelved until this week end when I read an interview[2] in the Guardian Week End Review section on Michael Moorcock, a prolific author of Science Fiction. On a good day Mr Moorcock can apparently produce 15 000 words of breathless prose and once authored a complete book in three days.

On the basis that as each year passes telecoms seems to more closely resemble science fiction I decided I should not be a wimp and get down to writing.

A few phone calls to some of my previous coauthors established that they were less than enthusiastic at joining me again in a technical publishing venture and asked why didn't I do this one on my own.

So at this point I would like to acknowledge with thanks my previous coauthors Mr Mike Fitch, now with British Telecom Research at Martlesham Heath, Mr David Ogley, now working in the psychiatric profession, an industry increasingly closely aligned with our own and my ex-codirector and business partner Roger Belcher with whom miraculously I am still on talking terms after nearly thirty years of close technical and commercial cooperation.

Roger's crucial role at RTT and on all previous book projects has been to spot and correct gross errors of a technical nature.

His absence as an arbiter of this particular work is therefore alarming but I am assuming that after thirty years of working on engineering projects in the industry I should be able to correct most mistakes on my own, an assumption that I hope does not prove to be misguided.

And rather than attempt this particular literary Everest completely solo and without the aid of extra oxygen I have called on the services of my other friend John Tysoe of the Mobile World to provide relevant financial analysis and market and business statistics that I hope will help integrate the technology and engineering story that we have to tell with the market and business dynamics of our industry today. Thanks are also due to my close friend and business

---

[1] Copies of these books can be bought either from Amazon or American Book Exchange http://www. abebooks.co.uk/.

[2] http://www.guardian.co.uk/books/2011/feb/04/michael-moorcock-hari-kunzru Hari Kunzru Interview with Michael Moorcock, Guardian 4 February 2011.

colleague Jane Zweig who has been patiently helpful at explaining why the US continues to be bafflingly different from the rest of the world. Thanks also to Tony Hay for unearthing useful articles on battery technology from the British Library.

John Liffen and Tilly Blyth in the telecommunications and computing curatorial department at the Science Museum have also been a great source of inspiration. In particular they have demonstrated that the Museum has case study resources that document with great detail the reasons why some innovations succeed and some fail. These resources have direct relevance to the formulation of present day industrial strategy in the telecommunications industry.

I would also like to thank the GSM Association for permission to reproduce some of the study work undertaken for them in 1997, Peregrine Semiconductor, Ethertronics, Antenova, Avago and Quintel for their permission to use specific product data references and the team at Nokia Siemens Networks who under Harri Holma's guidance produce such excellent technical books on LTE that this particular work does not aspire to emulate.

Other thanks are due to my long-suffering family, Liz, Tom and Hannah, though they are onerously familiar with my tendency to engage in projects that are interesting but seem to make little apparent financial sense. I have also found myself drawing on my father's work and experience gained over a fifty-year career in telecommunications engineering. Some of his knowledge can be found in the Newnes Telecommunications Pocket Book that he coauthored with his business colleague Ted Edis, published posthumously in 1992.

And finally my thanks to Mark Hammond, Susan Barclay, Sandra Grayson and the production team at John Wiley who I am anticipating will be their usual helpful selves in guiding me towards the production of a final manuscript that passes market muster.

I did suggest that approaching the writing of this work as a homage to Mr Moorcock might result in a work that sold in the tens of thousands thereby making us all exceedingly rich.

Mark pointed out that on the basis of my previous publishing history this outcome, though much to be wished for, was unfortunately exceedingly unlikely.

Still, it's always good to start these projects with a healthy dose of enthusiasm, though I have just noticed that it has taken me 45 minutes to write the 1000 words of this initial opening piece, which suggests that 15 000 words a day might be overoptimistic. But onwards and upwards, a job begun is a job half done – with 149 000 words to go how can that possibly be true?

Good heavens it's tea time already.

Geoff Varrall
4.00 Monday 4 February 2011

1

Introduction

1.1 Differentiating Technology and Engineering Innovation

This book is about technology and engineering analysed in the context of the impact of invention and innovation on the political, social and economic dynamics of the global telecommunications industry, tracing the transition and transformation that has occurred particularly since fixed and mobile broadband has become such a dominant deliverable. The occasional unapologetic excursion is made into the more distant past as and when relevant and valid.

The subject focus is technical but the ambition is to make the content relevant and accessible to as wide an audience as possible. It is common to find engineers who have an absorbing interest in the humanities. It is less common to find managers with a humanities background developing a similar passion for engineering. This is possibly because some engineering books can be dull to read. Our mission is to try and remedy this disconnect between two disciplines that in reality are closely interrelated.

Another myth that we attempt to dispel is the notion that somehow technology change is occurring faster today than in the past. This is a false flattery. Technology change viewed through the prism of the present appears to be tumultuous but is more accurately considered as part of a continuing process of transition. The past remains surprisingly relevant, can inform present judgement and should be used to help forecast the future. This is our excuse or rather, our reason for raiding the Science Museum archives and picture collection that are referenced in most subsequent chapters.

The derivation of the word science is from the Latin 'Scientia' meaning knowledge. The word physics is from the Greek 'Physis' meaning nature. A recurring narrative of this book is that informed decision making is contingent on studying knowledge and well-evidenced opinion from the past. However, any decision taken has to obey the fundamental laws of physics. Our understanding of those laws changes over time. A present example is the science of the very small, how the behaviour of materials change when constructed at molecular scale. The ability to understand the potential usefulness of a physical property is a skill that can transform the world. Newcomen's observation of steam and Edison's observation of the behaviour of a carbon filament in a vacuum are two prior examples. Paul Dirac's work on quantum mechanics is a near contemporary example and the work of Stephen Hawking on string and particle theory a contemporary example. The study of 'prior art' is often today the

Making Telecoms Work: From Technical Innovation to Commercial Success, First Edition. Geoff Varrall.
© 2012 John Wiley & Sons, Ltd. Published 2012 by John Wiley & Sons, Ltd.

job of the patent attorney, but would be more beneficially encompassed as an engineering rather than legalistic discipline.

The discipline of economics has always been the fundamental tool of analysis used to quantify and qualify business risk and opportunity. Our argument is that technology economics and engineering economics are a discrete subset of economic analysis, individually deserving of independent analysis and often overlooked or underappreciated.

Frustratingly, technology and engineering are often considered as interchangeable terms. Although closely coupled they are in practice distinct and separate entities. The word 'technology' is derived from the Greek word *'techne'* meaning craft, which the ancient Greeks considered represented the mechanical arts. 'Engineering' comes from the Latin word *'ingeniare', from which modern English derives the word 'ingenious'*.

The difference in practical terms can be illustrated by considering how Archimedes fortified his adopted home town in the Siege of Syracuse in 214 to 212 BC. Under attack from the Romans, Archimedes had a number of enabling *technologies* available to him including ropes and pulleys. These technologies were then combined together with *ingenuity* into throwing machines known as ballistas, from which the term ballistics is derived. Engineering in effect was the process through which usefulness or value was realised from several separate component technologies. More prosaically, Bran Ferren, the computer scientist, hit the nail on the head by defining technology as' stuff that doesn't work yet'.[1]

1.2   Differentiating Invention and Innovation

Similarly, the words invention and innovation are often used interchangeably but are also distinct and separate though closely linked. An invention will typically be an outcome from a 'mechanical art' and will result in an enabling technology. An innovation will generally be a novel use of that enabling technology and can therefore be categorised as engineering.

In the modern world the words technology and invention can be correctly and broadly applied to almost any type of mechanical component. The specification and performance of these components is usually described and prescribed in a technical standard. For example a modern mobile phone is a collection of component technologies, each with a discrete purpose but a common objective, to send or receive voice or data over a radio channel. Each component technology will have had engineering (ingenuity) applied to it to ensure the component behaves as expected and required.

1.3   The Role of Standards, Regulation and Competition Policy

The standards documents describe the expected performance and behaviour of the functions of the complete product. Engineering is applied to ensure the complete product behaves in accordance with these standards.

The cost and realisable value of a wireless network is heavily influenced by the characteristics of the radio channels used for communication. This is because wireless communications have a unique ability to interfere with one another. In the US in the 1920s for example, competing radio broadcasters progressively increased transmitted power close to the point where no

---

[1] http://qotd.me/q2007-01-29.html.

one could receive anything clearly. The Federal Communications Commission was established in 1934 to impose order on chaos and the basic principles of radio and TV regulation became established in national and later international law. As a result, spectral policy today has a direct impact on radio engineering cost and complexity.

If standards making and spectral policy are focused on achieving improvements in technical efficiency then the result will generally be an improvement in commercial efficiency. An improvement in commercial efficiency can be translated into improved returns to shareholders and/or some form of associated social gain. This topic would fill several books on its own but it is useful to reflect on two important trends.

Harmonised and mandatory standards have been crucial in delivering R and D scale efficiency more or less across the whole history of telecommunications. Within mobile communications for example, the rapid expansion of GSM over the past twenty years would almost certainly not have happened without robust pan-European and later global standards support. More recent experiments with technology neutrality in which an assumption is made that 'the market can decide' have been less successful. Markets are determined by short-term financial pressures. These are inconsistent with the longer-term goals that are or should be implicit in the standards-making process.

1.4   Mobile Broadband Auction Values – Spectral Costs and Liabilities and Impact on Operator Balance Sheets

In parallel, spectral policy has been increasingly determined by an assumption that the market is capable of making a rational assessment of spectral value with that value realisable through an auction process. Superficially that process would have appeared to be significantly successful, but in practice the 1995/1996 US PCS auctions produced the world's first cellular bankruptcies (Next Wave and Pocket Communications), the 2002 UMTS auctions emasculated BT, left France Telecom excessively geared and brought Sonera and KPN to the brink of financial collapse.

The UK UMTS auction in 2002 is an example of this law of unintended consequences. The auction raised £22.5 billion none of which was spent on anything relevant to ICT. It was assumed that the investment would be treated as a sunk cost and not be passed on to the customer but in practice the licenses have added £1.56 to every bill for every month for 20 years. Digital Britain is being financed from old ladies' telephone bills.[2]

The impact on operator balance sheets was equally catastrophic. Taking T Mobile and Orange together their *pro forma* return on investment including spectral license cost shows cumulative investment increasing from 70 billion euros in 2001 to just under $100 billion in 2008 with a cumulative cash flow of $20 billion euros. The merger of the two entities into Everything Everywhere may delay the point at which a write down is made but does not disguise the net destruction of industry value that has occurred.

This translates in the relatively short term into a destruction of shareholder value. In the longer term the impact is more insidious. One of the easiest and fastest ways to cut costs either with or without a merger is to reduce research and development spending. The problem with this

---

[2] My thanks to John Tysoe of The Mobile World for these insights, which as always are both colourful and acutely observed.

is that it becomes harder to realise value from present and future spectral and network assets. Thus, a short-term three- to five-month gain for a national treasury translates into a longer-term loss of industrial capability that may have a negative impact lasting thirty to fifty years.

In the US, the 700-MHz auctions between 2005 and 2007 allowed the two largest incumbent operators to squeeze out smaller competitors and resulted in a payment to the US Treasury of $20 billion dollars. The band plan is, however, technically inefficient. In particular, it is proving difficult to achieve the scale economy needed to develop products that can access all channels within the band. As a result, companies have paid for spectrum and built networks and then been frustrated by a lack of market competitive performance competitive user equipment. Some entities have not even got as far as building networks because no one can supply the filters needed to mitigate interference created by or suffered by spectrally adjacent user communities, in this example TV at the bottom of the band and public safety radio at the top of the band.

In theory the regulatory process exists to anticipate and avoid this type pf problem but in practice a doctrine of *caveat emptor* has been applied. If entities are prepared to bid billions of dollars for unusable spectrum, so be it. It is up to those entities to have performed sufficient technical due diligence to arrive at an informed view as to whether the spectrum being auctioned is an asset or a liability. However, a failure to bid and win spectrum typically results in a short term-hit on the share value of the bidding entity. Therefore, there is a perverse incentive to bid and win spectrum. In the US case the opportunity to return the US market to a duopoly was an associated incentive but resulted in an outcome directly contrary to US competition policy.

1.5   TV and Broadcasting and Mobile Broadband Regulation

Regulatory objectives have changed over time. TV and broadcasting regulation may have its origins in spectral management but competition policy and the policing of content have become progressively more dominant. In telecommunications the original purpose of regulation was to act as a proxy for competition, protecting consumers from the potential use of market power by landline monopolies or later by their privatised equivalents.

Cellular operators in the 1980s had to meet spectral mask requirements in terms of output power and occupied radio bandwidth, but were otherwise largely unregulated. The assumption was that market regulation would inhibit network investment during a period in which return on investment was largely an unknown quantity.

Over the past thirty years this has changed substantially and regulatory powers now extend across pricing, service quality and increasingly social and environmental responsibility, potentially the first step towards more universal service obligations. This is in some ways an understandable response to a profitable industry enjoying revenue growth that significantly outpaced increases in operational cost. Whether this will continue to be the case depends on future technology and engineering innovation.

There is also arguably a need for a closer coupling of standards and regulatory policy across previously separate technology sectors. The transition to digital technologies both in terms of the mechanisms used to encode and decode voice, image and data, the storage of that information and the transmission of that information across a communications channel has resulted in a degree of technology convergence that has not as yet been fully realised as a commercial opportunity or reflected adequately in the standards and regulatory domain.

For example, the same encoding and decoding schemes are used irrespective of whether the end product is moved across a two-way radio network, a cellular network, a broadcast network, a geostationary or medium or low-earth orbit network and/or across cable, copper of fibre terrestrial networks. Similarly cable, copper, fibre and all forms of wireless communication increasingly use common multiplexing methods that get information onto the carrier using a mix of orthogonal[3] or semiorthogonal phase, amplitude- and frequency-modulation techniques.

1.6 Technology Convergence as a Precursor of Market Convergence?

Despite the scale of this technology convergence, these industry sectors remain singularly independent of each other despite being increasingly interdependent. This is a curious anomaly, probably best explained by the observation that commercial exigency drives commercial convergence rather than technology opportunity. The adoption of neutral host networks covered in the third part of this book is an example of this process at work.

The technology case for a neutral network managed by one entity but accessed by many has been obvious for some time. It is simply an extension of the fundamental principle of multiplexing gain. The concept is, however, only recently gaining market traction.

The reason for this is that data demand is increasing faster than technology capacity. Technology capacity is substantially different from network capacity. As demand increases across any delivery domain, wireless, cable, copper or fibre, it is technically possible to increase network capacity simply by building more infrastructure. Existing capacity may also be underutilised, in which case life is even easier. However, if infrastructure is fully loaded additional capacity is only financially justifiable if revenues are growing at least as fast as demand or at least growing sufficiently fast to cover increased operational cost and provide a return on capital investment.

Technology can of course change this equation by increasing bandwidth efficiency. The transition to digital for example yielded a step function gain in bandwidth efficiency that translated into profitability, but as systems are run ever closer to fundamental noise floors it becomes progressively harder to realise additional efficiency gain.

This is problematic for all sectors of the industry but particularly challenging for wireless where noise floors are determined by a combination of propagation loss and user to user interference. Present mobile broadband networks provide a dramatic illustration.

1.7 Mobile Broadband Traffic Growth Forecasts and the Related Impact on Industry Profitability

In a study undertaken last year[4] we forecast that over the next five years traffic volumes would grow by a factor of 30, from 3 to 90 exabytes.[5] On present tariff trends we projected revenue to grow by a factor of 3. On the basis of the data available since then, particularly the faster

---

[3] Mathematically expressed, orthogonality is when the inner product of two functions within the signal is zero. An orthogonal signal is one in which the phase of the signal is arranged to ensure that, in theory at least, no interference is created into an adjacent signal.

[4] LTE User equipment, network efficiency and value Published 1 September 2010 available as a free download from http://www.makingtelecomswork.com/resources.html.

[5] An exabyte is a million terabytes. These forecasts including tariff trends are drawn from industry sources and modelling and forecasting undertaken by RTT and The Mobile World.

than expected uptake of smart phones, we have revised our 2015 figure to 144 exabytes (see graphs in Chapter 4).

Irrespective of whether the final figure is 90 exabytes or 150 exabytes or some larger number, the projected efficiency savings from new technology such as LTE are insufficient to bridge the gap between volume growth and value growth.

New solutions and/or new income streams are therefore required in order to enable the sustainable growth of the mobile broadband industry. This includes technology and engineering innovation and specifically for wireless, a combination of materials innovation, RF component innovation, algorithmic innovation and radio engineering innovation. Market innovation to date seems to have been focused on producing ever more impenetrable tariffs and selling smart-phone insurance that users may or may not need.

1.8   Radio versus Copper, Cable and Fibre – Comparative Economics

Radio engineering is a fascinating discipline but is best analysed in a broader context in which technical and commercial viability is benchmarked against other delivery options, copper, cable and fibre. The comparative economics are complex and can yield idiosyncratic and counterintuitive results. For example, if a large file is not time sensitive then the most efficient delivery method, at least in terms of cost, may be to put the data on a memory stick and post it.

In fact, the mechanics of the postal service are quite relevant when analysing network economics and the end-to-end economics of delivering voice and data over wireless, fibre, cable and copper networks.

Part of the efficiency gain in telecommunications has traditionally been achieved by multiplexing users over either single or multiple delivery channels to achieve what is commonly known as 'trunking gain'. Telephone lines for example started using frequency multiplexing experimentally in the 1920s and time division multiplexing became increasingly prevalent from the 1960s onwards. Trunked radio systems and first-generation cellular radio in the 1980s and 1990s were frequency multiplexed, second-generation cellular radio systems were frequency and time multiplexed, third-generation cellular systems can also be spatially multiplexed over the radio part of the end-to-end channel. Optical systems today can also be frequency and/or time multiplexed.

Simplistically, trunking gain is achieved because most users only need resources for a relatively short time and therefore have no need for a dedicated channel. However, conversational voice is sensitive to delay and delay variability. Historically these end-to-end channels have therefore been deterministic with known and closely controlled end-to-end delay and delay variability characteristics. The transition from voice-dominant communication to a mix of voice and data, some of which is not time sensitive has allowed for additional multiplexing in which traffic is segregated into typically four traffic classes, conversational (voice and video calling), streaming, interactive and best effort. In theory at least, this allows for a significant increase in radio and network bandwidth utilisation that should translate into lower cost per bit delivered. This is the basis on which IP network efficiency gain assumptions are traditionally based.

However, reference to our post-office analogy suggests the cost/efficiency equation is more complex. A letter or packet posted first class stands at least a chance of arriving the next day. A second-class letter will take longer to arrive as it will be stored at any point where there is not enough delivery bandwidth, lorries, aeroplanes or bicycles, for the onward journey. In theory, this means that the delivery bandwidth can be loaded at a point closer to 100% utilisation, thus reducing cost.

This is true but fails to take into account the real-estate cost of storing those letters and packets in some dark corner of a sorting office and managing their timely reintroduction in to the delivery path. Depending on how these costs are calculated it can be argued that a second-class letter or packet probably costs the postal service rather more to deliver than a first-class letter or packet with of course less revenue to cover the extra cost. An end-to-end packetised channel is the post-office equivalent of having first-, second-, third- and fourth-class stamps on every letter and packet. The routers have to inspect each packet to determine priority and store lower priority packets as needed. Packet storage needs fast memory, which is expensive and energy hungry and constitutes a directly associated cost that needs to be factored in to the end-to-end cost equation. If there is insufficient storage bandwidth available the packets will have to be discarded or other time-sensitive offered traffic will need to be throttled back at source. This introduces additional signalling load that in itself absorbs available bandwidth and may compromise end-user experience. As end-user experience degrades, churn rates and customer support costs increase.

On the radio part of the network, the additional address overhead needed to discriminate between delay-tolerant and delay-sensitive traffic combined with the signalling bandwidth needed to maintain a reasonably optimum trade off between storage bandwidth and delivery bandwidth utilisation comes directly off the radio link budget and will therefore show up as a capacity loss or coverage loss, effectively a network density capital and operational cost. The user's data duty cycle (time between battery recharge) will reduce as well.

These caveats aside, multiplexing in its many forms does produce an efficiency gain, but this will not always be quite as much as claimed by the vendor community. The efficiency gain can be substantial if we are tolerant to delay and delay variability or can access delivery bandwidth in periods of low demand. For example our business offices are situated in a residential area from which most people commute every day. Day-time contention rates on our ADSL lines are low, evening contention rates are high, which suits us just fine.

1.9   Standardised Description Frameworks – OSI Seven-Layer Model as a Market and Business Descriptor

All of the above suggests a need for a standardised description framework that can be used as the basis for analysing how different functions in a network interact together and the coupled effects that determined delivery cost and delivery value.

Such a framework already exists and has been in use since 1983.[6] Known as the Open Systems Interconnection Model, this seven-layer structure remains as relevant today as it has ever been and encapsulates the topics addressed in all subsequent parts and chapters in this book. (See Table 1.1.)

There are two immediately obvious ways to approach an analysis using this model, to start at the top and work downwards or to start at the bottom and work upwards. Most industry

---

[6] http://www.tcpipguide.com/free/t_HistoryoftheOSIReferenceModel.htm.

Table 1.1   OSI seven-layer model

Description		Example protocol or layer function
Layer 7	Application Layer	Windows or Android
Layer 6	Presentation Layer	HTML or XML
Layer 5	Session Layer	Reservation protocols
Layer 4	Transport Layer	TCP prioritisation protocols
Layer 3	Network Layer	IP address protocols
Layer 2	Data Link Layer	ATM and Ethernet Medium Access Control
Layer 1	Physical layer	Fibre, cable, copper, wireless including free-space optical, infrared, ultraviolet and radio frequency

analysts would err towards starting at the top and working down on the basis that the application layer has the most direct impact on the customer experience and therefore should be regarded as the prime enabler. This assumes that we are working in an industry that is customer led rather than technology led, but this is at odds with historical evidence that suggests that the economics of telecommunications supply and demand are directly dependent on fundamental innovations and inventions at the physical-layer level. The innovations in the lower layers are generally, though not exclusively, hardware innovations, the innovations in the upper layers are generally, though not exclusively, software innovations.

This functional separation determines the structure of this book, with Part 1 reviewing user equipment hardware, Part 2 reviewing user equipment software, Part 3 reviewing network equipment hardware and Part 4 reviewing network equipment software. In terms of analysis methodology we are firmly in the bottom-up brigade.

Readers with long memories might recall that this is uncannily similar to the structure used in '3G Handset and Network Design' and we would point out that this is entirely deliberate.

The business of technology forecasting should in practice be a relatively precise science, but even if imperfect can be used to calibrate future efforts. '3 G Handset and Network Design' still provides a perfectly adequate primer to many of today's technology trends but lacked an ambition to integrate the technology and engineering story with economic analysis.

1.10   Technology and Engineering Economics – Regional Shifts and Related Influence on the Design and Supply Chain, RF Component Suppliers and the Operator Community

Which brings us back towards our start point in which we said that we aim to make the case that technology and engineering economics are separate disciplines that deserve separate analysis. In this context we will find that technology and engineering capacity is also determined by the gravitational effect of large emerging markets that absorb R and D bandwidth that would previously have been available elsewhere.

Ten years ago Jim O Neill, a partner at Goldman Sachs, wrote a paper entitled Building Better Global Economics and coined the term BRIC identifying the four countries, Brazil, Russia, India and China that combined large internal markets with high growth potential.

Over subsequent years these countries have not only been tracked as an entity by analysts but have also developed political and economic links that aim to leverage their combined economic power. In 2010 China invited South Africa to become a member so the acronym is now officially BRICS.

The BRICS are benchmarked in terms of geography and demography and economic growth rate. All are markets in which large telecommunication investments are presently being made, but of the four China stands out as having the largest number of internet users and the largest number of mobile phones that combined together suggest that China is already by far the world's largest single potential market for mobile broadband access. In terms of mobile connections China and India together are now five times the size of the US market. China's present dominance is illustrated in Figure 1.1.[7]

Figure 1.2 shows China Mobile's particular leverage.

In turn, this has had a dramatic effect on the local vendor community. Over a five-year period between 2006 and 2011, Huawei has moved from being a subscale (in relative terms) infrastructure provider with an $8.5 billion turnover to being number two globally with $28 billion dollars of revenue and 110 000 employees across 150 nationalities.[8] A similar pull-through effect has created a local development and manufacturing community both in China and Taiwan that has challenged and will continue to challenge Nokia both at the lower, mid and upper end of the user equipment market. HTC in Taiwan, for example, at the time of

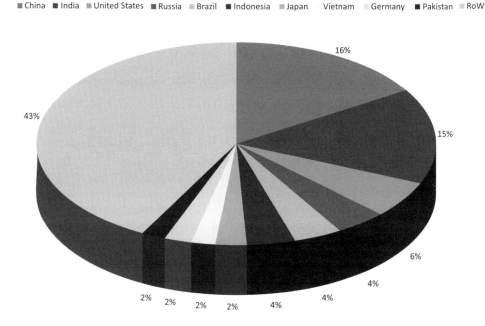

Figure 1.1   Mobile connections by country Q1 2011. Reproduced by permission of The Mobile World.

---

[7] All market statistics and modelling in this book are sourced from *The Mobile World*.

[8] http://www.cambridgewireless.co.uk/docs/FWIC%202011_Edward%20Zhou.pdf.

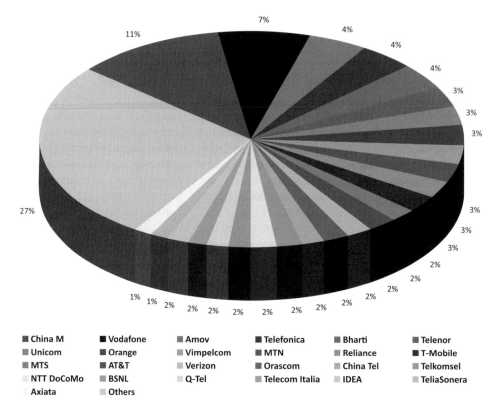

Figure 1.2   World mobile market, connections by operator Q1 2011. Reproduced by permission of The Mobile World.

writing has a substantially higher market capitalisation than Nokia and Nokia has just posted a quarterly loss, both would have been unthinkable five years ago. All the more galling for Nokia as HTC has invested substantially less than Nokia in R and D.

If you are an RF component supplier, this new vendor community, the China market, the China operator community and China Mobile as the largest operator in China are hard to ignore.[9] However, China has a unique combination of cellular radio band allocations and home-grown technology standards that need to be supported. This implies R and D risk and cost.

Vodafone and Telefonica have similar but different requirements but their leverage over the RF component and handset vendor supply chain is now less than it was both in terms of volume and value. The US is different again both in terms of band allocation and technology mix with the added complication that the two largest operators, AT and T and Verizon, do not have external markets to provide additional scale.

As a result, even large Tier 1 operators are finding it difficult to source price-competitive performance competitive user equipment for their local market customers. For example, as stated earlier AT and T and Verizon invested close to $20 billion in the 700-MHz spectrum

---

[9] This is particularly true of venture-capital-backed businesses who regard a China Mobile order number as a trophy asset.

and more than as much again in associated 700-MHz network hardware and software – a $50 billion investment. The band requires a unique mix of RF components, filters, switches, oscillators, RF amplifiers, low-noise amplifiers and RF amplifiers and associated matching components.

These are typically 50 cent devices. The RF component industry is highly fragmented and works on margins that are significantly lower than other component sectors. Servicing a subscale customer therefore implies an insupportable opportunity cost. The end result is that a $50 billion investment can be compromised and/or invalidated by an inability to source a few 50 cent components.

The critical enablers for RF performance both in user equipment and radio network equipment can be summarised as materials innovation, for example gallium arsenide, or more recently gallium nitride, in RF power amplifiers, component innovation, for example SAW or FBAR filters, and algorithmic innovation, for example adaptive matching techniques or at higher layers of the OSI model adaptive traffic management and scheduling.

Historically, these inventions and innovations have tended to originate from the West Coast of the US or Japan or Germany or Sweden or Finland or the UK or France to a greater or lesser extent.

China, however, is producing 450 000 university graduate engineers a year with India not far behind. India already has a major footprint in global ICT and China has an increasingly global footprint in the provision of telecommunications equipment. Companies such as Huawei and ZTE have large internal markets and can increasingly exploit local intellectual resources to develop in-house intellectual property value that can be leveraged into external markets.

A CTO from a European vendor pointed out that partly because of present exchange-rate anomalies, average engineering wage rates in mainland China are one tenth of European levels with a second level of subcontractors earning an order of magnitude less again. Combine this with preferential access to capital and it becomes obvious that Chinese vendors are well positioned to take market share from the European and US vendor community.

In the past, Europe and the US may have been able to realise a measure of market protection through aggressive management of the standards process and associated use of litigation to defend intellectual property value

However, this incurs a cost that at a practical engineering level translates into hundreds of engineers attending hundreds of international standards meetings that generate thousands of pages of standards documents that only the authors can interpret. This is probably no longer a sustainable business model. The progressively more active engagement of Chinese and Indian vendors in telecommunication standards work is a symptom of this transition in present progress.

And that summarises the underlying narrative that will be explored in the next twenty chapters. Technology economics and engineering economics are distinct and separate disciplines that are amenable to analysis and can yield significant insight into future competitive differentiation and competitive advantage opportunities. Accurate historic data sets are available that can be used to measure the rate of technology change. These can be used to underpin forward forecasts, but have to be qualified against any external factors that have undergone significant change.

So, for example, it could be argued that the spectral auction process has taken probably at least $200 billion dollars of cash out of the mobile telecommunications sector of the telecommunications industry. This has resulted in a decrease in available R and D resource.

In parallel, new markets have emerged, specifically China. These are having a gravitational effect on already diminished and diluted global technology and engineering capability. This will create a vacuum that will increasingly become filled by locally sourced resource. China's present investment in telecommunications engineering education and training closely mirrors the educational policy adopted by the Finnish government in the late 1980s and 1990s. The availability of well-trained highly competent hardware and software engineers was an essential ingredient in Nokia's transition from small niche player to global market leader in mobile phones, a position now increasingly under threat.

1.11   Apple as an Example of Technology-Led Market Innovation

Some of the ebb and flow of competitive advantage and opportunity can be directly ascribed to technology innovation. The Apple I Pod proposition is built on the availability of low-cost solid-state memory, the Apple I Phone and I Pad proposition is based on capacitive touch-screen interactive display technology. Both are examples of component-level innovation transforming user behaviour, a technology opportunity that Nokia failed to capitalise on early enough.

Apple also successfully managed to add content value and software value to the hardware proposition. Quite who has done what when and who owns what is, however, always subject to legal interpretation. Nokia sued Apple for patent infringement and won a royalty agreement in June 2011 worth just under 5% of the value of each iPhone sold plus a one-off payment of 800 million euros.

This corroborates our thesis that a technology and engineering opportunity analysis and ownership analysis should be the starting point not the end point of the strategic planning process.

And fortuitously, the methodology is beguilingly simple. To forecast three to five months ahead look three to five months behind, to forecast three to five years ahead look three to five years behind, to forecast thirty to fifty years ahead look thirty to fifty years behind.

It might seem absurd to contemplate a thirty- to fifty-year planning cycle, but in reality many technology cycles in telecommunications are of this order of magnitude. This is why standards or regulatory policy if determined by short-term three- to five-year political interest can destroy shareholder, stakeholder and social value. This is why short-term return on investment expectations, three- to five-year or three- to five-month or three- to five-day horizons can destroy longer term competitive capability.

Additionally, it is an inescapable fact that global scale is now more or less an exclusive precondition of commercial success in the telecommunications sector. Arguably only one country is large enough to have sufficient economies of scale to sustain any kind of nationally specific telecommunications technology policy. Previous attempts to protect local markets have, however, proved disastrous, the impact of nationally specific Japanese standards on the Japanese vendor community being a relatively recent example.

'Making Telecoms Work – from technical innovation to commercial success' is an all embracing title for an all-embracing attempt to bridge the traditional gap between engineering and the business planning process.

It is an absolute truth that hugely ambitious commercial plans can be invalidated by apparently trivial technical details. This brings us to Chapter 2.

Part One

User Hardware

2

Physical Layer Connectivity

2.1   Differentiating Guided and Unguided Media

This chapter explores the dynamics that determine the deployment economics of copper, cable, fibre and wireless networks both in terms of bandwidth efficiency and power efficiency and explores some of the metrics that are presently used for comparison and could potentially be used in the future, for example cost per bit, bits per Hz and joules per bit.

All of these physical layers use a combination of frequency or time multiplexing. Cable copper and wireless all use a combination of nonorthogonal or orthogonal frequency, phase or amplitude modulation. These techniques are also becoming, and will become more widely, applied in the optical domain.

Telecommunication delivery options can be divided into guided and unguided media. Guided media includes fibre optic, coaxial cable and copper or aluminium twisted pair. Unguided media include free-space optics, fixed wireless (one-way and two-way), portable wireless and mobile wireless.

A first-order summary of the available options and their comparative merits can be presented as shown in Table 2.1.

Self-evidently guided media cannot provide mobility in any meaningful sense of the word unless walking in a circle with a cable in your hand happens to be acceptable. What is immediately apparent is that all these delivery options represent work in progress and are all moving forward in terms of supported data rates and user functionality.

2.2   The Transfer of Bandwidth from Broadcasting to Mobile Broadband

Fifteen years ago this prompted Professor Nicholas Negroponte to suggest that as the capacity of guided media increased over time it would become steadily less sensible to use wireless broadcast as a way of communicating with static users – the 'Negroponte switch'. This and similar thinking encouraged the regulatory community to transfer bandwidth from terrestrial broadcasting to mobile broadband. Fortuitously for national treasuries this created lucrative spectral auction opportunities that in some, but certainly not all, cases also yielded longer-term shareholder value for the bidding entities. However, even taking this redistribution into

Making Telecoms Work: From Technical Innovation to Commercial Success, First Edition. Geoff Varrall.
© 2012 John Wiley & Sons, Ltd. Published 2012 by John Wiley & Sons, Ltd.

Table 2.1 Guided and unguided media

Guided media			Unguided media			
				Fixed wireless one		
Fibre optic	Coaxial cable	Twisted pair	Free space optics	way/two way	Portable wireless	Mobile wireless
50-year-old upstart Highest bandwidth, lowest attenuation and lowest specified bit error rate of all delivery options, can potentially deliver terabits of data	50-year-old upstart Can convey a larger volume of data than copper of the order of 200 Mbps but does not have the legacy reach of copper in most developed markets	150-year-old grand old man of telecommunica-tions can now support 1 Mbps over short distances (100 metres) – a fully amortised sunk investment	30-year-old upstart, high data rates except in dense fog	100 years of history, still a long way to go	100 years of history, still a long way to go	100 years of history, still a long way to go Has to adapt to very variable channel conditions This reduces capacity and incurs a higher bit error rate but lots of user functionality

account, if an analysis was done on the amount of video and audio consumed over wireless and a comparison made with the amount of video and audio delivered uniquely over an end-to-end guided media delivery route then it would be obvious that this switch has not occurred in any meaningful way.

Partly this is because terrestrial TV has far from disappeared. As we shall see later in Chapter 15 both the European and Asian terrestrial broadcasting community and US terrestrial broadcasters are actively developing portable TV standards that by definition guided media cannot match. Additionally in the mobile domain there are now nearly six billion mobile users who expect to consume media on the move or at least in an untethered nomadic environment so what has actually happened is that consumption has increased in all domains rather than transferred from one domain to another.

Technology has therefore not taken us in the direction expected by the visionary pundits fifteen years ago. One explanation for this is that over the past ten years guided and unguided media data rates have, by and large, increased in parallel. However, our ability to manage and mitigate and to an extent exploit the variability of radio channels, particularly bidirectional radio channels with two-way signalling capability, has increased rather faster.

2.3 The Cost of Propagation Loss and Impact of OFDM

All types of guided transmission media suffer propagation loss to a greater or lesser degree, but that loss is predictable and largely nonvariant over several orders of time scale. For example, a one-kilometre length of fibre, cable or copper will have defined loss characteristics that will stay stable over seconds, hours, days, months and years. Unguided media also suffer propagation loss but the loss varies over time. Portable or mobile wireless in particular will be subject to slow and fast fading that can cause the received signal to fluctuate by an order of 20 dB or more.

Various mechanisms have been developed over time to mitigate these effects. First-generation cellular systems for example used a combination of handover algorithms and power control to respond to changing channel conditions. Second-generation cellular systems had (and still have) a combination of frequency-domain (frequency-hopping) and time-domain (channel-coding) averaging techniques that are used to provide additional performance gain over and above first-generation systems. Digital compression of voice, images and video also increased the user's perception of performance gain.

Third-generation systems retain all of the above techniques but add in a FFT transform at the physical layer. This has been enabled by an increase in DSP capability that has allowed us to process signals orthogonally in both the time and frequency domain. The technique is known generically as orthogonal frequency division multiple access (OFDMA). The DSP capability is needed both to perform the transform but also to deliver sufficient linearity in the RX (receive) and TX (transmit) path to preserve the AM (amplitude-modulated) characteristics of the signal waveform that has an intrinsically higher peak-to-average ratio than other waveforms.

OFDM techniques are used in guided media; for example on copper access ADSL and VDSL lines. OFDMA techniques are used in unguided media, for example terrestrial broadcasting, WiFi and mobile broadband to deliver higher data rates and throughput consistency. The consistency comes from the ability of OFDMA to mitigate or at least average out the slow and fast fading effects of the radio (or to a lesser extent free-space optical) channel. OFDMA flattens the difference between guided and unguided media. The technique is particularly effective in systems with a return signalling channel that includes mobile broadband. Because unguided media offers more deployment flexibility and is faster to deploy and because it offers more user functionality it has captured more traffic than the pundits of fifteen years ago expected or predicted.

This in turn has enabled the mobile broadband industry to transfer relatively large amounts of relatively heavily compressed data across a highly variable radio channel in a way that would have been hard to imagine 15 years ago.

2.4   Competition or Collaboration?

This helps to develop the discussion as to whether technology convergence is a precursor to market and business convergence. Different parts of the telecoms industry today remain quite separate from each other. Although the mobile cellular and terrestrial broadcast industry have site shared for thirty years the two industries remain separate because an adversarial auction process reduces the commercial incentive to collaborate.

Similarly, mobile phone calls travel over copper and fibre networks for at least part of their journey with routing that is closely coupled and managed at a technical level. It could be argued that this would suggest a more closely linked commercial relationship would be beneficial. Competition policy, however, is pulling in the opposite direction. Satellite networks also tend to be treated separately both in terms of standards policy, spectral policy and regulatory policy yet technically they are becoming more closely integrated with other delivery options. Cable networks are similar. Technical similarity is increasing over time but commercial interest remains distinct and separate – a competitive rather than collaborative delivery model.

The reason for this is that standards making, regulatory policy and competition policy, responds to change but does not anticipate change. Pure economists will argue that collaborative delivery models are incompatible with the principles of open market efficiency and that a market-led model will be more efficient at technology anticipation. However, this is an obvious Catch 22[1] situation. Market entities may or may not be good at anticipating change. Indeed, an ambition of this book is to try and point out how technology forecasting can be improved over time, which in itself implies a need for improvement. However, even if there are unambiguous indications that technology convergence is happening at a particular rate it may be hard or impossible to respond because regulatory and competition policy remains based on historic rather than contemporary or forward-looking technology and engineering models. Even this is not quite true. Regulatory policy appears often to be based on a misinterpretation of past precedent. The past can only inform the future if accurately analysed.

Thus, there is a need for industry and regulators and economists to understand the technology and engineering change process. Because the regulatory and competition policy response only starts when the changed market position becomes evident then by default the subsequent policies fail to match technology requirements that by then will have changed substantially.

None of this needs to happen because fortuitously technology is amenable to rigorous analysis and can be shown to behave in predictable ways with a predictable influence on

---

[1] Catch 22 – a situation in which a desired outcome or solution is impossible to attain because of a set of inherently illogical rules or conditions. Catch 22 authored by Joseph Heller, first published in 1961.

market and business outcomes. Two engineering methodologies can be used as proxies for this prediction process.

2.5   The Smith Chart as a Descriptor of Technology Economics, Vector Analysis and Moore's Law

The Smith Chart developed by Phillip H Smith[2] and introduced in 1939 has been in use now by RF engineers for over 70 years as a way of finding solutions to problems in transmission lines and matching circuits, for example finding the inductance, capacitance and resistance values needed to realise an optimum noise match on the receive path of a mobile phone through the low-noise amplifier and the optimum power match from the RF power amplifier in the other direction.

While you cannot read a profit and loss account from a Smith Chart it effectively describes the potential efficiency of a bidirectional communications function taking into account multiple influencing parameters – a direct analogy of what we need to do when assessing technology economics in a real-life rather than theoretic context.

Noise matching is a proxy of how well a technology performs against competitive technologies; power matching is a proxy expression of how much potential added value can be realised from a technology in terms of its ability to match to market needs. If a power amplifier is poorly matched, power will be reflected from the output stages ahead of the device. If a technology is poorly matched to a market need, the value of that technology will be dissipated in the distribution channel or be reflected in high product return rates or service/subscriber churn.[3]

Conveniently, this can then be measured by using vector analysis. A vector can be simply described as having direction and magnitude, but can be imagined in several dimensions.

The direction can be either ahead or left and right or up and down or the same and all of these will have a rate of change that can be ascribed to it. This is useful when analysing economic trends but particularly useful when analysing technology economics because the direction and magnitude will remain relatively stable over time.

Probably the most often quoted example of this is Moore's[4] Law that observed and quantified the scaling effect of semiconductor processes linked to a parallel though not necessarily linear increase in design complexity. This simple and largely sustainable statement has informed business modelling in the semiconductor, computing and consumer electronics industry certainly up to the present day and probably remains valid for the foreseeable future.

Occasionally the impact can be even more dramatic. In solid-state memory for example the scaling effect combined with improved compression techniques (the MP3 standard) and a short period of over capacity in supply led to an extraordinary decrease in cost per bit for stored data with very low associated power drain. In parallel-miniature hard-disk technologies were being introduced.

---

[2] http://www.tmworld.com/article/323044-How_does_a_Smith_chart_work_.php.

[3] Churn is the measure of how many subscribers decide to leave a network because of poor technical or commercial service.

[4] Gordon Moore was director for research and development at Fairchild Semiconductor and published an article in Electronics magazine on 19 April 1965 that pointed out that the number of components on a given area of semiconductor wafer increased at a rate of roughly a factor of two per year.

Apple Inc. were smart enough to spot and then ride this wave with the iPod introduced in November 2001, initially using miniature hard disks then from 2005 transitioning to solid-state memory. Technology had created a new market with a new market champion. The development of resistive, capacitive and multitouch interactive displays similarly developed new markets that Apple have been astoundingly successful at exploiting. Note that the algorithmic innovation that is behind the Apple Touch Screen is often seen as the secret sauce (and source) of the success of the iPhone and iPad but this algorithmic innovation would not have been possible without fundamental materials innovation.

2.6 Innovation Domains, Enabling Technologies and their Impact on the Cost of Delivery

But this is a digression; displays are looked at again literally in Chapter 5, whereas this chapter addresses the innovation domains that impact cost of delivery. These can be summarised as materials innovation, component, packaging and interconnect innovation, system innovation and algorithmic innovation including adaptive techniques that mitigate and to an extent exploit the variability implicit in guided and particularly unguided media. Materials innovation includes the innovative use of existing materials, including natural materials, and/or the discovery of new materials that may be natural or man-made.[5]

As an example, in the 1880s in the USA Thomas Alva Edison was experimenting with different types of filament to try and create a light bulb that would last more than a few hours, deciding by a process of trial and error and observation that the best option was a carbon filament operating in a near vacuum.

Edison, however, also famously observed and questioned why an uneven blackening occurred within the glass bulb particularly near one terminal of the filament. Adding an extra electrode within the bulb he observed that a negative charge flowed from the filament to the electrode and that the current from the hot filament increased with increasing voltage. Although the phenomenon that came to be known as thermionic emission had been observed a few years earlier by Frederick Guthrie in Britain, Edison was the first to measure it. The effect was then explained and given a name (the electron) by JJ Thomson in 1897 but was referred to for a number of years as the Edison effect. Edison was also witnessing the effect of photonic energy. This could be regarded as a light-bulb moment in more senses than one.

The British physicist John Ambrose Fleming working for the British Wireless Telegraphy company then discovered that the Edison effect could be used to detect radio waves and produced the first two element vacuum tube known as the diode, patented in November 1904 and shown in Figure 2.1.

Three years later the concept was improved upon by the American Lee de Forest by adding an additional grid between the anode and cathode to produce a device, the triode valve that could either amplify or oscillate. This is shown in Figure 2.2.

The telecoms industry now had an enabling component technology that meant that tuned circuits could be built that would be capable of concentrating transmission energy within a discrete radio channel, the foundation of the radio broadcasting industry and two-way radio communication.

---

[5] It might be argued that man-made materials are fundamentally combinations of natural ingredients.

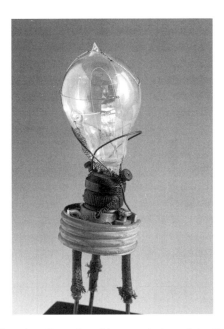

Figure 2.1 Fleming's diode valve. Reproduced by permission of the Science Museum/Science and Society Picture Library SSPL 10 311 209.

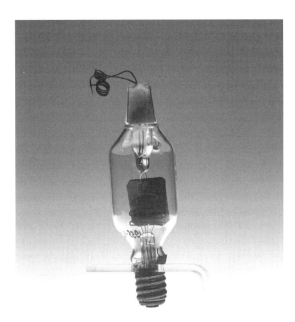

Figure 2.2 Lee de Forest's triode valve. Reproduced by permission of the Science Museum/SSPL 10 324 050.

Figure 2.3   The EF 50 valve – an example of the evolution of component technology over time. Photo Jacob Roschy, Radio Museum.[6]

This provided the basis for the beginning of medium-wave and short-wave radio broadcasting, for example in the UK the formation of the BBC in 1922 and fourteen years later the first VHF television broadcasts from Alexandra Palace in London.

Television created a demand for television receivers that in turn created a demand for valves and tuned circuits capable of working at VHF with sufficient dynamic range/gain control and receive sensitivity.

In the late 1930s, NV Philips of Holland had designed a new high-gain, low-capacitance series of all-glass valves, the EF50 series shown in Figure 2.3. These were used to produce televisions that would work more effectively in areas with weak reception. This valve was also used in radar products and two-way radio systems. As such it has been described as 'the valve that won the war', which is debatable but at least illustrates how the application of engineering development effort allows an enabling technology to improve over time.

As valves became smaller and more efficient both in terms of audio amplification and RF (radio frequency) amplification they could also be packed more closely together but this required a parallel development in interconnection technology, the printed circuit board. An example is illustrated in Figure 2.4.

The vacuum tube therefore enabled a mass market to develop in broadcast receivers, transformed military and civilian mobile communications and transformed the economics of telephone networks. Valves still live on today in high-end audio and guitar amplifiers and can be sourced either as original components or replaced with newly built modern equivalents.[7]

---

[6] http://www.radiomuseum.org/.
[7] http://www.oldtechnology.net/components.html.

Figure 2.4 1942 Radio with a printed circuit board. Reproduced by permission of the Science Museum/SSPL 10 439 336.

Similarly, valve-based two-way radios can still be found operating in some markets or in the hands of enthusiastic collectors.

However, although valve performance continued to improve there was a general recognition by the middle of the 1930s that a lower-cost way to generate, amplify and switch signal energy would be needed to achieve a step function reduction in delivery cost both in wireline and wireless networks. The answer was to find a way to harness the unique properties of semiconductors, materials that can act as a conductor and an insulator.

In 1876 a German physicist Ferdinand Braun,[8] a 24-year-old graduate of the University of Berlin demonstrated a rectification effect that could be created at the point of contact between metals and certain crystal materials, in effect a semiconductor device.

Just over 70 years later in 1947 John Bardeen and Walter Brattain built the first transistor, a device with a gate that could be used to control electron flow. Eleven years later in 1958 Jack Kilby invented the integrated circuit and in 1961 Fairchild Semiconductors produced the first planar transistor in which components could be etched directly onto a semiconductor substrate. Fifty years later these devices remain at the heart of every electronics and telecommunications product.

So, we have a historical time span of one hundred and seventy years from the 1840s to the present that takes us from Cooke and Wheatstone's telegraph to the modern internet all based on the mechanics of moving electrons around or more accurately harnessing and transferring the releasing the power of electrons in electronic circuits.

---

[8] http://www.computerhistory.org/semiconductor/timeline/1874-semiconductor.html.

2.6.1   Graphene – The New Enabler?

At some stage substitutes for silicon may also become available. Graphene is a possible candidate. Graphene is a form of carbon constructed as a flat layer of carbon atoms packed into a two-dimensional honeycomb arrangement. Its thinness means that it is practically transparent. As a conductor of electricity it performs as well as copper, as a conductor of heat it outperforms all other known materials. At a molecular scale grapheme could potentially enable a new generation of ultrafast transistors. It might also be suitable for coating windows to collect solar energy and may enable a new generation of supercapacitors.

One of the more relevant announcements in terms of telecommunications was from IBM who on 13 June 2011 announced[9] the development and production of an integrated circuit made from graphene that performs the function of a broadband frequency mixer operating at frequencies up to 10 gigahertz. The main challenge had been to overcome the fabrication challenges of getting graphene to adhere to metals and oxides. IBM appears to have solved this. Rather nice as the announcement was timed to coincide with IBMs 100-year anniversary.

2.6.2   The Evolution of Photonics

Enabling technologies can be shown to have had a similar transformational effect in photonics – a parallel universe with an equally fundamental impact on the economics of modern telecommunication systems.

Guiding light by refraction, the underlying principal of fibre optic transmission was demonstrated by Daniel Colladon and Jacques Babinet in Paris in the early 1840s with early applications emerging in the medical profession to light up internal body cavities.

In 1880 Alexander Graham Bell demonstrated the photophone a device that transmitted voice signals over an optical beam, arguably the start of the free-space optical industry (see later in this chapter). The phonograph appeared in the same year. He had already invented the telephone patented in 1874. The gramophone followed in 1887.

Optical communications had of course been used for centuries – semaphore systems and beacons for example, but optical transmission as we know it today needed two enabling technologies, an efficient way of generating photons and a low-loss medium to carry that energy.

Albert Einstein had theorised about the principles of stimulated emission in 1917. Stimulated emission is when an already excited atom is bombarded by yet another photon, causing it to release that photon along with the photon that previously excited it. The emitted photon and bombarding photon emerge in phase and the same direction. In order to produce a steady source of radiation many atoms or molecules have to be pumped to a higher energy state.

A laser uses an arrangement of mirrors within a cavity that reflect light waves so that they reinforce one another. An excitable gas or liquid or solid in the cavity determines the wavelength of the resulting beam. The process of pumping introduces energy into the cavity that excites the atoms causing a population inversion.

A population inversion is a state in which there are more excited atoms than grounded atoms. This is stimulated emission. The released photons oscillate backwards and forwards

---

[9] http://www-03.ibm.com/press/us/en/pressrelease/34726.wss.

between the mirrors in the cavity causing other atoms to release more photons. One of the mirrors allows some of the released photons to escape the cavity to create the beam. This is either launched into free space (free-space optics, free-space photonics, optical wireless) or into fibre.

Charles Townes working at the Bell Laboratories during the Second World War and at Columbia University with Arthur Schawlow produced a device that used ammonia gas and microwave radiation that became known as the MASER, microwave amplification by stimulated emission of radiation. The device was used to amplify radio signals and as a detector for space research. Gordon Gould, one of Townes students at Columbia, was probably the first person to construct an optical laser in 1958, Ali Javan produced the first gas laser using helium neon in 1960, Robert Hall created the semiconductor injection laser in 1962 and Kumar Patel produced the first carbon dioxide laser in 1964.

In 1965 Charles Kao and George Hockham suggested that attenuation in fibre was partly caused by impurities that could be removed. This paved the way for low-loss high-purity silica glass-based optical transmission. An attenuation limit of 20 dB/km was achieved in 1970. At the time this represented a significant performance threshold. In 1986 David Payne at the University of Southampton and Emmanuel Desuvire at Bell Labs codeveloped the erbium-doped fibre amplifier. The combination of low-loss fibre, relatively low-cost lasers and efficient amplification meant that optical communication could compete with and then overhaul copper in terms of cost per bit.

However, what we are trying to establish are the *relative* technology economics of fibre against copper, cable and wireless – both in terms of relative cost as at today and the relative cost economics going forward – a simple question with a complicated answer.

A starting point is to differentiate between long-haul transmission, core transmission and local access. Long-haul can be across distances of thousands of kilometres at very high aggregated data rates; core transmission is typically over shorter distances but at very high aggregated data rates, both realise high multiplexing gain. Local access is typically from a local switch servicing individual premises and individual users within those premises at lower aggregated rates.

Some but not all of the technologies and techniques used in long-haul and core transmission translate across into the local access domain but per premises cost and per user cost becomes far more dominant than per bit cost and is particularly influenced by the material and installation cost of the fibre, the cost of transceivers and connectors and the cost of signal splitting or time or frequency multiplexing. These factors are less dominant in core transmission and long-haul systems.

Fibre is now capable of supporting data rates in the region of tens of Tbits[10] but only over short distances and capacity reduces with distance, a relationship generically described as optical Moore's law.[11]

Impairments in long-haul networks include noise, attenuation, nonlinear effects and dispersion including polarisation mode dispersion (PMD). These impairments are less of a problem in local access networks though PMD becomes a dominant constraint at data rates over and above 10 Gbit/s.

---

[10] A terabit is equivalent to 1000 gigabits.

[11] Optical Moore's law can also be applied to optical storage where storage versus density versus cost metrics apply rather than just capacity versus distance. This topic is revisited in later chapters.

Receiver noise is a product of the thermal noise introduced when the photocurrent generated by received photons is amplified. Noise is introduced due to the random motion of electrons. The photodetection process, the generation of a photocurrent in response to incident photons, also generates random electrons and therefore noise.

Attenuation reduces the optical signal-to-noise ratio, which in turn has an impact on the bit error rate, the proportion of incorrect (error) bits received. Optical transmission systems have a target to achieve less than one error bit for every trillion bits received (1 in 10^{12}). Fixed wireless systems by comparison have a target to achieve 1 in 10^{10} and mobile broadband networks 1 in 10^3, for reasons that will be examined later. A decrease in power can be offset by using optical amplifiers, but these also introduce noise. Multiplexers additionally introduce insertion loss.

The ultimate capacity of an optical fibre system can be calculated using Shannon's law and expressed as information spectral density (ISD). ISD in the optical domain can be shown to be of the order of 4 to 6 bits per Hz. However, fibre is a nonlinear transmission medium, which means that as the signal power increases the number of bits per Hz increases to a maximum value and then starts decreasing. Nonlinear effects include unwanted intermodulation between different channels in a wavelength division multiplexed system, the intermodulation will also modulate the refractive index of the channels, other effects include phase shifting in single channel systems (a function of the signal intensity) and scattering and reflected energy effects, not dissimilar to VSWR in radio transmission systems.

The more linear the system the higher the ISD. Even assuming an ISD of one bit/s/Hz there is over 100 THz of available bandwidth across the 1000 to 1700 nanometre optical spectrum, which would yield at least 100 T/bit/s/Hz of transmission capacity in a long-haul system, so capacity does not seem to be an immediate problem.

Similarly in the access network the shorter range means that little or no amplification is needed so the optical signal-to-noise ratio is higher and nonlinear effects are less severe due to the lower launch powers and shorter distances involved. ISD should therefore be higher than in long-haul systems.

The final impairment process is dispersion, which is either intermodal, chromatic or polarised.

Intermodal dispersion occurs when several modes or angles are allowed to travel along a light-carrying core. It only occurs in multimode fibre.

Chromatic dispersion occurs because the refractive index, which dictates the speed at which light travels down the fibre varies with wavelength and will result in pulse spreading in the time domain. This is known as material dispersion. There will also be dispersion known as waveguide dispersion, which is a function of a portion of the light wave at a given wavelength travelling in the cladding as well as the core, both media having a different refractive index.

Chromatic dispersion reduces with narrower line widths though device cost increases proportionately. Some wavelengths are less susceptible to chromatic dispersion. Additionally, high-speed equalisers can be used to compensate for the effect though these are computationally expensive particularly at higher data rates.

Polarisation mode dispersion (PMD) happens because light waves are made up of two orthogonal components, one travelling along the X plane and one along the Y plane. Imperfections in the fibre will mean that these two components will have a different refractive index, which means they will travel at different speeds. PMD also varies with temperature so compensation can be quite complex and expensive and for that reason is only usually deployed in long-haul systems.

Mitigation of all these impairments over time increases the distance product and indirectly the capacity product, which is also a function of the bit error rate or to use an internet term, 'good put'.

Capacity product increases are also due to improved fibre and to the use of laser sources that can generate a range of optical frequencies that are chosen on the basis of their propagation loss and dispersion characteristics, gallium arsenide laser sources for example working at wavelengths around 800 nanometres[12] (374 THz), indium gallium arsenide laser sources at 1300 nanometres (230 THz) and indium gallium arsenide phosphide laser sources at 1500 nanometres (199 THz).

These are either laser diodes or lower-cost LED (light-emitting diodes). Light emitting diodes are electroluminescent and emit incoherent light with a relatively wide spectral bandwidth of 30 to 60 nanometres. The LED devices are relatively inefficient typically only converting about 1% of their input power to launched power but similarly priced devices known as vertical cavity surface emitting lasers have better power and spectral properties. Both couple effectively into multimode fibre.[13]

Multimode fibre has a large light-carrying core of between 50 and 100 micrometres, commonly 62.5 micrometres in the US. The wider spectral bandwidth makes the optical signal more susceptible to dispersion as the light is dispersed into multiple paths as it travels along the path causing pulses to overlap over longer distances. Multimode is normally used at 850 or 1300 nanometres and works well for distances of two miles or less and can work adequately at lower data rates over ten miles or more. It is best suited to local area network applications with bit rates of 10 to 100 Mbit/s

Single-mode fibre is normally used at 1310 or 1550 nanometres and has a light-carrying core of 8 to 10 micrometres that largely eliminates the dispersion limitations of multimode, thus providing a significantly better capacity distance product. However, a narrow spectral width laser is needed that is typically two to four times the cost of a multimode light source.

For radio communication systems capacity generally increases with frequency but distance reduces or looking at it the other way round attenuation at longer wavelengths/lower frequencies is lower but so is capacity. So, for instance, the distance product for long wave radio is high, the capacity product is low.

In the optical domain a similar relationship exists in that longer wavelengths/lower frequencies exhibit lower propagation loss but also there are particular windows with minimum dispersion and minimum attenuation characteristics that determine the optical band plan as shown in Table 2.2. The windows are determined by two phenomena, Rayleigh scattering, which reduces as wavelength reduces and hydroxide anion absorption, also referred to as the water absorption peak, which occurs at 945 nm, 1240 nm and 1380 nm. Avoiding these peaks produces a window at 800 to 900 nm that has fairly high loss and is only ever used for short-distance transmission (usually with LEDs for cost reasons – see above). The second and third telecom windows are at 1300 nm and 1500 nm, but are subdivided as follows.

---

[12] A nanometre is a billionth of a metre. Fingernails grow a nanometre every second or 86 400 nanometres in a day http://www.sciencelearn.org.nz/Contexts/Nanoscience/Science-Ideas-and-Concepts/Nanometres-and-nanoscale.

[13] Light introduced into a fibre has to be guided by the core and has to be launched at a critical angle in order for transmission to occur. The critical angle is a function of the refractive indices of the core and the cladding.

Table 2.2   Optical band designations and relative frequencies and wavelengths

Band	Description	Frequency THz	Wavelength
O band	Original (second-generation window)	238 to 220	1260 to 1360 nm
E band	Extended	220 to 205	1360 to 1460 nm
S band	Short wavelength	205 to 195	1460 to 1530 nm
C band	Conventional ('erbium window') Third generation window	195 to 191	1530 to 1565 nm
L band	Long wavelength Third-generation window	191 to 185	1565 to 1625 nm
U band	Ultralong wavelength	185 to 179	1625 to 1675 nm

The second telecom window was originally used for long-haul transmission. Long-haul transmissions are now all based on the third window. C and L bands between them have over 11.5 THz of bandwidth available. In the longer term, evolved semiconductor optical amplifier and Raman amplifier technologies will allow additional bandwidth to be economically accessed.

As with radio communication there is a choice of channel spacing within each band. In the optical domain this is referred to as wavelength division multiplexing and can be used to increase the capacity–distance product. Similarly different modulation schemes can be deployed including coherent transmission (and detection).

Modulation was typically (is typically still in legacy systems) on off keying (OOK) with direct detection. The combination of these techniques can achieve data rates up to 10 Gbit/s per channel but any increase beyond that requires more complex modulation for example differential phase shift keying with coherent detection. Coherency requires an ability to control the phase of the modulated signal, which requires an optical phase-lock loop and then demodulation of the signal in the receiver, which requires coherent detection. Essentially, these are techniques now widely deployed in mobile broadband and point-to-point microwave systems and in copper and cable but now extended into the optical domain.

Wavelength division multiplexing (WDM) and/or dense wavelength division multiplexing (DWDM) is fairly self-explanatory in that it is based on modulating and detecting multiple optical channels within a discrete band of optical frequencies. These techniques have become more applicable as dispersion compensation techniques have improved.

In parallel, optical splitting techniques have been developed for the local access domain that have allowed a single fibre to be shared between multiple premises and multiple users. The splitting techniques are unpowered, hence the description passive optical network.

Laser sources are relatively narrowband devices, whereas optical receivers are wideband. In practical terms this means that wavelength demultiplexers need to be installed in user premises. This would be an issue for example for Open Reach in the UK as and when they decided to upgrade from their present GPON[14] network, to a WDM PON network as the demultiplexers are expensive and temperature sensitive.

A GPON network as presently implemented typically conforms to the ITU T G.984 standard and delivers asymmetric service based on 2.488 Gbps of downstream bandwidth and

---

[14] gigabit passive optical network.

1.244 Gbps of upstream bandwidth supporting time multiplexed voice, Ethernet and ATM. The physical reach of these networks is of the order of 20 km.

PTP networks, as the name implies, have dedicated point-to-point fibre for each user or at least for each end point, which means more fibre than a PON so the fibre cost is higher, though this is offset by not having to have wavelength multiplexers/demultiplexers. A PTP network uses more duct space, which can be a problem if the duct space is already full up.

PTP networks use a multiplexer that splits the signal into transmit and receive wavelengths, either 1310/1550 nm or 1310/1490 nm, supporting *symmetrical* services and specified to operate over 10 km.

Passive optical networks and point-to-point fibre access networks are therefore the two competing deployment options for local distribution (as opposed to long-haul and core transmission, which is also implicitly point-to-point).

As stated earlier, WDM systems are usually deployed in single-mode fibre and come in three densities, coarse, conventional and dense.

A *course system* typically provides 8 channels on a single-mode fibre across the entire frequency band between 1310 and 1550 nm, which contains a minimum dispersion window and a minimum attenuation window. The advantage is that transceiver designs are relatively inexpensive.

A *conventional system* provides eight channels in the third transmission window at 1550 nm.

A *dense system* packs more channels into the same bandwidth, for example 40 channels at 100 GHz spacing, 80 channels with 50 GHz spacing or 100 channels at 25 GHZ spacing (ultradense wavelength division multiplexing).

In the 1990s increases in data rates were achieved by adopting techniques that mitigated impairments such as dispersion (dispersion compensation) and by implementing forward error detection (covered in Chapter 3). Further increases in bit rates in local access optical networks are therefore dependent on the development of low-cost high-performance temperature-independent optical components, photonic integrated circuits and DSP processing capability, for example for channel equalisation.

Delivery cost economics are not, however, just a function of data throughput but are determined by how efficiently that throughput can be used. Core transmission and long-haul transmission systems support large numbers of simultaneous users and can achieve substantial multiplexing gain.[15] In a local network only a small number of users or a single user may be supported and the dominant cost becomes the transceiver equipment, fibre material cost and installation cost.

So for example shorter distances and lower data rates ($<$10 Gbit/s) mean lower-cost Fabry–Perot[16] light sources can be used rather than DFB lasers. Fabry–Perot lasers are less temperature sensitive making them more suitable for use in customer premises equipment.

Cost has also reduced as fibre quality has improved, which in turn is due to improvements in manufacturing technique. For example, hydroxide ions (OH–) absorb particular wavelengths.

---

[15] Multiplexing gain is studied in Chapter 2.

[16] Fabry–Perot laser diodes are a type of diode laser combined with a mirror system. They are generally slower and noisier than distributed feedback (DFB) lasers. DFB lasers use diffraction gratings rather than mirrors to create resonance and oscillation in the cavity. They produce a finer range of wavelengths than a Fabry–Perot cavity.

Reducing OH ion content is allowing wider bandwidths to be supported allowing up to 16 wavelengths to be supported on a single fibre pair. Making fibre that bends easily in a relatively tight radius and the use of simpler lower-cost connector and terminating devices have helped make domestic fibre to the home more economic.

The choice of technologies within the access network include point-to-point Ethernet, wavelength division multiplexing passive optical networks (WDM PON) and time division multiplexed passive optical networks (TDM PON).

Point-to-point Ethernet delivers dedicated bandwidth to individual users, TDM PON provides shared bandwidth up to 64 users, WDM provides dedicated bandwidth to each user by allocating a specific wavelength. As the performance of photonic integrated circuits and temperature-independent optical components improve, all three options become more efficient.

Both TDM and WDM systems require demultiplexers and space needs to be found within splitter sites to accommodate these. In practice, physical access costs remain a dominant cost component. Allowing telecom operators to string fibre on telegraph/utility poles, a proposal presently being actively pursued in some markets, would substantially change fibre access economics though the data pipe still needs to be coupled to the user over the final few metres into the home or office.

So first let's look at the present and future evolution of optical delivery systems.

Fibre began as single wavelength then progressed to multiple wavelengths and more recently has progressed to multimode, three generations of optical technology evolution.

Single mode is used for backhaul and can support data rates of 40 g/bits or above. It has a loss of about 0.37 dB per km at 1310 nanometres (229 THz). It is expensive when compared with multimode fibre and requires expensive connectors and exotic laser diodes.

Multiplexing in the frequency and time domain, for example means that a single optical fibre pair can support 160 channels at 10 Gbit/s or 40 Gbit/s over distances of 4000 km. As stated earlier these long-distance links are cost efficient because a high level of traffic aggregation can be achieved.

Multimode is used in final-mile applications. In a multimode fibre, if the angle of the light entering the fibre is changed then that light 'channel' will follow a certain reflective path. This means it will have travelled a different distance to other light channels introduced at a different angle and will therefore arrive at a different time at the receiver.

This is the basis for time division multiplexing and allows for up to 64 'channels' to be supported. Multimode fibre is relatively low cost when compared to single-mode fibre and does not require expensive high-precision connectors. However, it has a relatively high propagation loss, about 0.8 dB per km at 1310 nanometres (229 THz) and is relatively low bandwidth when compared with single mode.

Newer multimode fibre also supports shorter-wavelength transmission at 1510 nanometres (199 THz). Attenuation losses reduce to about 0.23 dB per km, but chromatic dispersion is worse.

Multimode fibre is used extensively in passive optical networks. The ITU-T G983 specification defines the *downstream* at 1490 nm (213 THz) at 622 Mbits/s, a 155 Mbits *upstream* at 1310 nanometres (229 THz) and an RF video overlay at 1550 nanometres (193 THz).[17]

---

[17] http://www.precisionphotonics.com/converter.asp Frequency is equal to the speed of light divided by wavelength, with the speed of light defined as 299 792 458 metres per second.

The 18 billion dollar FioS Optical network being deployed by Verizon in the US is an example of a passive optical network based on a combination of multimode and single-mode fibre transport.

Even allowing for multiplexing on the multimode drop, these networks offer super fast downloads of recorded content including high-definition video, multiple (200 or more) standard TV channels and high-definition TV. Peak rates are capped per user but can be 45 Mbit/s and could theoretically be much higher.

Both multimode and single-mode fibre have substantial untapped bandwidth potential that will be realised as the cost performance ratio of optical components improves over time.

So, for example, Fabry–Perot lasers remain cost effective but have a wide spectral line that makes them unsuitable for long-distance transmission (the effects of chromatic dispersion) and dense WDM (requiring close frequency spacing). The devices are, however, temperature tolerant, which makes them suitable for CPE (customer premises equipment) applications.

Distributed feedback lasers have a wide spectral range and narrow line width, of the order of a few hundred MHz. When temperature stabilised they are good for both long-haul and DWDM systems but are much more expensive. They can also be wavelength tuned over several nanometres. Vertical cavity resonators are cheaper and now perform very similarly to DFB lasers.

Other enabling technologies already mentioned include modulators that control optical phase as well as amplitude, optical receive filters and photonic integrated circuits that combine previously discrete components such as transmitters modulators and receivers into a single chip, though some standards work needs to be done to ensure sufficient economies of scale are realised to cover R and D and manufacturing investment. The standardisation of 100 Gbit/s transceivers by the Internetworking Forum (OIF) is one example.

All these trends will result in a steady decrease in cost per bit (as throughput rates increase) and capital cost (as fibre cost and component cost reduces over time). Installation cost remains more of a problem given that competitive access options have either been fully amortised (copper) or written down (cable).

Delivery cost economics are therefore determined by the amount of legacy investment and physical reach of these competitive delivery options and the topology and demography serviced by those delivery systems.

Service asymmetry is also a factor and cost per bit is only a valid comparison metric if the bits can be used efficiently, for example multiplexed over a sufficient numbers of users with capacity closely matched to user uplink and downlink requirements.

Some first-order comparison can be achieved by analysing cost delivery economics in terms of addressable user density differentiated across five application domains, dense urban, urban, suburban, rural and remote rural as shown in Table 2.3.

Self-evidently only wireless can deliver mobile or portable service, but as we will see later there is a delivery cost/benefit trade off in terms of signalling and coding overhead.

With fixed service provision there is a choice of wireless that could be either radio frequency or free-space optical with point-to-multipoint offering a lower cost per subscriber but lower bit rate when compared with point-to-point.

Fibre has a higher capacity/distance product than cable, which has a higher capacity/distance product than copper. However, cable and copper networks may have been amortised over a long period and installation costs have long since disappeared into the noise floor.

Table 2.3   Dense urban to remote rural teledensity

	Dense urban	Urban	Suburban	Rural	Remote rural
Fibre	⬆				
Cable					
Copper					
Wireless fixed point-to-point	Per bit cost reduces as data rates increase but per subscriber costs also increase				
Wireless fixed point-to-multipoint					
Wireless portable					
Wireless mobile					
	Investment cost per subscriber increases with distance and as user density decreases Bit rate also reduces with distance Therefore, per subscriber cost increases with distance and decreasing user density				
	⬅				

The per bit cost benefits of fibre can only be realised as fiscal gain if the data rate can be matched efficiently to individual and group subscriber needs. The investment cost per subscriber increases with distance and as user density reduces. Bit rate decreases with distance for all the delivery options. The combination of these factors means that capitalisation and operational cost per subscriber increases with distance and decreasing user density.

Inconveniently, individual income and household income also tends to be lower in for example rural and deep rural areas, which compounds this effect. Providing a dedicated wavelength division multiplexed multimode fibre to your elderly aunt in Rosyth whose nearest neighbour is ten kilometres distant is not an economically attractive proposition unless an existing copper run exists that can be easily upgraded. If no copper run exists then point-to-point wireless may be the only feasible option. In practice, the most economic mix ends up being a combination of several or all options.

2.6.3   Free-Space Optics

Free-space optical hardware systems also known as free-space photonics or optical wireless systems have been in existence since the 1970s, initially for military applications but now increasingly promoted for use as alternatives to point-to-point microwave systems.

The systems can support data rates of between 100 Mbps and 2.5 Gbps over distances of several kilometres but are sensitive to propagation impairments including fog, absorption, scattering, physical obstructions, building sway and scintillation. Longer-term use of wavelength division multiplexing techniques is claimed to be capable of increasing this to 10 Gbps.

High-quality photodiode hardware is available for the 800-nanometre band and 1550 nanometres. A risk of eye damage limits the power in the lower band so 1550 nanometres is arguably operationally better. Additionally, longer wavelengths have lower photon energies. A 1550-nanometre photon has half the energy of an 800-nanometre photon, which means that for the same amount of energy (watts of power) a beam of 1550 nm has twice the number of photons.

2.6.4   Infrared Optical Systems

Most of us of course use at least one optical wireless system every day when we turn a television or audio system on or off. Infrared remotes have been around now for 25 years. They require line of site and have limited range of at most thirty feet but seem to work well enough to keep us sitting on the sofa which is their main function.

Infrared controllers get easily confused by sunlight and strong fluorescent lighting. To achieve selectivity infrared receivers usually work at around 980 nanometres with filters on the receive path that block out light at other wavelengths. Table 2.4 sets these frequencies and wavelengths in the context the range of light waves (colours) visible to the human eye.

2.7   Cable Performance Benchmarks

Cable data rates are lower than fibre but in common with fibre are increasing over time. The increase in data rate has primarily been delivered through the adoption of higher-order

Table 2.4 Visible wavelengths

Colour	Frequency	Wavelength
violet	668–789 THz	380–450 nm
blue	631–668 THz	450–475 nm
cyan	606–630 THz	476–495 nm
green	526–606 THz	495–570 nm
yellow	508–526 THz	570–590 nm
orange	484–508 THz	590–620 nm
red	400–484 THz	620–750 nm

modulation schemes. The downstream frequencies used in cable are typically from 42 to 850 MHz, with an upstream between 5 and 42 MHz in the US and between 5 and 65 MHz in Europe. Table 2.5 shows how data rates have increased over the past ten years over three generations of DOCSIS (data over cable service interface specification).

DOCSIS 3.0/Eurodocsis is needed to support a credible offering of high-definition TV and is designed to be IPV6 compatible.[18] The higher data rates are achieved partly by higher-order modulation schemes and partly by channel bonding. Eurodocsis is essentially a redistribution of the uplink/downlink budget to support the 8-MHz digital TV channel bandwidths and data rates needed to meet European DVB-C broadcast requirements.

Table 2.6 shows DVB-C modulation options and data rates. Present standardisation efforts are being directed towards producing a new iteration of DVB-C known as DVB-C2.

2.8 Hybrid Fibre Coaxial Systems

In a hybrid fibre coaxial system, the optically modulated signal at a fibre node is downconverted to an RF modulated signal that begins at 50 MHz but extends to 550 to 1000 MHz. The reverse path is either between 5 and 42 MHz (ATSC in the US) or 5 to 65 MHz (DVB C in the rest of the world). The downstream is either carrying 6-MHz ATSC channels or 8-MHz DVB C channels.

Table 2.5 Three generations of DOCSIS cable specification

DOCSIS	DOWNSTREAM	UPSTREAM
DOCSIS 1.0	38 Mbit/s	9 Mbit/s
DOCSIS 2	38 Mbit/s	27 Mbit/s
EURODOCSIS 2/DVB C	51 Mbits	9 Mbit/s
DOCSIS 3.0	160 Mbit/s	120 Mbit/s

---

[18] IPV6 is addressed in a later chapter.

Table 2.6 Example modulation options and data rates for an 8-MHz digital TV cable channel

Modulation	16 QAM	32 QAM	64 QAM	128 QAM	256 QAM
Data rate Mbit/s	25.64	32.05	38.47	44.88	51.29

2.9 The DVB-S Satellite Alternative

HFC networks are directly competitive with satellite networks. Satellites are power-budget limited. For this reason, DVB S satellites use QPSK or 8 PSK modulation. These modulation schemes have less amplitude modulation than QAM schemes and therefore support more power-efficient RF power amplifiers.

Present standardisation efforts are being directed towards producing a new iteration of DVB-S, known as DVB-S2 based on using 16 and 32 PSK modulation combined with adaptive variable source coding (AVC).

Table 2.7 shows the data rates needed for present SDTV and HDTV transmission and the possible data rates achievable using AVC coding. High-definition TV occupies over four times the bandwidth of a standard definition transmission. Compression schemes such as AVC provide a partial solution to this but increase error and jitter sensitivity. Similar constraints apply to MPEG4 coding schemes. A 36-MHz satellite transponder capable of supporting 25 standard definition TV programs will be hard pressed to support more than 5 or 6 HDTV channels.

2.10 Terrestrial TV

Terrestrial TV has a significant advantage over satellite in that it is potentially neither power nor bandwidth limited. The need to coshare spectrum with analogue transmissions has required digital TV to transmit at suboptimum power levels, but this constraint disappears once analogue transmissions are switched off.

There are obvious opportunities to increase signal strengths over and above present analogue TV transmissions by increasing the number of DVB transmission sites. Cellular operators are a natural partner in this process. This opens up the possibility of supporting a wider range of portable and mobile devices including the use of these devices within buildings.

It could of course be argued that in building portable and mobile device applications can also be serviced by a combination of fibre, cable or copper access combined with WiFi. However, it would be economically hard for any alternative delivery options (including mobile broadband

Table 2.7 DVB-S Data rates

Definition	Data rate	Data rate with AVC coding
SDTV	4.4 Mbit/s	2.2 Mbit/s
HDTV	18 Mbit/s	9 Mbit/s

and/or possible future combined WiFi/mobile broadband networks) to match the ability of terrestrial TV to deliver a 20-Mbit or 30-Mbit downlink to devices irrespective of where or how they are being used.

Discussions are ongoing as to how much of the UHF bandwidth terrestrial broadcasters retain after the analogue to digital switchover is complete. Even if the cellular industry took 100 MHz of digital dividend spectrum there would still be 300 MHz available to support a broad range of national and local standard and high-definition TV multiplex options.

Terrestrial TV therefore has a significant role to play in the standard and high-definition TV delivery process, including the delivery of these services to portable and mobile devices. We revisit terrestrial broadcasting both in terms of physical layer evolution and network topology in Chapter 15.

2.11   Copper Access – ADSL and VDSL Evolution

As stated earlier, optical transmission systems have a target to achieve less than one error bit for every trillion bits sent and received (1 in 10^{12}). Copper access has a target to achieve one error bit for every 10 billion bits sent or received (1 in 10^{10}), wireless systems have a target to achieve one error bit for every 1000 bits sent or received (1 in 10^{3}).

However, it is not just the error rate but the distribution of the error rate that matters. If errors occur in bursts then coding techniques will be needed to mitigate the effect this will have on throughput efficiency, a topic we cover in more detail in the next chapter.

Error bursts are bursts that are inherently more likely in unguided media and less likely in guided media though error coding is typically still used in guided media to minimise the impact of error rates.

Most voice conversations and data exchanges or web or internet access sessions travel over copper cables at some point in their end-to-end journey including of course mobile to land line voice or data exchanges.

In the early years of telecommunications copper cable was just a cable made of copper.

In 1880 Oliver Heaviside came up with the idea of coaxial cables.

Coaxial cables have an inner conductor surrounded by a flexible tubular insulating layer and an outer shield. The derivation of coaxial refers to the geometric relationship between the inner conductor and outer shield. The advantage of coaxial cabling is that the electromagnetic field carrying the signal is guided through the space between the inner and outer conductors. This means that the cable is resilient to external electromagnetic interference and can be laid next to metal objects, for example gutters and metal conduit. The cable acts as an efficient waveguide at higher (RF) frequencies. Coaxial cable is more expensive than fibre on a per bit basis, cheaper than some types of fibre on a per metre basis but more expensive than copper particularly on a per metre basis. The most common applications today include the connection between televisions and a roof-mounted dish or antenna and/or cable TV connectivity.

Twisted-pair cables were invented by Alexander Graham Bell in 1881. They are either shielded (STP) or unshielded and both conductors are floating and balanced with respect to ground and neither is connected even partially to ground.

The problem with copper pairs as a lower-cost alternative to coaxial cable is that they suffer from crosstalk. This is mitigated by twisting the cables at different twist rates. They can then be bundled together without the need for shielding. This is known as an unshielded twisted pair or UTP. Unshielded means the twisted pairs are unshielded from each other rather than the outside world.

UTP cables are used for indoor telephone wiring, for example for phone or Ethernet connections, and often grouped into sets of 25 pairs within an outer sheath.

Outdoor telephone cables can contain hundreds or thousands of pairs divided into bundles – the cables within each bundle have different twist rates and the bundles are then bundled together again using different twist rates.

Twisted-pair copper cables are repurposed for data by adding a microfilter that ring fences the 3.4 kHz of voice bandwidth, leaving the higher frequencies to be used as a series of multiple carriers in the frequency and time domain. These higher frequencies suffer more aggressive impairments including far-end crosstalk (FEXT), near-end crosstalk (NEXT) and impulse noise.

Impulse noise is typically of the order of a few microseconds and is caused by power-line transients or electromechanical switches or current surges in electrical equipment. Impulse related error rates can be reduced by forward error correction.

Far-end crosstalk occurs when signals from the far end of a twisted-pair couple with the weak received signals from the far end of another twisted pair. This is a dominant impairment in ADSL. Near-end crosstalk occurs when signals transmitted towards the far end couple with weak signals originating from the far end.

A telecoms engineer from fifty years ago would be surprised and certainly impressed by what can now be squeezed along two pieces of copper wire. Many of the techniques used are, however, based on prior art used in legacy systems and close examination of the components used would reveal more similarities than differences.

So, for example, the challenge today is to achieve higher data rates (capacity gain) by using higher frequencies in the twisted-pair local loop. The challenge in a legacy network has traditionally been to realise more voice capacity, for example on the coaxial circuits feeding the local loop. A 9.5-mm coaxial cable carrying 2700 voice channels would typically have a 6-dB per 2 kilometre propagation loss at 300 kHz and a 40-dB loss at 12.5 MHz. This required (still requires today in legacy networks) attenuation equalisation and phase equalisation.[19]

Attenuation equalisation is needed to avoid strong signals swamping weak signals. Phase equalisation is needed to mitigate the effects of group delay caused either by the propagation medium or by filters or repeater amplifiers. Amplifiers also of course introduce noise and create intermodulation. As with coaxial cable, the capacity–distance product for a twisted pair is a function of wire size and data rate. A DSL (digital subscriber line) example is shown in Table 2.8.

The length and the quality of the local loop determine the upper limit of the frequencies that can be used, which in turn increases the capacity. Distance can be increased by using repeater amplifiers.

Apparent capacity gain is also achieved by configuring the upstream/downstream to be asymmetric in the downstream direction, matching bandwidth availability to an assumed asymmetric traffic demand. Hence, the term asymmetric digital subscriber line.

The advantage of asymmetry was first observed by Joseph Leichleder, a scientist at the Bell Core laboratories in 1989 when working on options for delivering TV over the local loop

---

[19] A full treatment of these topics can be found in Newnes Telecommunications Pocket Book, by EA Edis and JE Varrall, ISBN 0-7506-0307-0, pages 29 to 31, 57, 58, 85 and 158.

Table 2.8 Wire size and distance

Wire size	Data rate	Distance
0.5 mm	1.5–2 Mbps	5.5 km
0.4 mm	1.5–2 Mbps	4.6 km
0.5 mm	6.1 Mbps	3.7 km
0.4 mm	6.1 Mbps	2.7 km

to compete with cable TV. HFC networks proved to be technically and commercially more efficient than ADSL and quickly became dominant in this application domain.

In retrospect this hardly mattered, because ADSL proved to be just what was needed for low-cost internet access. Trials in 1996 led on to implementation from 1998 onwards. The ADSL 1standard G992.1 ratified by the ITU was finalised in 1999 with the option of deploying on an ISDN line from 120 kHz upwards or a POTS[20] line from 25 kHz to allow for a microfilter to protect existing voice services between 300 Hz and 4 kHz (If implemented with a legacy voice service).

An ISDN 2B+D line supports two 64-k 'bearer voice channels and a 16-k data channel. The majority of ADSL implementation in a majority of countries is on the legacy POTS network.

The ADSL upstream bandwidth on a shared POTS line with legacy voice extends from 25.875 to 138 kHz, which is divided into discrete multitones (frequency subcarriers) set at a channel spacing of 4.3125 kHz. Each tone carries information bits from a bin. A bin, as the name implies, is a destination for data bits that are then mapped onto available subcarriers. The mapping can be selective in order to avoid impaired subcarriers or otherwise unavailable bandwidth. Upstream ADSL1 has up to 32 bins (32 times 4.3125 kHz = 138 kHz). Downstream bandwidth is from 138 kHz to 1104 kHz supporting 256 tones at 4.3125 kHz spacing mapped to 256 bins.

Headline data rates are 1.8 Mbits/s on the upstream and 12 Mbits/s on the downstream but the practical capacity/distance constraints reduce this to 8.448 Mbps at 9000 feet, 2.048 Mbps at 16 000 feet or 1.544 Mbps at 18 000 feet.

The ADSL 2 standard G992.3 was ratified in 2002 and is similar to ADSL 1 except that the downstream can be extended below 138 kHz to provide an enhanced upstream of 3.5 Mbps. The ADSL2+ standard G.992.5 was ratified in 2003 with upstream bandwidth extended to 12 MHz to support 512 subcarriers. The headline data rates are 2.3 Mbps for the upstream and 24 Mbps for the downstream. VDSL1 (very high bit rate digital subscriber line) was ratified in 2003 but the more comprehensive G993.2 VDSL2 standard was completed in 2006. The various iterations of both standards are described as profiles. VDSL1 was specified for 8- and 12-MHz operational bandwidths extended in VDSL to include frequencies up to 17 and 30 MHz. The profile number denotes the bandwidth and the letter denotes the power that is determined by whether or not the VDSL service overlaps other services. For example, a reduced power is needed in some circumstances to mitigate crosscoupling effects. The 30-MHz option also supports wider channel spacing, 8.625 kHz rather than 4.3125 kHz.

Table 2.9 lists the profile and power outputs and the subsequent table shows the ADSL and VDSL band plan.

[20] plain old telephone service.

Table 2.9 VDSL 2 profiles

Profile	8a	8b	8c	8d	12a	12b	17a	30a
Bandwidth (MHz)	8.832	8.832	8.5	8.832	12	12	17.664	30
Number of subcarriers (tones)	2048	2048	1972	2048	2783	2783	4096	3479
Carrier spacing	4.3125	4.3125	4.3125	4.3125	4.3125	4.3125	4.3125	8.625
Line power dBm	+17.5	+20.5	+11.5	+14.5	+14.5	+14.5	+14.5	+14.5

Higher power, for example over +20 dBm, will increase power consumption as the line driver will require a higher bias current to deliver the higher voltages needed on the line. The higher power outputs also require more linearity, which is harder to achieve over wider operational bandwidths. For example, the 8b option above with +20.5 dBm of power is downstream limited to 8.5 MHz operational bandwidth. The wider bandwidths will require higher sampling rates, which create DAC/ADC performance challenges that will be compounded by the additional clock cycles needed for the IFFT/FFT transform as the number of tones increase.

But the real problem will be the crosstalk created as new high-bandwidth services are introduced that have to coshare with legacy services. The potential for crossinterference is highlighted in Table 2.10.

There are effectively two constraint mechanisms – the fundamental propagation constraints of the medium and the unwanted crosstalk between services cosharing the medium.

Note that the downstream data rates for VDSL derive from submultiples of SONET (synchronous optical network) and SDH (synchronous digital hierarchy) data rates that presumably would imply that 155.52 Mbps would be the next benchmark to aim for. VDSL is, however, presently configured as symmetric bandwidth.

These higher data rate channels are susceptible to channel time dispersion (CDT) and frequency-dependent noise (FDN) and will degrade the performance of other legacy services sharing the medium. There are, however, various techniques that can be applied to mitigate these effects. These can be illustrated by tracing through the processing steps in the transmit/receive chain.

In the transmitter an error-correction code such as Reed–Solomon is used to encode the high-speed data stream and then extracted into lower-speed data substreams that are then placed on each subcarrier using quadrature amplitude modulation. The number of carriers to be used is determined during bit loading and then transformed into a time domain signal using an inverse discrete Fourier transform (IDFT). A cyclic extension is added to mitigate channel dispersion effects.

In the receiver the time-domain signal is equalised and a discrete Fourier transform brings the signal back into the frequency domain. Frequency equalisation is then performed on each carrier and then demodulated and passed on to the error decoder.

Splitting the high data rate stream into lower bit rate streams reduces the problem caused by ISI and the combination of a time-domain equaliser and frequency-domain equaliser provides channel throughput gain with a relatively minimal processing overhead. Sub carriers with low signal-to-noise ratios can be avoided or bit loading can be reduced. Countries like Japan with legacy ISDN for example will end up not carrying ADSL or VDSL on lower-frequency subcarriers.

Table 2.10   Band plan for ADSL and VDSL[21]

Upstream										
Down-stream										
Voice	300 Hz to 4 kHz	Micro-filter								
ISDN	300 Hz to 125 kHz									
ADSL1 1999 standard		25.875 kHz to 138 kHz	138 kHz to 1104 kHz							
ADSL 2			As ADSL 1 but can be extended below 138 kHz to provide more upstream bandwidth							
ADSL2+	Extended to 12 MHz									
VDSL 1			138 kHz to 3.75 MHz	3.75 to 5.2 MHz	5.2 to 8.5 MHz	8.5 to 12 MHz	12 to 16.7 MHz	16.7 to 17.6 MHz		
VDSL2		25 kHz to 138 kHz	138 kHz to 3.75 MHz	3.7 MHz to 5.2 MHz	5.2 MHz to 8.5 MHz	8.5 MHz to 21.567 MHz				21.5 MHz to 30 MHz

There will also be interference between upstream and downstream bandwidth that needs to be accommodated with guard bands although in theory the subcarriers should be orthogonal to each other and therefore should not interfere with each other.

This, however, depends on the length of the loop. A connection from a home or business to a fibre node a few hundred metres away will behave differently to a medium or longer loop

---

[21] http://www.eetimes.com/design/communications-design/4009317/DMT-Solves-UTP-Challenges-in-VDSL-Designs/.

Table 2.11 Headline data rates and capacity/distance ratios

Standard	Headline data rates		Capacity/distance through 24 AWG (0.511 mm) twisted pair
	Upstream	Downstream	
ISDN	2B +D 2 by 64 kbps +16 K data		Compatible with voice local loop
ADSL1	**1.8 Mbps**	**12 Mbps**	1.544 Mbps at 18 000 feet
			2.048 Mbps at 16 000 feet
			8.448 Mbps at 9000 feet
ADSL2	**3.5 Mbps**	**12 Mbps**	
ADSL2+	**3.3 Mbps**	**24 Mbps**	
VDSL1	**2.3 Mbps**	**51.4 Mbps**	12.96 Mbps at 4500 feet
			25.82 Mbps at 3000 feet
			51.84 Mbps at 1000 feet
VDSL 2	**100 Mbps**	**100 Mbps**	As VDSL 1 though shorter distances for higher data rates

application of three or four kilo feet (1–1.2 km). The longer loop will require additional echo cancellation and time-domain equalisation and the higher frequencies in the loop will be more highly attenuated. This is shown in Table 2.11.

So there are two fundamental performance considerations. How does the individual point-to-point link behave and what does it do to other user's cosharing the same medium. Take for example a live bundle with a pair count of 25 twisted pairs. Once the first DSL circuit is added the amount of crosstalk will drastically increase and increase further with each additional DSL circuit. There may also be T1[22] and E1 signals on the line. The spectral density of these services may make them incompatible with DSL circuits.

This becomes problematic when cable bundles are shared between telecommunication companies and imply a need for some form of spectral policing that in turn implies a need to measure power spectral density across all utilised frequencies.

So in many respects copper is similar to wireless. As data rates have increased power spectral density has increased and this is increasingly causing user-to-user interference that has to be actively managed. The use of OFDM allows an individual user's bandwidth to be subdivided down into frequency subcarriers that are orthogonal to each other, or in other words theoretically at least do not interfere with each other. Each individual subcarrier has a relatively low symbol rate that minimises problems with intersymbol interference.

Different users are, however, not orthogonal to one another and the result is an increasing amount of crosstalk. This complicates access economics. Supplying a high data rate service to an individual user, for example a VDSL2 connection has a direct cost, for example a VDSL line card and modem. This cost can be easily accounted. However, implementing the service will have a direct impact on all other users' cosharing the delivery medium. This cost is not easily accountable and the users who suffer service degradation may be supported by another service provider.

---

[22] A T1 circuit in North America, Japan or Korea is a full duplex 1.544 Mbit/s circuit typically carrying 24 64-kbit channels. E1 is the equivalent used in the rest of the world with a line data rate of 2.048 Mbit/s downstream and 2.048 Mbit/s upstream split into 32 time slots.

2.12   The Copper Conundrum – the Disconnect between Competition Policy and Technical Reality

Thus, we have a conundrum. The copper access network has been progressively unbundled on the basis that allowing competing providers to share access bandwidth encourages market efficiency. Unfortunately, this has coincided with technology innovation that makes this cosharing less rather than more efficient unless firm spectral policies are developed and enforced.

The assumption is often made that OFDM provides additional user-to-user interference resilience. This is not actually the case. What actually happens is that the modem will avoid using subcarriers that are badly impaired. There is therefore a direct physical layer capacity cost suffered by other users whenever a new high data rate service is deployed in a local loop. This is physical layer contention loss expressed in the frequency domain. There will also be MAC layer contention loss expressed in the time domain, but that's a topic for Chapter 3.

In terms of access economics copper has the advantage over fibre in that it is usually already deployed and fully amortised. The problem is that the income from a user is not directly coupled to the cost that will be borne by other users in the form of a degraded service offering. The problem has been exacerbated by competition policy that has encouraged service unbundling in the belief that this will improve market efficiency and a relaxed regulatory environment that has failed to take into account potential spectral interference issues.

Thus, theoretical improvements in market efficiency are being dissipated by the technical inefficiency that is consequent on the competition policy that was introduced on the assumption that overall delivery efficiency would improve. This is not necessarily the case and a strong argument could be made that artificially created competition in the local loop will prove to be an expensive experiment that will fail in the longer term.

My good friend David Cooper of Hillebrand and Partners[23] reflected on the conundrum of unbundling the local loop, to the delight of free-market regulators but with adverse technical effects

He commented as follows

> At the micro economic level, entrepreneurs necessarily have to align economics with technology if they want to remain competitive.
>   Governments and regulators can afford to misalign economics and technology, and there is no control mechanism that limits this.

Perhaps there should be.

The problem is compounded when the misalignment is supported by economic theory that transparently fails to observe contrary evidence that market forces do not necessarily result in technical and/or commercial efficiency.

2.13   OFDM in Wireless – A Similar Story?

Nicholas Negroponte based his 'switch thesis' on the assumption that wireless, which in this context we are describing as a transmission at a radio frequency, is bandwidth limited. This is of course fundamentally untrue. Wireless systems are power limited but have almost limitless

---

[23] http://www.hillebrand-partners.com/.

spectrum available to them. For example, today we have wireless systems implemented at long wave, medium wave, short wave and UHF and microwave frequencies. WiFi chip sets are presently under development that have 2.4 GHz, 5 GHz and 60 GHz transceiver capability and all three frequencies (though particularly 60 GHz where there is an oxygen absorption peak) have high frequency reuse potential. Mobile broadband networks around the world have the best part of 1 GHz of bandwidth allocated to them, not all in the same place or at the same time admittedly and the bandwidth can be hard to access due to component limitations but essentially the amount of bandwidth available dwarfs the 30 MHz available to us in the copper local loop.

The assumed constraints of wireless in terms of free-space loss attenuation are also open to question. A microwave dish with a one-degree bandwidth communicating with another microwave dish with a one-degree bandwidth will have >50 dB of gain relative to an isotropic radiator. Antenna directional selectivity can therefore be translated directly into range gain. Similarly the use of multiple antennas and multiple paths between a transmitter and receiver can be translated into capacity gain or at least a gain in peak data rates for individual users. Wireless could be described as a guided media operating in free space.

Several fundamental trends have been ever present in radio-frequency transmission. A radio transceiver is composed of a set of building blocks each of which use a subset of passive and active components. A crystal or possibly in the future a MEMS device produces a frequency reference that is multiplied up and combined with a phase-locked loop to produce a frequency reference that can be changed and stable. On the transmit path a carrier can be produced and information placed on the carrier using a combination of phase, frequency and amplitude modulation. The signal is then amplified and filtered using surface acoustic wave or bulk acoustic resonator filters and then applied to an antenna. In a users device this signal can be in the order of 10 milliwatts (+10 dB), 100 milliwatts (+20 dB) or a watt (+30 dB) or more. You will probably remember that one watt is equal to 1 joule (J) of energy per second.

On the receive path the received signal will be several orders of magnitude smaller.

Table 2.12 shows some levels typically encountered in radio systems. A picowatt is one million millionth of a watt – no wonder we use a logarithmic notation to describe this stuff.

Table 2.12   Typical received signal levels

Received signal in picowatts[24]	dBm
1	−90
0.1	−100
0.01	−110
0.001	−120
0.0001	−130
0.00001	−140
0.000001	−150

---

[24] Conversion tables for dBM and a comprehensive treatment of RF transceiver design can be found in The Mobile Radio Servicing Handbook Belcher, Fitch, Ogley, Varrall, ISBN 0-434 92187-4, first published in 1989 by Heinemann Newnes.

To put this in context a modern mobile phone with a GPS receiver is expected to recover a signal below −150 dBm in immediate proximity to a transmit signal of the order of 27 or 28 dB, a dynamic range of nearly 180 dB. This signal has to be recovered and filtered and amplified through a low-noise amplifier without being compromised by oscillator noise or other sources of internal or external interference.

Over time there has been a steady improvement in component performance, a topic revisited in more detail in Chapter 5. This includes active device performance, amplifiers for example and passive devices, filters and discrete devices such as capacitors, inductors and resistors, the modest workhorses of every electrical system.

As individual component performance has improved together with packaging design it has been possible to extract more performance at circuit and system level in terms of stability, selectivity, stability and DC to RF and RF to DC conversion efficiency.

This has provided the basis for moving to higher frequencies. For example, the availability of low-cost high-quality FR4 printed circuit board material was an essential precondition for realising 800- and 900-MHz transceivers for first-generation mobile phones manufactured in the mid-1980s.

The move to higher frequencies allowed for more voice channels to be supported but also allowed wider channel spacing to be realised to support higher per user peak data rates.

In first-generation cellular systems channel spacing was either 30 kHz (US) or 25 kHz (Europe and rest of the world). Second-generation GSM systems relaxed the channel spacing to 200 kHz, which made network planning a bit easier and relaxed some of the RF component specifications in the RF front end of a mobile phone. The 200 kHz was then multiplexed in to 8 time slots with each time slot optimised for a 12 kbps voice path. Data was not a predominant design requirement but as demand increased, time slots were concatenated together to produce higher data per user data rates. Third-generation systems increased channel bandwidth either to 1.25 MHz (CDMA in the US) or 5 MHz (UMTS in Europe and Asia)

First-generation and third-generation networks achieve user-to-user and channel-to-channel selectivity in the frequency domain (5-MHz channels), time domain (time slots) and code domain. A full treatment of how code domain selectivity is achieved can be found in Chapter 4 of 3 G Handset and Hardware Design[25] and will not be repeated here. All we need to know is that as data rates increase code domain selectivity becomes progressively less efficient and achieves progressively less baseband gain.

LTE takes a different approach with channel bandwidths that scale from 1.4 MHz (as a multiple of the 200 kHz GSM channel raster) to 5, 10, 15 or 20 MHz channels. Advanced LTE will take this further still with channel bonding to give per user channel bandwidths of up to 100 MHz. As with ADSL and VDSL, higher per user data rates create intersymbol interference problems and time-domain equalisation becomes computationally expensive and introduces too much delay.

ISI is mitigated in LTE by dividing the carrier into orthogonal frequency subcarriers. As with VDSL, user bits can be mapped on to frequency subcarriers that have favourable channel conditions at particular moments in time.

[25] 3G Handset and Network Design, Belcher and Varrall, John Wiley and Son ISBN 0-471-22936-9, first published 2003.

We could at this point go in to an extensive explanation of how OFDM is implemented in modern mobile broadband systems, but this has already been done in very excellent depth by Harri Holma and Antti Toskala in LTE for UMTS Evolution.[26]

Rather than attempt to repeat something that has been already done very well we use the rest of this chapter to reflect on the general merits (and demits) of OFDM applied at RF frequencies and use WiFi rather than mobile broadband as an example.

The specifics of realising OFDM in a mobile broadband transceiver are addressed in Chapter 5. All we need to know at the moment is that the LTE transform takes a channel that can be anything between 1.4 and 20 MHz and produces subcarriers at 15 kHz that are grouped together as 12 subcarriers to create a resource block with a bandwidth of 180 kHz, very similar to GSMs carrier bandwidth.

The theory (and largely the practice) is that the high per user data rates that are achievable using wider channel bandwidths of 5 MHz or more can be delivered without high levels of ISI, which would require computationally intensive time-domain equalisation. The equalisation would also introduce additional delay.

In the earlier section on ADSL and VDSL we highlighted that higher per user data rates are achieved by adding an additional process into the transmit and receive path by moving from the frequency domain to the time domain and back again.

The combination of time-domain and frequency-domain processing at the system level is already widely implemented – so for example in GSM and third-generation networks gain is achieved both in the frequency and time domain.

A time domain to frequency domain transform in an OFDM system, however, has the express purpose of increasing per user data rates. To a lesser extent it should also increase average data rates for all users, but this depends on the efficiency and effectiveness of contention protocols and queuing and allocation algorithms (treated in Chapter 3).

Transforms are the process of changing something into something else. Transforms can be unidirectional and irreversible. Setting fire to something is an example of a unidirectional and irreversible transform. More usefully, transforms can be bi-directional and reversible. AC/DC and DC conversion in electrical engineering is one of the most ubiquitous examples of a day-to-day use of transforms (hence the use of the word transformer). Our prime interest though is in the use and application of transforms in communications engineering.

Transforms change the representation of a signal without changing or destroying its information content. They change the way the signal is viewed. The purpose of the transform is to make a signal easier to process. Typically, this means making the signal easier to compress and easier to send. 'Easier' implies an ability to process the signal at a lower power and/or to process the signal using less storage or (radio and network) transmission bandwidth.

2.13.1   Transforms in Historical Context

Transforms can be described and implemented mathematically and exploit techniques that we take for granted but are the result of several thousand years of mathematical progress.

Maths (applied maths) helps us to do things we want and need to do. The real-life signals that need to be transformed in communications all start out and end as analogue waveforms

---

[26] LTE for UMTS Evolution to LTE Advanced, Second Edition, John Wiley, ISBN 978-0-470-66000-3.

(sound waves, light waves and radio waves) so an understanding of waveform geometry as a specific part of mathematical theory is particularly important to us.

Information in radio systems is modulated onto radio waves. The radio waves are sinusoidal but carry information that is modulated by changing the phase, frequency and/or amplitude of the (composite) carrier. We express phase, amplitude and frequency in terms of radians (the speed and angle of rotation of the phase vector). These waveforms can be described mathematically and manipulated mathematically. Mathematical manipulation and the efficiency of the mathematical manipulation process are the basis for modern digital radio communication systems.

The history of maths and the contribution that individual mathematicians have made to waveform and transform theory goes something like this:

15 000 BC – the start of organised agriculture involves the counting of goats and other mobile objects of value. This is arithmetic, the science of numbers. Farmers need to work out the size of their fields. This is geometry – the science of shapes, from the Greek geo meaning earth and metro to measure. Then structures need to be built. This requires trigonometry – the science of angles. The amount of land and the fertility of the land available determines how many goats you can support.

2000 BC – the Egyptians produce a numeral system to support trade and agriculture

1900–1600 BC – the Babylonians introduce the base 60 system of applied mathematics still used today with 360 degrees in a circle and 60 minutes in an hour and 60 seconds in a minute.

640–546 BC – Thales of Miletus predicts the lunar eclipse of 585 BC and proves a number of properties of triangles including vertical angles being equal and an angle inscribed in a semicircle being a right angle.

572 BC – Pythagoras of Samos produces all those useful equations for triangles and circles and shows how complex shapes can be described in terms of simple straight lines and circles, an early example of a transform.

427–347 BC – Plato founds his academy in Athens (387 BC) and encourages mathematicians such as Theaetetus and Eudoxus, the exhaustion man, to work on geometry and proportional theory.

300 BC – Euclid produces 'the bible of mathematics', 13 books and 456 propositions on plane and solid geometry and number theory including prime numbers, the infinitude of the primes. Euclid's axioms include the concept of Euclidean distance, the straight-line distance between two signal vectors that still serves today as a starting point for many mathematicians. Note that 'distance' can be related to many performance metrics in present communication systems including system sensitivity and selectivity.

287–212 BC – Archimedes equates the area of a circle with a triangle and works out the value of Pi using inscribed and circumscribed polygons as well as that business with the bath.

A bit of a gap due to the Romans being more interested in building roads than maths.

430–501 AD – Tsu Ch'ung-chih – the famous Chinese mathematician defines Pi more accurately.

529 AD – the closing of the Alexandrian library marks a shift towards Arabia in terms of mathematical innovation, including the evolution of algebra, the art of calculating with unknown quantities represented by letters after the arabic al gebr/al jabr meaning to equalise.

900 AD – Al Khumar Rizmi works on algorithmic approaches to problem solving. The word algorithm is named after him and is used to describe the step-by-step iterative process for solving problems using a predetermined set of rules within a finite number of steps.

1452–1519 – Leonardo Da Vinci and the arrival of Renaissance man including;

1501–1576 – Gerolamo Cardamo of Milan works on the mathematics of probability and the application of negative numbers used today in IQ quadrature modulation.

1550–1617 – John Napier and Henry Briggs (1561–1631) work on logarithms.

1596–1650 – Rene Descartes – works on merging algebra and Euclidean geometry, giving us the Cartesian coordinates still used as the basis for modern constellation diagrams.

1601–1665 – Fermat and his theorem. Pierre De Fermat of Toulouse works on analytic geometry and number theory and differential calculus, determining the rate of change of a quantity. He produces irritating problems for other mathematicians such as why a cube cannot be divided into two cubes.

1642–1726 – Isaac Newton develops a geometric approach for calculating Pi as well as that business with the apple and gravity and his work on optical theory.

1707–1783 – Leonard Euler proves that complex numbers, the extension of a real number by the inclusion of an imaginary number, can be related directly to the real trigonometric functions of sines and cosines – the basis of the Fourier base and Fourier transform used in all present OFDM radio systems.

1768–1830 – Jean Baptiste Joseph Fourier narrowly avoids being guillotined in the French revolution and produces a paper in 1807 on the use of sinusoids to represent temperature distribution. He uses the technique in a study on heat propagation that finally gets published in 1822, the Theorie Analytique De La. Chaleur. In the meantime he claims that any continuous periodic signal can be represented as the sum of properly chosen sinusoidal waves. Challenged by Lagrange (1736–1813) and Laplace (1749–1827), Lagrange asserts that sinusoids cannot be used to describe signals with corners, for example square waves. Lagrange is (sort of) right but in practice over time it is shown that sinusoids can be used to the point where the difference between the two has zero energy (the Gibbs effect). This becomes the basis for translating signals from the time domain to the frequency domain and back again, taking simple elements in the frequency domain and combining them to create complex signals in the time domain, the inverse discrete Fourier transform used in OFDM signal synthesis and decomposing complex time-domain signals into simple frequency components, the discrete Fourier transform used in OFDM signal analysis. Unfortunately the process involves a lot of calculations and proves to be laborious. The theory is understood, the practice is impractical.

1777–1855 – Carl Friedrich Gauss develops the theory of complex numbers following on from Euler and works out simpler and faster ways of doing Fourier transforms. He forgets to tell anyone how it's done. His work on the fast Fourier transform is published after his death in neo-Latin and remains unread by anyone who could possibly understand it. He also produces a body of work on non-Euclidean geometry, triangles with more than 180 degrees in their angles, and work on statistics (Gaussian distribution) which was to prove fundamental to our later understanding of white noise and filter design.

1845–1918 – Georg Cantor works on exotic number theory and transcendental and irrational numbers. (A number is transcendental if it is not the solution of any polynomial equation with integer coefficients. A number is irrational if there is no limit to the number of its decimal places. Pi and Euler's constant are examples of irrational numbers though bizarrely when added together they become rational.)

1887–1920 – Srinivasa Ramanujan produces ever more accurate approximations of Pi.

1862–1943 – David Hilbert works on algebraic number theory and geometry including the functional analysis of space and distance, the concept of vectors as linear sequences of complex numbers and functions and related work on harmonic and functional analysis. Hilbert is rather overshadowed by Einstein (1879–1955) but his contributions to modern communications, signal processing and vector theory remain very significant. He has a transform named after him, the Hilbert transform used in wideband 90 degree phase shifts.

The above represents a selective but fundamental body of work that modern living mathematicians can draw on to produce transforms optimised to perform specific tasks.

In 1965, 110 years after the death of Carl Friedrich Gauss, JW Cooley at the IBM laboratories working with John W Tukey of Princetown University published a paper that showed how the Fourier series could be reduced recursively by decimating in time or frequency, an approach similar to Gauss whose work at the time was unknown to them. This became the basis for the fast Fourier transform and is often now cited as the genesis of modern digital signal processing.

The Cooley–Tukey algorithm inspired a whole generation of mathematicians to produce fast and faster implementations of the Fourier transform that trade accuracy against the amount of multiplication needed against the number of additions needed against the amount of memory space needed against the number of clock cycles needed against the processing delay and power limitations of the host device. Examples include the Guo and Burrus wavelet based FFT in 1996, the Edelman fast multipole method (based on the compressibility of the Fourier matrix rather than the compressibility of the data), and other options like the Good

Thomas algorithm, the Rader Brenner algorithm, the Bruun algorithm and the Winograd algorithm.

Some of the techniques (for example Guo and Burrus) draw on work done by Benoit Mandelbrot. Mandelbrot's work on fractal geometry has been translated into Basque, Brazilian, Bulgarian, Chinese, Czech, Italian, Portuguese, Rumanian and Spanish so is unlikely to share the fate of Heinrich Gauss's work on the FFT.

Fractals have particularly compelling advantages in image processing, particularly where natural objects (trees and leaves) need to be encoded. We revisit this topic when we look at fractal compression schemes as an extension of existing DCT-based image compression (Mandelbrot maths in mobile phones) in Chapter 9.

Benoit Mandelbrot's published work in 2004 with RL Hudson applied the concepts of fractal geometry and fractal transforms to the evaluation of risk in financial markets – the science of fiscal transforms. A profit and loss statement, balance sheets and share price movements are mathematical expressions and are therefore amenable to fiscal mathematical trend and pattern analysis. Fractal transforms are particularly good at exposing patterns and coupling effects in fiscal systems.

Back in the engineering rather than fiscal domain, fractal transforms are increasingly relevant to the modelling of offered traffic behaviour in multimedia networks. We revisit this topic in Chapter 12.

Like Cooley and Tukey, Mandelbrot's work continues to inspire mathematicians to produce faster and more effective transforms both for signal-processing and image-compression applications.

Note in passing how pure maths (maths for its own sake) and applied maths requires certain infrastructural conditions in order to flourish and deliver political, social and economic value. Plato's academy and the Alexandrian Library (387 BC to AD 529), the growth of Islam after 650 AD, the Gutenberg Press in 1450, the postal service that allowed Newton to correspond with his peers and more recently the internet are examples of changes that have facilitated mathematical progress.

2.13.2 *How Fourier Transforms Deliver Performance Differentiation*

Present radio systems including DAB/DVB-T and DRM, WiFi and LTE HSOPA all depend on the Fourier transform in order to support relatively high data rates.

As bit rates have increased over the radio physical layer it has become progressively more attractive to use orthogonal frequency division multiplexing (OFDM) to improve resilience to intersymbol interference. An early example in terms of a European specification is the DAB standard (1991) but OFDM is now a fundamental part of local area WiFi systems and used in LTE mobile broadband networks. Let us briefly review the differentiation achievable through optimised implementation of the transforms and their associated iterative algorithms.

Note that once upon a time many years ago I was advised by a publisher (not John Wiley) that every equation used in a book halved the number of books sold. Due to the fact that my understanding of higher level maths is at best tenuous I have enthusiastically embraced this advice ever since. So what follow is an explanation of the Fourier transform that eschews maths and/or diagrams (that I find difficult to draw and not helpful).

2.13.3 Characteristics of the Fourier Transform

Fourier's fundamental assertion that has made him famous through time is simply that any complex signal viewed in the frequency domain can be understood as a composite of a number of simple sinusoidal signals that have been added together.

To place this into our specific context of interest (OFDM), his theorem led to the understanding that a composite waveform can be described as a mix of sine and cosine waves (90 degrees apart) that also have specific phase and amplitude characteristics. More specifically, signal points in the frequency domain can be combined to create complex signals in the time domain (the inverse Fourier transform) and complex time-domain signals can be resolved back into specific signal points in the frequency domain (the Fourier transform).

If the process is applied to a discrete waveform sample (as in an OFDM signal burst of a defined and repetitive length), then it is described as a discrete Fourier transform (DFT) or inverse discrete Fourier transform (IDFT).

The overall purpose of the Fourier transform in the context of implementing an OFDM transmitter and receiver is to take a complex time-varying waveform and express it as a tractable mathematical formula. Tractable implies an ability to either compact the formula itself and/or the data used in the formula – the basis for the algorithms used in fast Fourier transforms.

An IDFT synthesis equation on the transmit path multiplies the discrete frequency domain by a sinusoid and sums over the appropriate time-domain section (a 3.2-microsecond burst in WiFi).

A DFT analysis equation on the receive path multiplies the time-domain signal by a sinusoid and sums over the appropriate time domain section (3.2-microsecond burst).

If it was a continuous signal (which it isn't) the transform would integrate rather than sum.

2.13.4 The IDFT and DFT as Used in a WiFi Transceiver

Take a WiFi transceiver as an example. In essence, a WiFi transceiver is based on a mix of seventeenth century (Cartesian) and nineteenth-century (Fourier) mathematical principles. The modulation can either be BPSK, QPSK, 16 QAM or 64 QAM. Using a 16-QAM signal as an example, 4 data bits are mapped onto a 16-QAM constellation (based on Cartesian coordinates).

The position of the symbol that relates specifically to the 4 bits is described as a complex number with two values, one of which describes the amplitude and the second of which describes the phase. The phase coordinate has a numerator j after it which can be $+j$ or $-j$. The j numerator denotes whether the symbol is mapped to a positive or negative phase shift. These two values are then put into a bin and the process is repeated 52 times and then another 12 times using zeros as a value to make up 64 complex number bins.

So, for example, in the 802.11a implementation of WiFi at 5 GHz the operational bandwidth can be anything between 100 and 455 MHz depending on the country. This compares with the 83 MHz allocated at 2.4 GHz for 802.11 b and g. This means that 802.11 a has anything up to 19 nonoverlapping channels of 20 MHz compared with the three nonoverlapping channels at 2.4 GHz.

Table 2.13   802.11 a data rates, modulation options and coding schemes

Data rate	6/9 Mbps	12/18 Mbps	24/36 Mbps	48/54 Mbps
Modulation	BPSK	QPSK	16 QAM	64 QAM
Coding	$\frac{1}{2}$, 2/3, 3/4			

The channel spacing of 20 MHz supports an occupied bandwidth (-3 dB bandwidth) of 16.6 MHz that is subdivided down into 64 subcarriers spaced 312.5 kHz apart. The 64 subcarriers include 48 carriers for data, 4 pilot carriers and 12 virtual subcarriers that perform a guard band function. The 52 'useful' subcarriers are equally spaced around the centre frequency of each channel. The four pilot signal carriers (-21, -7, 7 and 21) are used to model and track frequency offsets. The modulation options and associated gross data rates are shown in Table 2.13 – the coding rate options are also shown.[27]

A 54 Mbps OFDM carrier using 64 QAM has a symbol rate of 4 microseconds. The symbol length in single carrier 64 QAM would be 50 nanoseconds. The guard interval is 0.8 microseconds; the data burst is 3.2 microseconds long.

The job of the IDFT (inverse discrete Fourier transform) is to do the maths to calculate the sine and cosine waveform values needed to create the composite time-domain sine and cosine waveforms that will represent these 64 frequency subcarriers each of which has a specific phase and amplitude. The back end of the waveform is copied to the front in order to provide a guard period and the waveform is clocked out at 20 MHz to form the 3.2-microsecond modulated burst with the 0.8-microsecond guard period.

In the receiver, a DFT is applied to the composite time-domain waveform and (all being well), the waveform across the burst is resolved into 64 frequency subcarriers each with the phase and amplitude information needed to decide which 4 bits were originally modulated onto the signal. The process exploits a number of convenient properties of sinusoidal waveforms.

2.13.5   Properties of a Sinusoid

A cosine /sine function has a period of 360 degrees.

The cosine curve is the same shape as the sine curve but displaced by 90 degrees.

Providing the system is linear, a sinusoidal input results in a sinusoidal output.

The average of the values described by any true sinusoidal waveform is zero if averaged over an even number of cycles.

Multiplying $\cos(x)$ by $\sin(x)$ produces a third waveform that is smaller in amplitude and at a higher frequency, but that will sum to zero when averaged out.

The cosine (real) computation only measures the cosine component in the time series at a particular frequency. The sine (imaginary) computation only measures the sine component in the time series.

The amplitude and phase of the signal can change but the frequency and wave shape remain the same.

---

[27] A full treatment of modulation and error coding can be found in Data Over Radio, Varrall and Belcher, ISBN 0-930633-14-8, first published by Quantum Publishing 1992.

The periodic nature of sinusoids provides the basis for implementing orthogonal signals (signals that do not interfere with each other) in either the time or frequency domain.

Being sinusoidal, the subcarriers can be arranged such that the peak signal energy of one subcarrier coincides with the minimum energy in the adjacent subcarrier (a property well known to designers of FFSK modems in the 1980s). This property provides the basis for OFDM signal orthogonality.

If you want to find out whether a time series contains a sine or cosine component, create a cosine function at frequency F and a sine function at frequency F, multiply by the cosine function (which we will call real F (I) and multiply by the sine function, which we will call imaginary F (Q). Both should be nonzero values but any modulation applied will show up as sine and cosine offsets, or in other words, the IQ-modulated waveform.

2.13.6   The IDFT and DFT in Wide-Area OFDM (LTE)

Wide-area OFDM, of which LTE is one example, implies a specific need to have a relatively slow symbol rate with an extended guard band (to deal with the delay spread on the channel) combined with the ability to support high data rates.

Keeping the symbol rate down requires the implementation of a more complex FFT and/or higher-order modulation, for example the option of a 256-point FFT and/or 64 QAM. Note that these radio systems in common with WiFi are two way so require both an inverse discrete Fourier transform to achieve the waveform synthesis on the transmit path and the discrete Fourier transform to achieve signal analysis on the receive path.

As such, two-way OFDM radio systems are much harder to implement than receive-only systems (DAB/DVB broadcasting and DRM). As bit rates increase over time it will be necessary to implement higher-order FFTs. This will place a premium on the techniques needed to reduce the complexity/calculation overheads implicit in these schemes. We will need faster more efficient transforms. Note that these schemes all depend on either the compressibility of the transform itself or the compressibility of the data that in turn depends on the symmetry and essentially repetitive nature of sinusoids and their associated input number series.

Note also how compression techniques become more important as FFT complexity increases. A 32-point FFT is typically ten times faster than a standard FT. A 4096-point FFT (used for example in a DVB H receiver) is typically one thousand times faster.

2.13.7   The Impact of Transforms on DSP and Microcontroller
Architectures – Maths in the Microcontroller

We said earlier that the efficient implementation of fast Fourier transforms (and transforms in general) imply a trade off between multiplication, division, addition, subtraction, memory space, bus width, bit width and clock cycle bandwidth. These trade offs are in turn dependent on the architecture of the host device.

Devices using parallel processing (super scalar devices capable of operating multiple parallel instructions) have become increasingly useful for all sorts of tasks including the FFT, DCT and related filtering, convolution, correlation, decimation and interpolation routines.

Hardware optimisation includes the integration of hardware repeat loops for filtering and FFT routines and the task-specific optimisation of fixed-point, floating-point and saturation arithmetic – the maths of the microcontroller. Although FFT algorithms are a commodity item in most DSP libraries there are still plenty of future performance optimisation opportunities particularly as increasingly complex higher-order FFTs are introduced into next-generation radio systems.

2.13.8   How Fourier Transforms Add Value

In the context of OFDM, Fourier transforms add value by improving the performance of higher bit rate radio systems both in local area and wide-area radio networks. The 'cost' is additional clock cycle/processor bandwidth but this in turn helps sell silicon. The ownership of the intellectual value that has been built around the FFT, in particular the compression/discard algorithms that speed up the transform will also become increasingly significant as higher-order FFTs are deployed over the next three to five years.

The efficiency of these algorithms and their mapping to optimised DSP and microcontroller architectures will translate directly into mobile phone and radio system performance including power budgets and supportable bit rates.

At the beginning of this section we stated that wireless systems are not bandwidth limited but power limited. Data rates can be increased by increasing transmit power but this degrades receive sensitivity internally within the user's transceiver and externally increases the noise floor visible to all other users. And pragmatically we only have a finite amount of transmit power available in a portable/mobile device, which is determined by battery capacity.

2.13.9   Power Drain and Battery Capacity

Table 2.14 shows typical battery densities.

Lithium ion batteries using a liquid electrolyte deliver better energy density (120 Wh/kg) but also have a relatively high self-discharge rate. Lithium metal batteries, using a manganese compound, deliver about 140 Wh/kg and a low self-discharge rate, about 2% per month compared to 8% per month for lithium ion and 20% per month for lithium polymer.

Lithium thin film promises very high energy density by volume (1800 Wh/l). However, delivering good through-life performance remains a nontrivial task. These very high density batteries like to hold on to their power, they have high internal resistance. This is not inherently well suited to devices that have to produce energy in bursts on demand (we return to this topic again in Chapter 5).

Table 2.14   Battery density comparisons

	Ni-Cads	NiMh	Lithium Ion	Zinc Air	Lithium Thin Film
Wh/kg	60	90	120	210	600
Wh/l	175	235	280	210	1800

Additionally, most of the techniques required to achieve high per user peak data rates reduce DC to RF conversion efficiency because they require more linearity. This applies particularly to RF power amplifiers. LTE mitigates this to some extent by deploying a single-carrier implementation of OFDM on the uplink. This reduces the amplitude of the modulated waveform but requires a discrete Fourier transform and an inverse fast Fourier transform on the transmit path and a fast Fourier transform, a maximum-likelihood sequence equaliser and inverse discrete Fourier transform to be performed in the base station receive path. It moves the agony around rather than removing it.

2.14   Chapter Summary

All physical layer options have benefited from improved component performance combined with an increase in DSP capability.

This has allowed for a wider range of frequencies to be used more aggressively both in the optical and RF domain and over copper cable and twisted-pair networks and peak data rates have therefore largely increased in parallel for all delivery options.

Fibre clearly offers the most bandwidth but is only economic when that bandwidth can be shared amongst multiple users or if a user has relatively extreme bandwidth requirements and the willingness to pay for that bandwidth.

Copper still has a considerable way to go in terms of peak data rate capability and has the advantage of using a delivery medium whose cost has been thoroughly amortised.

Wireless at RF frequencies is not bandwidth limited and can potentially match cable and copper delivery bandwidth but only if DC to RF and RF to DC conversion efficiency can be improved as a route to realising acceptable data duty cycles (time between battery recharge). Joules per bit rather than bits per Hz will be the big challenge going forward.

The options for meeting this particular challenge are discussed in Chapter 5.

Copper access and radio access networks presently have a problem in that while peak per user data rates have increased significantly, this has to an extent been achieved by degrading the average throughput rates of other users. We explore this in the following chapter.

3

Interrelationship of the Physical Layer with Other Layers of the OSI Model

3.1 MAC Layer and Physical Layer Relationships

Having stated unequivocally in Chapter 2 that we would leave the mysteries of the LTE physical layer in the capable hands of Harri Holma and Antti Toskala it is necessary for us to take a look at the differences between guided and unguided media at the physical layer and how these differences impact on the MAC (medium access control) layer with LTE[1] as a particular and specific example.

The MAC layer, OSI Layer 2 as the name implies describes the functional protocols that manage ingress and egress of traffic across each of the physical layer (Layer 1) delivery options. In terms of LTE this requires us to look at the specific challenges of delivering mobile broadband access over a particularly wide range of operational and channel conditions. The offered traffic mix may also be highly variable.

The big difference between guided media, coaxial cable, copper twisted-pair and fibre and unguided media (free-space optics and RF) is that unguided media channel conditions change faster and are harder to predict, though they can be measured and managed. Signals will typically travel along multiple reflected paths that will produce differential time delays at the receiver. Phase reversals may also have occurred at each reflection point.

This can also be observed in point-to-point microwave links where ground proximity within what is known as the Fresnel zone causes phase cancellation at the receiver. Early long-distance microwave links over the sea for example exhibited phase cancellation when the tide reached a particular height.

Slow and fast fading and time dispersion are a dominant characteristic of mobile broadband connectivity. Time dispersion becomes a particular challenge at higher data rates.

---

[1] LTE stands for long-term evolution and is the generic term used by the industry to describe the next generation of OFDM-based cellular radio networks.

Making Telecoms Work: From Technical Innovation to Commercial Success, First Edition. Geoff Varrall.
© 2012 John Wiley & Sons, Ltd. Published 2012 by John Wiley & Sons, Ltd.

One of the narratives of this book is that our ability to process signals in both the frequency and time domain has translated/transformed this challenge into an opportunity by increasing the user experience value realisable from mobile broadband access faster than other access media. This is different from what has happened in the past. In the past, any increase in data rate in wireless systems has been matched by an equivalent increase in the data rates achievable over fibre, cable or copper. This is no longer the case. The business of mobile broadband is being transformed by transforms.

The extent of the transformation, however, is as yet unclear – this is work in progress. For example, we can show that present LTE devices can deliver high peak data rates on a single-user basis but power efficiency has not improved on a *pro rata* basis. Sending data faster may result in a system efficiency improvement but this is not the same as a power-efficiency improvement. Power efficiency for the single-user experience has to improve in order to realise added value. This could be achieved by increasing network density but this increases capital and operational cost, so the user will need to be charged more for access.

Similarly, the high per user data rates have an associated cost that is inflicted on other users. If the other users either do not notice or do not mind then that is fine. If they do notice and/or mind then this is effectively a loss of added value and an increase in opportunity cost. It is pointless increasing single-user value if multiple-user value decreases by an equivalent or greater amount

So there are two questions that we need to try and answer in this chapter.

*Can OFDM combined with enhanced functionality at the MAC layer and higher layers of the protocol stack deliver a significant improvement in system efficiency **and** power efficiency and can this in turn be translated into additional **single-user value**?* To do this we need to take a view of how transformative the transforms that we use are really going to be in terms of realisable single-user added value and how effectively contention and scheduling algorithms arbitrate between the single- and multiuser experience.

*Can OFDM combined with enhanced functionality at the MAC layer and higher layers of the protocol stack be translated into **multiple-user value**?* One argument made here is that OFDM systems are more resilient to user to user interference but this has yet to be proven in practice within fully loaded real-life networks. The real answer to this question as above is more likely to be found in contention and scheduling algorithms, both of which we explore.

3.2   OFDM and the Transformative Power of Transforms

In Chapter 2 we reviewed the Fourier transform and its role in processing OFDM signal waveforms. Essentially, the difference at the physical layer between 3G mobile broadband systems and LTE '4G' systems is that Fourier transforms have replaced Hadamard–Rademacher–Walsh transforms.

So the first thing we need to try and do is quantify the difference between Hadamard–Rademacher–Walsh transforms and Fourier transforms.[2]

---

[2] We should also be covering Hilbert transforms and their role in IQ modulation but for reasons of space and brevity we recommend that you download this excellent tutorial http://www.complextoreal.com/tcomplex.htm. A less comprehensive but broader treatment can be found in the August 2011 RTT Technology Topic Transforms that can be downloaded here http://www.rttonline.com/techtopics.html.

Fast Fourier transforms are already used in 3G systems and indeed in 2G systems at the application layer (layer 7) to source code voice, audio, image and video. Almost all compression algorithms use an FFT at some stage to achieve bandwidth efficiency, exploiting the ability of the Fourier transform to represent complex signals viewed in the frequency domain as a composite of simple sinusoidal signals.

In 4G radio systems we user Fourier transforms at the physical layer (layer1) to translate a symbol stream on to discrete frequency subcarriers. The effect is to provide additional frequency diversity and time diversity in that the symbols are further apart in time than the original bit stream. We do this so that we can support higher single-user data rates.

However, present cellular systems make wide use of the Hadamard transform both in the channel encoding/decoding process (convolution and block coding) and code division multiple access (CDMA). Channel coding provides time diversity. CDMA provides frequency diversity (by spreading the original data signal across a wider bandwidth). Does Fourier really deliver significant additional performance gain?

To answer this we need to consider the history of matrix maths.

3.2.1   The Origins of Matrix Maths and the Hadamard Transform

Matrix maths is the science of putting numbers in boxes or rather, arranging numbers in rows and columns (a horizontal and vertical matrix). The numbers can be binary or nonbinary.

The Sumerian abacus came into use around 2700 to 2300 BC providing a mechanism for tabling successive columns of successive orders of magnitude. The Babylonians in the third or fourth century BC used matrices on clay tablets, a forerunner of the counting tables used by the Romans. Counting tables were simply tables with a ridged edge. The tables contained sand and the sand would be divided up into squares to help in counting and calculation.

Matrix concepts were developed by the Chinese in the Han Dynasty between 200 and 100 BC to solve linear equations. This implied an understanding that matrices exhibit certain properties, later described as 'determinants' that are revealed when the numbers in the matrix are added or multiplied either vertically by row, horizontally or diagonally.

This understanding was documented in 'Nine Chapters on the Mathematic Art' put together around 100 AD and was the product of a period now often described as 'The First Golden Age of Chinese Mathematics' (the second golden age was in the thirteenth and fourteenth century).

Historically, it is important to realise that the development of matrix theory in China was contemporaneous with the work of Archimedes on curves and circles and Pi between 287 and 212 BC.

Thus, the origins of the Fourier transform and the Hadamard transform can both be traced back to the pre-Christian era. As with the science of geometry, it took over 1000 years before much else meaningful happened in matrix theory.

In 1683, the Japanese mathematician Takakazu Seki wrote the 'Method of solving dissimulated problems' that precisely described how the earlier Chinese matrix methods had been constructed.

At the same time, in Europe, Gottfried Leibniz was producing work on determinants that he called 'resultants'. This became the basis for a body of work to which mathematicians like Cramer, Maclaurin, Bezout and Laplace all contributed throughout the eighteenth century.

The term 'determinant' was used by Gauss in 1801 in his 'Disquisitiones arithmeticae' side by side with his work on the coefficients of quadratic forms in rectangular arrays and matrix multiplication. This became the basis for a body of work to which other mathematicians such as Cauchy (who proved that every real symmetric matrix is diagonisable), Sturm, Cayley and Eisenstein contributed in the first half of the nineteenth century.

However, it was JJ Sylvester who is first credited with using the term 'matrix' in 1850. Sylvester defined a matrix to be 'an oblong arrangement of terms which could be used to discover various 'determinants' from the square arrays contained within it.' This work was developed in his snappily titled 1867 paper 'Thoughts on Inverse Orthogonal Matrices, Simultaneous Sign-successions and Tesselated Pavements in two or more colours, with applications to Newton's Rule, Ornamental Tile Work and the Theory of Numbers'. This paper established a new level of understanding about pattern behaviour in matrices (using ornamental tiles as an example). In between times, Sylvester gave maths tuition to Florence Nightingale who in turn revolutionised health care through her understanding and application of the statistical analysis of illness and associated treatment options, yet another example of applied maths.

At which point we can introduce Jacques Hadamard. Jacques Hadamard and his family survived the Prussian siege of Paris in 1870 by eating elephant meat. Hadamard obtained his doctorate in 1892 with a thesis on analytic theory and related work on determinant equality, the property that allows matrices to be used as a reversible transform. Hadamard also produced pioneering work on boundary-value problems and functional analysis and is generally seen as 'the founding father' of modern coding theory.

Hadamard's work was developed by Hans Rademacher, particularly in the area of orthogonal functions now known as Rademacher functions that appeared in a paper published in 1922 and was the forerunner of pioneering work in analytic number theory. Hans Rademacher's work was contemporaneous with the work of Joseph Leonard Walsh known as 'Joe' to his friends. This included a publication in 1923 on orthogonal expansions, later called 'Walsh functions'. Joe Walsh became a full professor at Harvard in 1935 and produced pioneering work on the relationship of maths and discrete harmonic analysis.

3.2.2   The Hadamard Matrix and the Hadamard Transform

A transform changes something into something else. The process is most useful when it is reversible/bidirectional and the purpose is generally to make a particular process easier to achieve.

In our context of interest, we want to take a string of numbers and rearrange or redistribute the number string in rows and columns so that they are more easily processed. In other words, a Hadamard transform is a transform that exploits the properties of a Hadamard matrix in the same way that a Fourier transform exploits the properties of the Fourier number series (the ability to describe waveforms as a summation of sines and cosines).

3.2.3   The Properties of a Hadamard Matrix

Hadamard matrices possess a number of useful properties. Hadamard matrices are symmetric, which means that specific rows can be matched to specific columns. Hadamard matrices are orthogonal, which means that the binary product between any two rows equals zero. The binary product is simply the result of multiplying all the components of two vectors, in this case rows, together and adding the results. We will see why this is useful later.

This means that if you compare any two rows, they match in exactly $N/2$ places and differ in exactly $N/2$ places so the 'distance' between them is exactly $N/2$. We explain why 'distance' is useful later.

Exactly half of the places that match are +1s and the other half are −1s. Exactly half of the places that differ are (−1+1) pairs and exactly half are (+1–1) pairs (the symmetric properties of the matrix). You can turn a Hadamard matrix upside down (reverse the +1s and −1s) and it will still work.

The matrix has the property of sequency. The sequence number of each row is the product of the number of transitions from +1 to −1 in that row. A row's sequence number is called its sequency because it measures the number of zero crossings in a given interval. Each row has its own unique sequency value that is separate from its natural order (the row number).

The Hadamard transform can therefore be correctly described as a sequency transform that is directly analogous to describing the Fourier transform as a frequency transform. In other words, given that the rows in a Hadamard matrix are orthogonal, the Hadamard transform can be used to decompose any signal into its constituent Hadamard components. In this sense it works just like a Fourier transform but with the components based on sequency rather than frequency.

As with the Fourier transform, a number of these components can be discarded without destroying the original data so the Hadamard transform can be used for compression.

The Hadamard transform can also be used for error correction. A much-quoted example is the use of Hadamard transforms to code the pictures coming back from the visits to the moon in the 1960s and the Mariner and Voyager missions to Mars. The pictures were produced by taking three black and white pictures in turn through red, green and blue filters. Each picture was considered as a thousand by thousand matrix of black and white pixels and graded on a scale of 1–16 according to its greyness (white is 1, black is 16). These grades were then used to choose a codeword in an eight error correction code based on a Hadamard matrix of order 32. The codeword was transmitted to earth and then error corrected.

It was this practical experience with applied Hadamard transforms that led on to the use of Hadamard transforms in present-generation cellular systems including GSM and CDMA.

These use convolutional and block coding to increase the distance between a 0 and a 1 or rather a −1 and a +1. The process is described in more detail below. Note that channel coding is a distinct and separate though related process to code division multiple access. Both processes exploit the properties of the Hadamard matrix.

Channel coding produces 'coding gain' and code division multiple access produces 'spreading gain'. Coding gain can be of the order of 10 dB or so and spreading gain in the order of 20 dB for lower user data rates. Together they show the extent to which the Hadamard transform contributes to the link budget of legacy cellular radio systems. OFDMA systems have to improve on this benchmark on all performance metrics not purely on single per user peak data rates.

3.2.4 Differences Between the FFT and the FHT

The FFT is best at modelling curves and sinusoidal waveforms. The hardest curve to model with a Fourier transform is a step function, also known as a square wave, where the edges of the waveform exhibit a theoretically infinite number of sinusoids. In practice, these can be approximated but it is the 'Achilles heel' of the Fourier transform (often described as 'the Gibbs effect').

The FHT is best at capturing square waves. The hardest curve to model with a Hadamard transform is a basic sine/cosine curve. This is intuitively consistent with matrix theory – describing square waveforms by putting numbers into squares.

Hadamard transforms when implemented as fast Walsh Hadamard transforms use only additions and subtractions and are therefore computationally efficient.

Fourier transforms require many multiplications and are slow and expensive to execute. Fast Fourier transforms employ imperfect 'twiddle factors' so trade accuracy against complexity and 'convergence delay'.

The magnitude of an FFT is invariant to phase shifts in the signal. This is not true in the case of the FHT because a circular shift in one row of the Hadamard matrix does not leave it orthogonal in other rows. This is the Achilles heel of the FHT and is a weakness that underpins the ultimate limitations of CDMA in terms of error performance and susceptibility to AM/PM distortion.

However, with this proviso, the Hadamard transform has been and remains a fundamental part of the signal processing chain in present mobile phones both in terms of its application in discrete processes such as channel coding and code division multiplexing but also in a support role to other processes.

The fact that it is simpler to execute and has different but complementary properties makes it a useful companion to the FFT and the two processes working together provide the basis for future performance gain. Having mastered the theory, let's examine how the Hadamard transform is applied in present radio systems.

3.2.5   Some Naming Issues

For brevity and in due deference to Hans Rademacher and Mr Walsh, we shall describe these codes as Hadamard codes used as a Hadamard transform.

As with the Fourier transform, the Hadamard transform computation can be decimated to speed up the computation process in which case it is known as a fast Hadamard transform (FHT).

In some ways, the FHT is easier to implement computationally as it does not require the 'twiddle factors' implicit in the FFT. This is because the discrete FFT is approximating and then describing a composite sinusoidal waveform (hence the twiddle factors). The FHT is describing square waves and therefore does not need additional approximation correction factors. The fast Hadamard transform (FHT) is used on the receive path. The inverse fast Hadamard transform (IFHT) is used on the transmit path.

As we have said, the FHT has properties that are distinct and different from the FFT (the fast Fourier transform) but are also complementary.

The combination of the two techniques, the FHT and FFT together, deliver a number of specific performance advantages. These benefits should include cost reduction, improved coverage, improved capacity and more consistent and flexible radio access connectivity.

3.2.6   How the FHT and FFT Deliver Cost Reduction

In the 1970s there was a consensus that it was going to be easier (cheaper) to filter in the time domain rather than the frequency domain and by implication, to process and filter in the digital domain rather than the analogue domain.

This started the process whereby channel spacing in cellular systems has relaxed from the 25 or 30 kHz used in first-generation systems to the 200 kHz used in GSM to the 1.25 MHz or 5 MHz systems used in present 3G networks. The process is taken further, for example in WiFi systems (20 MHz) and 3G systems (1.25 or 5 MHz) or advanced 3G systems (10,15,20 or 100 MHz). The objective is to reduce the cost of RF (radio frequency) filtering both in the handset and the base station.

3.2.7 The Need to Deliver Cost Reduction and Better Performance

However, user expectations of performance increase over time. User data rates in first-generation analogue systems were typically 1200 or 2400 bits per second, GSM data rates are tens of kilobits, 3G data rates are (supposed to be) hundreds of kilobits and have to compete in the longer term with local area WiFi systems delivering tens of megabits and personal area systems delivering hundreds of megabits (potentially WiFi at 60 GHz).

The performance of a radio system can be measured in terms of the radio system's sensitivity, selectivity and stability.

Sensitivity is the ability of the radio system to extract a wanted signal from the noise floor. Improved sensitivity translates into improved range and/or an ability to support higher data rates. In terms of the user experience, *sensitivity equals coverage and capacity.*

Selectivity is the ability of the radio system to extract a wanted signal in the presence of unwanted signals from other users. As with sensitivity, improved selectivity translates into improved range and capacity. However, by relaxing the RF channel spacing over time, we have thrown away some of the selectivity inherent in narrowband radio systems so have the need to replicate this in some other way. In parallel, users expect to receive voice and nonvoice services in parallel so we have the need to support multiple data streams per user, which implies a need to provide additional per user channel to channel selectivity.

Stability is the ability of the radio system to perform consistently over temperature over time, which in turn is dependent on the short- and long-term accuracy of the frequency and time reference used in the transceiver. The move to higher frequencies in the microwave band has increased the need for a more accurate frequency reference, but this has been offset by the relaxation in RF channel spacing. The combination of higher data rates and the need to deliver improved sensitivity and selectivity in the baseband processing sections of the transceiver has increased the need for a more accurate (and potentially expensive) time reference. As we shall see later, this in a sense is the Achilles heel of present CDMA systems (their inability to scale to much higher data rates without an inconveniently accurate time reference). In Chapter 2 we showed how OFDM shifts some of the hard work involved here back into the frequency domain. In terms of the user experience, stability therefore translates directly into user data rates *and* the consistency of the user experience.

3.3 The Role of Binary Arithmetic in Achieving Sensitivity, Selectivity and Stability

In 1937, Claude Shannon's MIT thesis 'A symbolic analysis of relay and switching circuits' helped to establish the modern science of using binary arithmetic in wireless (and wireline) communications. The science was consolidated by Richard Hamming in his work on error

Table 3.1 Coding distance – sensitivity

Coding distance – sensitivity
0–1

detection and correction codes (1950), digital filtering (1977), coding and information theory (1980), numerical analysis (1989) and probability (1991). Hamming formalised the concept of distance between binary numbers and binary number strings that is in practice the foundation of modern radio system design.

In digital radio systems, we take real-world analogue signals (voice, audio, video, image) and turn the analogue signals into a digital bit stream that is then mathematically manipulated to achieve the three 'wanted properties' – sensitivity, selectivity, stability.

Note that analogue comes from the Greek word meaning proportionate. Analogue implies that the output of a system should be directly proportionate (i.e. linear) to the input of the system and is continuously varying. To represent these waveforms digitally requires a sampling process that has to be sufficiently robust to ensure that analogue waveforms can be reconstructed accurately in the receiver. (Harry Nyquist 'Certain Factors affecting telegraph speed' 1924).

Taking this small but significant proviso into account binary numbers can be used to deliver sensitivity.

For example, considering Table 3.1, moving the 1 further away from a 0 implies an increase in *distance* which implies an increase in *sensitivity*.

The greater the *distance* between two strings of numbers (code streams in CDMA), the better the *selectivity* between users. The two codes shown in Table 3.2 differ in 10 places, which describes their 'hamming distance'[3] from each other.

If two code streams are identical (no distance between them) they can be used to lock on to each other, for example to provide a time reference from a base station to a handset or a handset to a base station. Longer strings of 0s and 1s will produce distinct spectral components in the frequency domain that can be used to provide a frequency reference. An example is given in Table 3.3.

We can also use binary numbers as a counting system. Interestingly, as we shall see later, if we start arranging 0s and 1s in a symmetric matrix of rows and columns, the binary product (sometimes known as the dot product) of the numbers in a column or row can be used to uniquely identify the position of that column or row. This is the property (described earlier) known as sequency and is the basis of many of the error coding and correction schemes used in present radio systems.

Table 3.2 Coding distance – selectivity

Coding distance – selectivity
01101011010010100
10011011101100010

---

[3] Named after Richard Hamming who published a seminal paper on error coding in 1920.

Table 3.3 Coding distance – stability

Coding distance – stability (code correlation)
01101011010010100
01101011010010100

3.3.1 Coding Distance and Bandwidth Gain

A first step to increasing the distance between a 0 and a 1 is to change a 0 into a -1 and a 1 into a $+1$. If we take either a -1 or a $+1$ and multiply by a series of -1s and $+1$s running at a faster rate then the bandwidth of the composite signal is expanded. The converse process applied in the receiver will take the composite 'wideband' signal and collapse the signal back to its original (data) bandwidth. This is the principle of spreading gain used in CDMA systems and is in many ways analogous to the bandwidth gain achieved in a wideband FM radio system.

3.3.2 An Example – The Barker Code Used in Legacy 802.11 b WiFi Systems

Basic 802.11 b WiFi systems provide an example. The original 802.11 b standard supports data rates of 1Mbit/ and 2 Mbits/s using either BPSK modulation (1 Mbit/s) or QPSK (2 Mbit/s). The data bits are multiplied with an 11-bit Barker sequence at a 1-MHz data rate that expands the data bandwidth of 2 MHz to an occupied channel bandwidth of 22 MHz, giving just over 10 dB of coding gain.

Barker sequences are named after RH Barker. The concept was introduced, in his 1953 paper on 'Group Synchronisation of Binary Digital Systems' read at the IEE in London. They were/are widely used in radar systems to help in distance estimation and were first used in low-cost commercial two-way radio systems in the first generation of digital cordless phones developed for the 902–908 MHz US ISM band. Note that when demodulated the codes can be sum error checked by converting to binary numbers as shown in Table 3.4.

The 11-bit Barker code used in 802.11 b is as follows shown in Table 3.5.

Table 3.4 Counting in binary

Counting in binary						
1	1	0	1	0	0	1
64	32	16	8	4	2	1

Table 3.5 Barker sequence

11-bit Barker sequence
$+1\ -1\ +1\ +1\ -1\ +1\ +1\ +1\ -1\ -1\ -1\ -1$

Table 3.6 Spreading/despreading codes

Input data bit	−1										
Spreading code	+1	−1	+1	+1	−1	+1	+1	+1	−1	−1	−1
Composite code	−1	+1	−1	−1	+1	−1	−1	−1	+1	+1	+1
Despreading code	+1	−1	+1	+1	−1	+1	+1	+1	−1	−1	−1
Output data bit	−1	−1	−1	−1	−1	−1	−1	−1	−1	−1	−1

If we take this spreading sequence and multiply it with an input data bit −1 and apply the rule that if the signs are different, the result is a −1, if the signs are the same the result is a +1 then we get the outputs shown in Table 3.6.

As you can see, the despreading code is the same as the spreading code.

Effectively we have answered the question 'is at a −1 or +1?' eleven times over and it is this that gives us the spreading gain.

3.3.3 Complementary Code Keying used in 802.11 b

However, if the WiFi data rate is increased to 11 Mbits per second, the spreading gain disappears.

In this case, the Barker code is replaced with 64 different 8-bit codes. The data bits are grouped into 6-bit symbols and each 6-bit symbol is mapped to one of the 64 codes. When the receiver demodulates the symbol bit stream, the 8 bits received should be one of the 64 8-bit code sequences which correspond to one of the 6-bit input data symbols. This is described as complementary code keying and is a good example of the use of sequency in the encode–decode process.

There is technically no spreading gain with this arrangement, though there is some (modest) coding gain due to the equality of distance between each of the 64 codes. The occupied bandwidth remains at 22 MHz.

The difficulty then arises as to how to manage higher user data rates. The answer with 802.11 b is to use an OFDM multiplex.

3.3.4 Walsh Codes used in IS95 CDMA/1EX EV

Present-generation wide-area cellular CDMA systems have to date not needed to support the higher data rates expected in local area systems, and for that reason have not to date needed to use an OFDM multiplex.

Code multiplexing and channel coding were therefore chosen to provide an acceptable compromise between implementation complexity and performance.

IS95 CDMA, the precursor of the CDMA2000 and 1XEV/DO system in use today, uses a 64 by 64 Hadamard matrix. This consists of 64 codes of length 64 of which code 0 is made up of all 1s and is used as a pilot and code 32 is made up of alternating 1s and 0s and is used for synchronisation. The other codes have their 0s and 1s, or rather −1s and +1s, arranged so that each of the codes is orthogonal to each other. Orthogonal in this context means that the codes are equally distinct from one another, or in other words do not interfere with each other as a product of the (FHT) transformation process. These codes are often referred to as

Walsh codes (named after Joseph Walsh) but are in practice based on the Hadamard matrix. Each code has 32 places where it is different from other codes. In other words each code has a Hamming distance of 32 from other codes in the matrix.

In the uplink, every six information bits are mapped to one of the 64 bit rows of the Hadamard matrix. The 64 bits in the row are substituted for the original 6 bits and the 64 bits are modulated on to the radio carrier using QPSK modulation. This is an inverse fast Hadamard transform.

The base station applies a fast Hadamard transform on every 64 received bits. Ideally only one of the resultant FHT coefficients will be nonzero. The nonzero value determines the row number that in turn determines the 6 bits originally sent. In other words, the process exploits the property of 'sequency' implicit in the Hadamard matrix.

Elegantly, the IFHT/FHT delivers some useful spreading gain ($64/6 = 10.75$ dB). It is also error tolerant. Given that the Hamming distance between each Hadamard code is 32, up to 15 bits can be errored per block of 64 without corrupting the 64 bits actually sent.

The only slight snag is that all users are cosharing the 64 codes and have to be separated from each other by unique scrambling codes (-1s and $+1$s running at the same rate as the data). The spreading codes deliver sensitivity, the scrambling codes deliver selectivity, the pilot and synchronisation codes deliver stability.

In the downlink, each row in the Hadamard matrix can be used to carry a unique channel to a unique user. Theoretically, this means 62 channels per 1.25 MHz of channel bandwidth (taking out the pilot and synchronisation channel). Every single information bit is replaced with the entire 64 bits of the users code (a 64/1 expansion). A data rate of 19.2 kbps therefore is spread to an outbound data rate of 1.2888 Mbps occupying a 1.25-MHz channel. As with the uplink, a scrambling code that is unique to the base station is also applied to provide base station to base station selectivity (actually a single code 'pseudonoise' sequence off set in time for each base station).

Later evolutions of IS95 have increased the matrix to 128 rather than 64 but the same principles apply. Either way, the CDMA multiplexing or channel coding have proved to be an effective format for exploiting the properties of the Hadamard matrix to deliver beneficial performance gains over simpler radio systems.

3.3.5   OVSF Codes in W-CDMA

The orthogonal variable spreading factor codes used in W-CDMA were originally conceived as a reordering of the Walsh codes used in IS95 CDMA with the added twist that user data rates could be changed every 10 milliseconds with users being moved between different lengths of spreading code (hence the 'variable' description used).

This is shown in Table 3.7. The section in bold indicates that this is a tree-structured code. The codes to the right are longer copies of the codes to the left. The bold segment denotes a branch of the tree stretching from left to right. At SF4 (which means spreading factor 4) four users can be supported on each of 4 codes at, say a theoretical data rate of 960 kbits. The code tree then extends rightwards to SF256 (not shown for reasons of space and sanity) which would theoretically support 256 users each with a data rate of 15 kbits/s.

As the data rate changes, potentially every frame (every 10 milliseconds), users can be moved to the left or right of the code tree. However, if a user is at SF4, no users can be on

Table 3.7 OVSF Codes in W-CDMA

SF4	SF8	SF16
		+1+1+1+1+1+1+1+1+1+1+1+1+1+1+1+1
	+1+1+1+1+1+1+1+1	
		+1+1+1+1+1+1+1+1-1-1-1-1-1-1-1-1
+1+1+1+1		
		+1+1+1+1-1-1-1-1+1+1+1+1-1-1-1-1
	+1+1+1+1-1-1-1-1	
		+1+1+1+1-1-1-1-1-1-1-1-1+1+1+1+1
		+1+1-1-1+1+1-1-1+1+1-1-1+1+1-1-1
	+1+1-1-1+1+1-1-1	
		+1+1-1-1+1+1-1-1-1-1+1+1-1-1+1+1
+1+1-1-1		
		+1+1-1-1-1-1+1+1+1+1-1-1-1-1+1+1
	+1+1-1-1-1-1+1+1	
		+1+1-1-1-1-1+1+1-1-1+1+1+1+1-1-1
		+1-1+1-1+1-1+1-1+1-1+1-1+1-1+1-1
	+1-1+1-1+1-1+1-1	
		+1-1+1-1+1-1+1-1-1+1-1+1-1+1-1+1
+1-1+1-1		
		+1-1+1-1-1+1-1+1+1-1+1-1-1+1-1+1
	+1--1+1--1--1+1-1+1	
		+1-1+1-1-1+1-1+1-1+1-1+1+1-1+1-1
		+1-1-1+1+1-1-1+1+1-1-1+1+1-1-1+1
+1-1-1+1	+1-1-1+1+1-1-1+1	
		+1-1-1+1+1-1-1+1-1+1+1-1-1+1+1-1
		+1-1-1+1-1+1+1-1+1-1-1+1-1+1+1-1
	+1-1-1+1-1+1+1-1	
		+1-1-1+1-1+1+1-1-1+1+1-1+1-1-1+1

codes to the right on the same branch. Similarly, if you have two users at SF8 or 4 users at SF16 on the same branch no users to the right on the same branch can be supported and so on rightwards across the branch.

A user at SF4 will have minimal spreading gain. A user at SF256 will have maximum spreading gain with a difference of just over 20 dB between the two extremes. As you would expect, this means that as a user's data rate increases, the spreading gain decreases. The occupied bandwidth (5 MHz in this case) remains the same.

The spreading codes are used with scrambling codes (long codes) with the scrambling codes providing user to user/channel to channel selectivity on the uplink and base station to base station selectivity on the downlink. Additional short codes are used for uplink and downlink synchronisation.

This is the original *Release 99* WCDMA radio layer code scheme. It provides a significant amount of flexibility both in terms of being able to support a wide range of variable (and potentially fast changing) data rates per user and a significant amount of flexibility in being able to support multiple data streams per user.

It does, however, require careful implementation both in terms of code planning and power planning. Although the variable spreading factor codes are orthogonal (hence their name orthogonal spreading factor codes), this orthogonality can be lost if code power is not carefully balanced or the nonlinearities inherent both in the radio system and the channel are not managed aggressively.

If orthogonality is compromised, unwanted error energy is projected across code channels that will then suffer from an unacceptably high error vector magnitude that in turn compromises sensitivity and selectivity that in turn compromises coverage (range) and capacity.

The mechanism for power management is an outer and inner control loop. The inner control loop also known as fast power control runs at 1500 Hz and if correctly implemented can be effective but in its own right absorbs a significant percentage of the available signal energy (about 20%).

HSPA aims to simplify this process and reduce some of the power control signalling and energy overhead by only using the SF16 part of the code tree and dispensing with fast power control. However, as you might have noticed, this takes away one of the desired properties of the OVSF code tree, which is the ability to support lots of users each with multiple simultaneous channels each at a variable data rate. In other words, much of the multiplexing capability of the OVSF code tree disappears if only SF16 is used.

The answer used in HSPA is to have a high-speed data-shared channel (HS-DSCH) that can be shared by multiple users with a MAC-driven access control based on 2 millisecond (and later 0.5 millisecond frames) that is not dissimilar to the contention based MAC used in present WiFi systems.

The challenge here is that the shared channel requires a new high-speed shared physical control channel (HSDPCCH). This control channel has to carry the channel quality indication (CQI) messages and acknowledgement/negative acknowledgement (ACK/NACK) messages that the MAC needs to decide on admission control and other factors, such as coding overhead, choice of modulation and transmission time interval.

This signalling is discontinuous, but when present increases the peak to average ratio of the transmitted (composite) signal and can also encounter relative timing issues with the other (dedicated) control channels.

If the high-speed control channel is not correctly detected, no communication takes place, which is potentially a little hazardous. The peak to average ratio can be accommodated by backing off the PA, but this has an impact on coverage (range).

In a sense, HSPA has exchanged the code-planning and power-planning challenges inherent in Release 99 WCDMA with code-sharing and power-sharing issues. This means that the RF performance of the handset and base station remains as a critical component of overall system performance.

Although some of the functional complexity at the PHY (physical) radio level has been moved to the MAC (medium access control) level, the effectiveness and efficiency of the MAC is dependent on the careful measurement and interpretation of the CQI and ACK/NACK responses.

The 7-dB step change in power that occurs when the CQI and/or ACK/NACK signalling is transmitted can trigger AM/PM distortion. This may cause phase errors that in turn will compromise CQI measurements or disrupt the ACK/NACK signalling.

This will probably be the determining factor limiting coverage and will probably require some conservative cell-geometry factors (signal versus noise rise across the cell) in order to maintain the signalling path (without which nothing else happens).

The requirement for a more complex and flexible multiplex can be met either by having a rather overcomplex code structure and/or an overcomplex MAC. Either or both can be problematic both from a radio planning and/or a handset/base station design perspective.

3.3.6   CDMA/OFDM Hybrids as a Solution

This seems to imply that something else needs to be done to allow these wide-area radio systems to deliver data rates that meet user's likely future data-rate expectations.

The options are either to increase cell density and/or to increase the sensitivity, selectivity and stability of the handsets and base stations – preferably both.

In 1XEV, (including the most recent 1XEV-DO Revision A) handset enhancements are based on implementing receive diversity and advanced equalisation. Base-station enhancements include implementing 4-branch receive diversity (two pairs of crosspolarised spatially separated antennas) and pilot interference cancellation – a mechanism for getting unwanted signal energy out of the receive path.

A similar evolution path exists for HSPA. Advanced receiver techniques were first specified in Release 6 together with an enhanced uplink (HSUPA). Release 7 included standardised multiple antenna/MIMO proposals.

The downlink evolution road map for 1XEV and HSPA does, however, mean that we will have a combination of CDMA and OFDM in at least two mainstream wide-area cellular standards, sometimes generically described as 'Super 3G'.

It is therefore useful to have an understanding of how a combination of *CDMA and OFDM* will work in terms of *signal processing task partitioning*.

Table 3.8 shows the transforms, or rather inverse transforms used in the transmit path of a hybrid CDMA/OFDM transceiver.

The job of the inverse FFT in source coding is to take the composite time-domain waveform from the quantised voice, audio, image and video samples and to transform them to the frequency domain. The transform makes it easier to separate out the entropy (information) and redundancy in the source signal. The bandwidth of the signal is compressed. We cover this process in more detail in Chapter 9.

The bit streams representing the voice, audio, image or video samples are then channel coded using convolution and block encoding. This expands the occupied bandwidth but increases the distance between the information bits (-1s and $+1$s). This is an inverse Hadamard transform. The bit stream is then 'covered' with a spreading code and scrambling code. This is another inverse Hadamard transform. This further expands the occupied bandwidth.

The bit stream is then transformed again using an IFFT to distribute the data stream across discrete frequency subcarriers. Note that the IFFT is imposing a set of time-domain waveforms on a series of frequency subcarriers (lots of sine/cosine calculations). The number of points used in the I FFT/FFT and the characterisation of the IFFT/FFT determines the number of

Table 3.8   FHT/ FFT task partitioning in future radio systems

Transmit path			
Source coding	Channel coding	Channel Multiplexing, Orthogonal spreading codes and scrambling codes	Frequency multiplexing orthogonal frequency division multiple access
Voice, audio, image, video	Convolution and block coding	CDMA	OFDM
Inverse FFT	Inverse FHT	Inverse FHT	Inverse FFT

subcarriers and their spacing and is the basis of the 'scaleable bandwidth' delivered in an LTE system

In the receiver, the OFDM demultiplex (an FFT), recovers the wanted symbol energy from the discrete frequency subcarriers. The benefit here is that the symbol rate per subcarrier will be divided down by the number of subcarriers. The more subcarriers, the slower the symbol rate. This reduces intersymbol interference and makes the next stage of the process easier.

If this is a combined OFDM/CDMA system, this next stage involves the combining of the symbol stream (assuming this is a diversity or MIMO system) and the despreading and descrambling of the signal using an FHT. The result should be some useful diversity gain (courtesy of the multiple receive paths), some spreading gain (courtesy of the spreading/despreading codes) and additional selectivity (courtesy of the scrambling codes). This should make the next stage of the process easier,

This next stage is channel decoding, usually or often implemented as a turbo coder (two convolution encode/decode paths running in parallel). The FHT produces some additional 'distance' that translates into coding gain.

And finally, the bit stream now finally recovered from the symbol stream is delivered to the source decoder where an FFT is applied (frequency-domain to time-domain transform) to recover or rather reconstruct or synthesise the original analogue waveform.

3.4   Summary

We have studied the properties of the Hadamard transform and its practical application in present CDMA cellular systems and early generation WiFi systems. We have showed how the IFHT/FHT is used in the code division multiplex/demultiplex and in channel encoding/decoding to deliver 'distance' which can be translated into coverage and capacity gain. We have reviewed how the IFFT/FFT is used to add in an OFDM multiplex to slow down the channel symbol rate as a (potentially) power-efficient mechanism for accommodating higher user data rates.

Note that all these processes are absorbing processor clock cycles with the express purpose of achieving a net gain in system performance that can be translated into supporting higher data rates at lower power (the end-user benefit).

The challenge is to deliver these benefits consistently given that user expectations increase over time and that physics tends to dictate the limits of any process that we are using to realise performance gain.

In the context of HSPA, it is a real challenge to implement the PHY and MAC sufficiently robustly to provide a realistic chance of achieving the theoretical gains. The same implementation issues apply to future iterations of 1XEV/DO. In practice, it always tends to be a combination of techniques that together deliver performance improvements that are translatable into a better more consistent single-user experience. But now we need to establish how or whether this translates into an experience gain for multiple users.

3.5   Contention Algorithms

The well-established principle used in wire line and wireless systems is that a bidirectional communications channel is established between two communicating entities, people or machines for example.

In a digital system, whether wireless or wire line, the channel can be a specific time slot within a frame and the time-domain space can be shared with other users. Multiple time slot channels can also carry multiple information streams between two discrete end points or multiple end points. These information streams as stated earlier may be variable rate or in other words asynchronous. They may also have varying degrees of asymmetry as a session progresses.

In many telecommunications networks, asynchronous and asymmetric traffic is mapped on to an ATM (asynchronous transfer mode) multiplex for at least part of the end-to-end journey. Time slots in an ATM network are allocated in units of ten milliseconds.

ATM and related traffic-shaping protocols are covered in substantial detail in Chapter 8 of '3G Handset and Network Design' so will not be repeated here. More usefully, we will review contention protocols used in WiFi and LTE.

As a reminder the LTE physical layer OFDM multiplex creates 15-kHz wide subcarriers that are grouped together as bundles of 12 subcarriers to make a 180-kHz resource block in the frequency domain that can then be resolved in the time domain in one-millisecond increments.

This allows the bit stream to be mapped to specific subcarriers that are experiencing favourable propagation conditions and/or to avoid subcarriers that are suffering from fast or slow fading. This increases throughput for a single user and theoretically at least should deliver a power efficiency gain as well. However, other users will potentially end up on a less favourable combination of subcarriers. The performance gain therefore comes with an opportunity cost attached for other users.

At the MAC layer contention algorithms have to arbitrate access between multiple users and manage at least four different traffic types, all with varying degrees of latency sensitivity. Voice, for example, is latency sensitive, best-effort data is not. Interactive and conversational traffic latency requirements fall between the two. Note that latency can and should be described as a first-order and second-order effect. The first-order effect is the amount of end-to-end delay. The second-order effect is how much that delay varies through a session. The theory is that the inherent variability of the channel itself is not noticeable to the user or at least not to an irritating degree.

In LTE, rather poetically, a stream of happy bits and sad bits are seeded into a best-effort data exchange. These bits describe the fullness of the buffer in the sending device. If the buffer is nearly full, sad bits will be sent to indicate that more bandwidth is needed in order to prevent buffer overflow with an implied potential loss of data. Happy bits suggest that some of the bandwidth presently allocated to that user could be reallocated to another user.

3.5.1   802.11 QOS and Prioritisation

LTE contention is therefore rather more proactive than present wire line or wireless systems in terms of real-time bandwidth reallocation, although work is still needed on realising power-efficient voice over IP.

Wire line ADSL provides a baseline model. With ADSL a number of users coshare line bandwidth with other users, the number being described as the contention ratio. Higher contention ratios will reduce single-user data rates but very high contention ratios will also reduce multiple-user throughput rates as there will be a substantial number of failed access and retry attempts that will translate into a loss of throughput. This can be both bandwidth inefficient and power inefficient.

Table 3.9   802.11 e QOS and prioritisation

Bandwidth management/Parametised QOS					
Traditional CSMA/CA DCF PCF/PC New					
EDCA	Extended data channel access Wireless media extension	Background	Best Effort (Replaces DCF)	Video TXOP Video transmission opportunity	Voice TXOP Voice transmission opportunity
HCCA	Hybrid Controlled Channel Access Point co-coordinator (PC) replaced with a hybrid coordinator (HC)	HCCA establishes 8 queues in the Access Point (AP) designated by their traffic specification			

ADSL is increasingly used for voice and real time video exchanges so the need to arbitrate between traffic with different latency requirements has increased over time.

WiFi contention protocols are similar to ADSL. The channel will be inherently less variable than wide-area LTE, though will still exhibit fast and slow fading and as with ADSL there is an end-user expectation that simultaneous voice and video calls can coexist with best-effort data both between two users and/or a multiple users including conversational conferencing.

802.11 has always traditionally been a connectionless contention-based access protocol. The traditional bandwidth allocation mechanisms, primarily the distributed coordination function and point coordination (PC) functions are not well adapted to and were never intended for time-bounded services such as real-time voice and/or video.

Table 3.9 shows the evolution of traffic shaping protocols developed by the work group under the task name 802.11 e.

The traditional bandwidth-allocation mechanisms used in 802.11 b (distributed coordination function and point coordination function using a point coordinator) are supplemented with two new protocols specified by the 802.11 e work groups based on traffic prioritisation (802.11 d). The new functions are EDCA – extended data channel access also known as the wireless media extension. This establishes four prioritisation levels, background, best effort (equivalent to existing DCF capabilities), video and voice. Video and voice data streams are given dedicated transmission opportunities known as TXOP. HCCA – hybrid controlled channel access, replaces or rather supplements the existing point coordination functions with the point coordinator replaced by a hybrid coordinator. The hybrid coordinator establishes eight queues in the access point ingress and egress ports, which can then treat individual data/traffic streams in accordance with their traffic specification (TSPEC).

3.5.2   802.11 f Handover

802.11 has always worked on the principle of 'break before make' rather than 'make before break'. Simply put, if acknowledgement packets stop coming back, the device will channel scan and look for another beacon. The time taken to do this is typically about 70 milliseconds.

If you were walking from access point to access point within an office using WiFi for voice this would be annoying. A fast-roaming study group is looking at reducing these roaming delays to less than 50 milliseconds. The question then arises as to the degree of mobility being supported and whether the base station/access point or the user's device should take the handover decision. Seamless 'make before break' handover protocols (used in cellular voice networks) imply substantial amounts of measurement reporting and a rework of the beacon structure (see 802.11 k below).

3.5.3   802.11 h Power Control

If measurement reporting is used then it makes sense to introduce power control. Power control improves the battery life/duty cycle of the user device and should generally help to reduce the noise floor, which in turn should deliver some capacity and coverage benefits. Power control, however, implies a rework of the beacon structure (see 802.11k below).

3.5.4   802.11 i Authentication and Encryption

802.11 i (ratified in June 2004) addresses the replacement of the existing (semisecure) authentication and encryption procedures known as wireline equivalent privacy (WEP) with WiFi protected access (WPA). This adds in the user authentication missing in WEP and makes it easier to implement SIM-based access – effectively bringing WiFi together with existing cellular authentication procedures. 802.i also describes a temporal key integrity protocol, a combination of WPA and AES, the American encryption standard for streamed media. The challenge here is to keep the configuration simple and to minimise any impact on header overheads and end-to-end latency budgets.

3.5.5   802.11 j Interworking

Originally established to address issues of 802.11a and Hiperlan interworking, additional work items include handover between 802.11 b, g and a and in the longer term, handover between WiFi and cellular (or alternative 802.16 /802.20 wide-area systems.)

3.5.6   802.11 k Measurement Reporting

802.11k measurement reporting introduces many of the techniques presently used in cellular (including GSM MAHO mobile-assisted handoff). Measurements collected and sent to the MIB Management Information Base) would include data rate, BER, SNR and a neighbour graph. One proposal is to use beacon compression to take out redundant information in persistent sessions and therefore release beacon bandwidth for measurement reporting. This would be known as a radio resource management beacon (RRM beacon).

3.5.7   802.11 n Stream Multiplexing

802.11n is intended as a protocol for managing multiple HDTV channel streams with additional space for simultaneous voice and data. The standard is going to mandate either the use of MIMO (multiple input/multiple output) techniques to get throughputs of up to 100 Mbit/s and/or the use of channel bonding. The headline data rate for two adjacent 'bonded' 40 MHz channels is 250 Mbit/s. The MAC overheads bring this down to about 175 Mbit/s.

3.5.8   802.11 s Mesh Networking

Finally (for the moment), 802.11s addresses mesh networking and *ad hoc* network protocols. This potentially brings WiFi into much more direct competition with Bluetooth-based personal area networks (PANS) and device access/device area networks (DANS). Mesh networking protocols may also facilitate a whole new generation of wearable WiFi products both for consumers and professional users.

3.6   The WiFi PHY and MAC Relationship

As a reminder, an 802.11 b and g access point will typically be configured in the 2.4-GHz ISM band to support 3 nonoverlapping 20 MHz radio channels with centre frequencies spaced either 30 MHz apart (Europe, Channels 1, 7 and 13) or 25 MHz apart (channels 1, 6 and 11 in the US).

Access points using the 5-GHz ISM band (802.11 a) will typically be configured to support up to 8 channels spaced 20 MHz apart (Europe Band 1 and US Band 1 and 2), with additional channels available in Band 2 in Europe and Band 3 in the US. Raw data rates of 54 Mbits/second are achievable in strong C/I conditions using high-level modulation and an OFDM multiplex to improve channel resilience.

Multiple channels capable of supporting multiple users cosharing a common 54 Mbps looks like a lot of bandwidth, but in practice there are significant MAC overheads and link budget constraints that result in substantially lower net throughput rates.

These MAC overheads increase when periodic two-way time-bounded services for example real-time voice and/or real-time voice and video need to be supported alongside best-effort services.

3.6.1   Voice and Video Frame Lengths and Arrival Rates

As shown in Table 3.10, voice frames are typically 92 bytes long and arrive every 20 milliseconds (the frame rate is determined by the syllabic rate). Video frames are 1464 bytes long and arrive every 40 milliseconds (assuming a 25 frame per second video rate). A 92-byte voice packet arriving every 20 milliseconds implies a voice data rate of 36.8 kbits/second ($92 \times 8 \times 50$). A 1464-byte video packet arriving every 40 milliseconds implies a video data rate of 292.5 kbits/second.

Table 3.10   Voice frame lengths, video frame lengths and arrival rates

Voice frames	92 bytes	Every 20 milliseconds
Video frames	1464 bytes	Every 40 milliseconds (25 fps)
Data frames	1500 bytes	
Fast data frames	3000 bytes	

A combined voice and video call would have a combined data rate of 329 kbits per second. This is, however, the rate to support unidirectional voice and video. Although people do not (generally) speak at the same time, the MAC layer has to provision bidirectional periodic bandwidth (known as voice and video transmission opportunities) so the bandwidth occupancy is effectively doubled to 73.6 kbits/second to support a bidirectional voice call, 585 kbits per second to support two way video and 658 kbits per second to support two-way voice and video. This suggests a capacity of 13 voice channels per Mbit, 1.7 video channels per Mbit or 1.5 voice and video channels per Mbit (2-way video calls).

3.6.2   Data Throughput – Distance and MAC Overheads

Data throughput is dependent on the modulation used and channel coding. In 802.11 a and g, between 48 and 54 Mbps of gross data rate is available if 64 QAM is used in a lightly coded (3/4) channel, but this is dependent on having a strong C/I (carrier to interference ratio). As the C/I worsens, the gross data rate reduces to 23/36 Mbps (16 QAM), then 12–18 Mbps (QPSK), then 6–9 Mbps (BPSK) and the channel coding overhead increases from 3/4 to 1/2 (one error protection bit for each data bit). This is shown in Table 3.11.

Although there are more channels available in 802.11 a, the propagation loss is higher and the net throughput therefore falls faster as a function of distance (though this also means that access points can be positioned closer together so channel reuse can be a bit more aggressive). In 802.11g, a request to send (RTS) and clear to send (CTS) message is needed if bandwidth is coshared with an 802.11 b transmitter (known as working in mixed mode). This produces a significant decrease in real throughput. The effect of distance in 802.11a and the impact of RTS/CTS overhead in 802.11g when used with 802.11 b is shown in Table 3.12. Real throughput rates in 802.11g and 802.11a quickly fall to levels that are not much higher and sometimes lower than standard 802.11b.

RTS/CTS is a poll and response algorithm and therefore implies a scheduling delay in addition to introducing a significant protocol and time/bandwidth overhead.

Note that 'mixed mode' actually implies two sets of MAC overheads.

First, the way the contention MAC is managed in 802.11b is different from 802.11a and g. Both use time slot back off but 802.11 b uses a 20-microsecond slot width and a and g use 9 microseconds. If 11b devices are interoperating with 11g devices then the 20-microsecond slot length must be used. This means that contention overheads will be higher.

Similarly with 11b devices, there is a choice of a long 192-microsecond and/or short 96-microsecond preamble. The OFDM preamble is 20 microseconds. In mixed mode, either the long or short 11b preamble will need to be used to support 11b devices. This means that preamble overheads will be higher.

Table 3.11   Data rates, modulation and coding in 802.11 a and g

Data rate	6/9 Mbps	12/18 Mbps	24/36 Mbps	48/54 Mbps
Modulation	BPSK	QPSK	16 QAM	64 QAM
Coding	1/2 or 2/3 or 3/4			

Table 3.12 Effect of distance on 802.11 a, b and g throughput, effect of mixed mode b and g signalling overhead on 802.11g throughput

Distance(ft)	802.11b Mbps	802.11a Mbps	802.11 g only	802.11 g mixed mode
10	5.8	24.7	24.7	11.8
50	5.8	19.8	24.7	11.8
100	5.8	12.4	19.8	10.6
150	5.8	4.9	12.4	8.0
200	3.7	0	4.9	4.1
250	1.6	0	1.6	1.6
300	0.9	0	0.9	0.9

Secondly, mixed mode now also implies that the MAC will be simultaneously loaded with time-bounded (periodic) and best-effort traffic. This will have a significant impact on throughput and capacity.

Taking these MAC overheads into account, Table 3.13 shows typical throughputs for TCP/IP best-effort data and/or UDP throughput. The table is from an Atheros White Paper and includes their proprietary bonded (40 MHz) channel solution giving a max throughput of 108 Mbps. Note that time-bounded services would normally use UDP (unacknowledged datagram protocol) rather than TCP (with transmission retries).

SIP (session-initiation protocol) places an additional bandwidth overhead on the UDP throughput of approximately 8 kilobytes every time a voice, video or voice and video session starts or is modified.

3.6.3   Mobility, Handover and Power Control Overheads

The above also excludes any measurement and signalling overheads introduced by the need to support mobility. As long as a user stays in one place then these overheads can be avoided. This may be/probably is the case for Skype laptop users but may not/probably will not be the case for people expecting to use WiFi voice from their mobile broadband device either at a public access hot spot or on a corporate, SOHO or home wireless LAN.

Mobility overheads include the need to do measurement reporting (802.11k), the use of measurement reporting to manage handover algorithms (802.11f) and/or the use of measurement reporting to manage per packet power control (802.11h).

Most transceivers can now collect RSSI (received signal strength indication) at a per packet level. Given that the channel is reciprocal (same RF channel on the uplink and downlink) it is

Table 3.13 MAC overheads when supporting TCP or UDP

		Max TCP	Max UDP
802.11 b	11 Mbps	5.9	7.1
g plus b	54 Mbps	14.4	19.5
g only	54 Mbps	24.4	30.5
a	54 Mbps	24.4	30.5
a turbo	108 Mbps	42.9	54.8

Table 3.14 Power control dynamic range in 802.11.
Reproduced by permission of California EDU

Power in dBm	Power in milliwatts/microwatts
20	100 milliwatts
17	50 milliwatts
15	30 milliwatts
13	20 milliwatts
7	5 milliwatts
0	1 milliwatt
−10	100 microwatts

easier to do measurement reporting with WiFi than it is with cellular (which uses a different radio channel on the uplink and downlink each with different propagation properties).

However, it is all very well collecting this information but then you have to decide what to do with it. There has not been much point up to now in doing power control with best-effort data. If the RX level is good then you just send the data faster using higher-order modulation and/or with reduced channel coding.

Voice, video and/or voice and video combined are, however, different in that they occupy periodic bandwidth with typically a longer (more persistent) session length. If the user is close to the base station then it is worth reducing power to a level at which packet error rates can be kept above the coding threshold of the channel encoder and the error threshold of the source decoder. Reducing the TX power helps reduce battery drain on the uplink but also tends to improve receive sensitivity so the downlink usually benefits as well but care needs to be taken to make sure the power control doesn't take up more bandwidth and/or power than it saves.

The range of RSSI measurement in a 802.11 transceiver is typically 60 dB. The range of power control is typically either 20 or 30 dBm as shown in Table 3.14. This is less than you find in a wide-area radio interface (35 dB for EDGE, 80 dB for 1XEV or Rel 99 UMTS) but still potentially useful. The usefulness depends (as with handover algorithms) on how mobile the user is likely to be.

Table 3.15 is taken from a University research project on mobility thresholds.[4] It assumes two users a certain distance apart with one of the users walking away from the other user at 1.5 m (metres) per second. The closer the distance between the two transceivers the faster the rate of change in terms of required output power. For example, at a distance of 9 m, walking 2 m away (just over a second of elapsed time) results in a 2-dB power change. At a distance of 70 m, a distance of 20 m has to be covered before a 2-dB step change occurs (13 seconds of elapsed time). It is therefore plausible that a per packet power control algorithm could be deployed that could be reasonably stable when used in this type of 'gentle mobility' application and that could yield some worthwhile power savings and related link budget benefits. From a PA design perspective, however, it does mean the operating point of the amplifier will be constantly changing and this in itself has an impact on error vector magnitude and harmonics.

---

[4] http://www.cs.colorado.edu/department/publications/reports/docs/CU-CS-934-02.pdf.

Table 3.15   Mobility thresholds

Transmit power in intervals of 2 dBm	Distance (m)	Difference (m)
0–2	7	
2–4	9	2
4–6	10	1
6–8	21	11
8–10	26	5
10–12	36	10
12–14	46	10
14–16	70	24
16–18	90	20

The ability to power control handsets and access points also provides an opportunity to op-timise the radio system in terms of channel reuse and coverage. This will become increasingly important if voice users become a significant percentage of the offered traffic mix.

3.6.4   Impact of Receive Sensitivity on the Link Budget

The link budget is a composite of the TX power, TX accuracy (typical error vector magnitude for an 802.11g transmitter should be <1.5% and <2% for 802.11a, but is often much worse), path loss and sensitivity. Sensitivity is a function of data rate. With 802.11 a and g higher data rates are achieved by using higher-order modulation, every time the modulation state is doubled (for example from BPSK to QPSK), another 3 dB is needed on the link budget, moving from 16 QAM to 64 QAM implies a 6-dB increase. In practice, the fact that on the TX side, EVM tends to get worse with higher-order modulation means that the real implementation losses are higher. Table 3.16 shows some typical receive sensitivity figures (and some claimed sensitivity figures) at different data rates.

3.6.5   Linearity Requirements

Table 3.17 compares three generations of cellular transceiver with a WiFi (802.11 a and g) transceiver in terms of peak to average ratio (PAR), peak to mean ratio, whether the radio channels are full or half-duplex and the power-control dynamic range.

The use of OFDM in 802.11 a and g delivers some significant benefits in terms of channel resilience and ISI performance (a constant and relatively low symbol rate) but the cost is a substantial envelope variation on the composite modulated waveform (although the example

Table 3.16   Typical 802.11 a and g and b receive sensitivity

Data rate	54 Mbps (a and g)	11 Mbps (b)	5.5 Mbps (b)	2 Mbps (b)	1 Mbps (b)
Typical sensitivity (dBm)	−75 dBm	−85 dBm	−88 dBm	−89 dBm	−92 dBm
Claimed max sensitivity					−101 dBm

Table 3.17 Linearity comparisons between cellular and WiFi (OFDM)

Generation	System	PAR (dB)	PMR (dB)	Duplex	Power control (dB)
1G	AMPS	0	0	Full	25 dB
	ETACS	0	0	Full	25 dB
	J TACS	0	0	Full	25 dB
2G	GSM	0	0	Half	30 dB
	PDC	3–5	>10	Half	30 dB
	US TDMA	3–5	>10	Half	30 dB
3G	EDGE	>3	10	Half/Full	35 dB
	1XEV	>5	10	Full	80 dB
	HSPA	>5	>10	Full	80 dB (Rel99)
WiFi	OFDM	17 dB	>20 dB	Half	30 dB

of 20 dB peak to mean is a worst-case condition with all 52 subcarriers lining up over a symbol period). This requires additional linearity from the PA, which is difficult to realise in a power-efficient manner.

In contrast, Bluetooth 2.0 EDR (which uses GFSK, four-phase DQPSK or optionally eight-phase DPSK) is arguably more power efficient.

GSM, PDC, TDMA and EDGE are described as half-duplex in that they don't transmit and receive at the same time (except theoretically for EDGE Class 13 through 18) but they still have an RF duplex separation between transmit and receive that translates directly into an improved sensitivity figure.

WiFi is half-duplex in that it uses the same RF channel that is time division duplexed to separate the uplink and downlink. This means the sensitivity will always be less than a full duplexed cellular system using separate RF channels for the uplink and downlink.

This matters because sensitivity is part of the link budget and the link budget determines coverage (range) and capacity. On this basis it could be argued that WiFi is not particularly spectrally efficient. The additional linearity needed also means it is not particularly power efficient when compared to legacy cellular or Bluetooth systems.

3.6.6   WiFi's Spectral-Efficiency and Power-Efficiency Limitations

So the WiFi PHY is arguably less spectrally efficient and less power efficient than cellular and probably less spectrally efficient and certainly less power efficient than Bluetooth.

The WiFi contention optimised MAC when used for connection orientated time-bounded traffic is arguably less efficient than existing connection optimised MACS used in cellular and Bluetooth voice applications.

3.6.7   Why WiFi for IP Voice and Video?

IP voice, IP video and IP voice and video are all potentially supportable on WiFi radio systems but require careful implementation in terms of PHY management (channel reuse) and MAC management (the cosharing of common bandwidth between time-bounded voice and best-effort data).

Whether WiFi is efficient or not when compared to other options is to an extent irrelevant if costs are sufficiently low to drive adoption though an inefficient PHY and MAC will always have a cost in terms of additional battery drain.

The addition of OFDM increases processing overhead in the receive chain (the cost of the receiver FFT) and processing overhead in the TX chain (the inverse FFT). The additional linearity implied by the envelope of the composite waveform also reduces TX power efficiency when compared to other radio systems.

However, OFDM is really the only way to realise data rates in the region of tens of Mbps (direct sequence spread spectrum starts to run into jitter problems at these higher speeds and you need the OFDM multiplex to slow the symbol rate down in order to control ISI and increase multipath resilience).

In the longer term, WiFi with MIMO (multiple input multiple output) is one way of getting speeds of the order of 100 Mbps or more (the other way is to deploy in the 60-GHz band or to use channel bonding at 5 GHz.

This is one of those circular arguments. The WiFi PHY and MAC were never designed to support a significant mix of time-bounded services. It is reasonable to assume that IP voice, and in the longer term IP video and IP voice and video will become a progressively more important part of the offered traffic mix (and by implication a more important part of offered traffic value). This implies that handsets and access points will need to support higher data rates.

Higher data rates are achieved by implementing mixed mode 802.11 b and g, which implies additional contention overhead. The connection based nature of voice and voice and video calls also adds contention overhead and signalling bandwidth load.

The fact that users might expect to walk around when using WiFi IP voice implies the need to manage mobility, which implies the need to introduce network-assisted handover, which implies the need to implement RSSI or packet error measurements. If you are doing RSSI or packet error measurement you may as well implement per packet power control, which will help to improve capacity and coverage (by lowering the overall noise floor and by improving sensitivity in the receiver). This in turn will help reduce some of the power budget issues. The higher data rates are needed partly because time bounded services absorb bandwidth but also because time-bounded services are more expensive to deliver in terms of PHY and MAC utilisation.

This all has to make sense to the network operator. Probably the most significant shift here will be the inclusion of network operator-specific browsers in the next generation of SuperSIM smart cards. Rather like existing application layer WiFi browser products, these will identify preferred networks, which of course means preferred networks from a network operator perspective. At this point, WiFi becomes a profit opportunity not a threat. But also the proposition must make sense to the end user and this implies a need for power efficiency.

Table 3.18 shows how power consumption is distributed in a lap top between the computer functions and WiFi connectivity functions, in this case, 802.11g.

Table 3.18   Power-consumption comparisons

Lap Top	70%
WLAN Card	9%
WLAN Host	21%
Total	100%

Table 3.19   RF power budget

TX	2 watts
RX	0.9 watts
Listen	0.8 watts
Sleep	40 milliwatts

Note this includes the RF power budget and directly related IP uplink and downlink packet processing overheads.

Table 3.19 shows the RF power budget.

The RF power consumed including TX and RX efficiency loss is a function of the duty cycle

An example is given in Table 3.20 with a duty cycle that yields an average power consumption of 1.54 watts.

It might be considered that 1.54 watts is inconsequential, but in practice there is a further 2.85 watts of associated CPU power totalling 4.389 watts. 4 watts represents a 20 to 30% reduction in battery duty cycle, four hours is reduced to three hours.

Note that the typical peak power requirement is 20 watts for a lap top, 10 to 15 watts for a larger than smart-phone device, 5 to 10 watts for a smart phone and sub-5 watts for a basic phone. The RF power budget (including related processor overheads) of a smart phone is likely to be higher than 30% due to the additional signalling load incurred by mobile applications.

Table 3.21 shows typical WiFi throughput against received signal level. The table also shows an Atheros[5] implementation where higher throughputs have been achieved at these levels, the difference being the improved sensitivity in the Atheros front end. Table 3.22 shows the relationship between the link budget and the path loss which will determine the coverage radius from the base station.

There are several points to make here. Theoretically, in a noise-limited channel a 3-dB increase in power or the equivalent increase in sensitivity translates into the error rate dropping a thousand fold. However, as can be seen in the industry standard example, between

Table 3.20   Assumed uplink duty cycles

TCP Uplink power consumption	
Duty cycle	$0.6 \times$ TX $+ 2 \times$ listen $+ 0.2 \times$ receive $\times 0.2 \times$ sleep
Watts	$0.6 \times 2 + 0.2 \times 0.8 + 0.2 \times 0.9 \times 0.2 \times 0.04$
Total	1.54 watts

Table 3.21   Sensitivity gains

Received signal level dBm	−85	−88	−89	−92	−93	−101	−105
Industry norm	11 Mbps	5.5 Mbps	2 Mbps	1 Mbps			
Data rate achieved							
Atheros					6 Mbps	1 Mbps	250 kbps

---

[5] Atheros is now owned by Qualcomm.

Table 3.22 Path loss and distance

Path loss	10 m	20 m	30 m	40 m	50	60	70
dB	86	100	107	113	117	121	124

−88 and −89 dBm the packet error rate triggers a change of modulation scheme which reduces the throughput from 5.5 to 2 Mbps. As with LTE, an increase in sensitivity in the user device will delay the point at which additional error coding and/or a lower-order modulation is needed.

But one of the areas of particular interest for cellular operators is how LTE compares with WiFi both in terms of peak data rate and the throughput power budget for local connectivity (a few metres from the base station). On a like for like basis assuming similar TX power levels in a similar channel (20 MHz at 2.4 GHz or 20 MHz at 2.6 GHz) and a similar distance from the base station then the peak data rate and average data throughputs should be similar.

LTE replaces the OFDM used on the WiFi link with an OFDM-related modulation and multiplexing scheme described as single carrier frequency division multiple access (SC-FDMA). This reduces the envelope variation on the modulated waveform that at least in theory should allow the TX path to be marginally more efficient. However, it can be said that this 'improvement' is only from an onerous 12 dB or 16× to an only slightly less onerous 10 dB or 10×.

The relative performance will also depend on whether LTE is used in TDD or FDD mode and what band is used for local connectivity, but the assumption would be 2.6 GHz versus 2.4 GHz or 5 GHz for the WiFi connection. An alternative is to use LTE TDD devices that have been developed for the China market. However, if operators are aspiring to use LTE as an alternative to WiFi for example in femtocells then any noticeable performance advantage will probably need to be realised by improving sensitivity and selectivity in the user device.

From a user-experience perspective the throughput at 2.4 or 2.6 GHz will be very closely determined by distance. It can be seen that in these small-cell environments user-equipment sensitivity very directly determines the user experience in terms of distance from in this case the WiFi access point.

Note that in many technical papers written both about WiFi and wide-area mobile broadband, an assumption is made that high peak data rates equate to improved energy efficiency. This may be the case sometimes but not always and is dependent on how well the power down/power up algorithms are implemented in the user equipment (a function of software latency), hardware memory effects, for example in RF power amplifiers, the efficiency of the fast memory needed to cache short data bursts and the bandwidth and characteristics of the channel over which the data is transferred. The bandwidth characteristics of the channel will also determine the amount of gain realisable from MIMO operation.

Wider channel spacing, for example the 20 MHz used in WiFi and small-cell LTE deployments deliver high peak rates but the noise floor proportionately increases with bandwidth. In a large noise-limited cell it is quite possible that the average data throughput with narrower channel spacing, for example 3 or 5 MHz could be significantly higher than a 20-MHz channel used in the same propagation conditions and the megabyte per watt hour metric may be lower. For certain, the gain achievable from MIMO in larger cells will be marginal at best and may be negative in many channel conditions due to long multipath delay and/or low signal-to-noise ratio.

It is laudable and understandable that operators should expect and want LTE to be more effective and efficient than all other forms of connectivity in all operational conditions but whether this is an achievable or fiscally sensible short-term ambition must be open to question.

To get LTE to work as well or preferably better than WiFi for local connectivity implies a relatively aggressive adoption of MIMO techniques in LTE user equipment. The question then to answer is how often users will be close enough to an LTE base station to realise a real benefit from the composite MIMO LTE channel and how many of those users would actually be better served from a WiFi channel. The answer is probably not a lot or at least not enough to warrant the additional cost and complexity and weight and size and power drain added to some or most LTE devices.

For certain, it would seem that MIMO will deliver very marginal gains in larger-diameter cells, for example anything larger than a picocell or femtocell and may have a negative impact on throughput in many wide-area access channel conditions. In terms of band allocations this would suggest MIMO might be useful at 2.6 GHz and above but less useful in lower bands and more or less pointless at or below 1 GHz. This is just as well given the problems of getting enough volume and distance to achieve effective spatial antenna separation in low band device designs.

Additionally, there are proposals in LTE Advanced to bond channels together. This could either be adjacent channel bonding or bonding two channels from different bands, which would imply a need for yet another RX/TX processing path.

The rather pointless marketing pursuit of headline data rates is therefore resulting in the industry investing R and D dollars that from an end-user-experience perspective would be better being spent at least initially on optimising SISO performance within the existing band plans. In other words, present R and D priorities are probably disproportionate to the proportionate user experience gain that can be achieved. Inefficiently focused R and D spending has an associated opportunity cost (indirect cost). The direct and indirect costs of MIMO investment (and or work on channel bonding) have to be recovered from *all* LTE devices.

In the longer term, WiFi semiconductor vendors including Atheros are working on triband solutions that combine 2.4-GHz, 5-GHz and 60-GHz connectivity, suggesting that this may be the best option for delivering high and ultrahigh data rates in local area network environments.

Note that some parts of the industry still regard WiFi as being competitive rather than complementary to cellular connectivity. The development of recent applications where mobile broadband is used to provide connectivity to multiple WiFi devices and hybrid WiFi and mobile broadband integration in high-end automotive applications suggests this misconception will lessen over time.

In-car automotive applications are a form of machine to machine communication, albeit highly mobile. On a historic basis, automotive two-way area connectivity has been relatively narrowband and often predominantly downlink biased, GPS-based navigation being one example. Adding WiFi to the in-car environment could change this. If mobile broadband devices are integrated at the manufacturing stage then they can be reasonably easily performance optimised with efficient external antennas. They are also connected to a generously dimensioned 12-volt supply.

Devices introduced at a later stage may be much less efficient. A present example would be people using hand-held smart phones to do in car navigation. These devices can also be connected to a 12-volt supply typically via the cigarette lighter (who uses these any more for the purpose for which they were intended?). This solves the duty-cycle problem but the loading on the network can be substantial, a composite of signalling load to handle high mobility and the penetration loss into the car, particularly severe with tinted windows (of the order of 6 to 10 dB). Applications where maps are progressively downloaded into the device will be particularly bandwidth hungry.

This is an example where a particular LTE device used in a particular LTE application, which may not be the originally intended application, has the potential to inflict serious collateral damage on the network. The damage (opportunity cost to other users) will be disproportionate to the realisable subscriber and application value. If the device additionally has poor sensitivity and selectivity the collateral damage will be greater.

3.7   LTE Scheduling Gain

Scheduling gain is just one of many adaptive techniques used in any and all contemporary mobile broadband networks. These adaptive techniques can of course be used to compensate for poor channel conditions, which can be viewed as a proxy for insufficient network investment and/or poor user equipment sensitivity and selectivity, a proxy for a failure to invest in user equipment performance optimisation. This is the glass half-empty view of the world.

Alternatively, adaptive techniques can be seen as a mechanism for achieving performance extension based on network density that has been dimensioned to deliver specific user experience quality metrics across a wide range of loading conditions and user equipment where a compromise between cost and performance optimisation has been successfully achieved. This is the glass half-full view of the world. In practice, user device receive sensitivity and selectivity improves scheduler efficiency, which in turn increases the user data duty cycle.

Improving scheduler efficiency also improves e node B throughput that translates directly into lower energy cost per subscriber session supported and faster more responsive applications (improved application and task latency). This increases user value. In parallel, scheduler efficiency gain reduces network cost including energy cost. However, both the increase in user value and the decrease in network cost per subscriber served needs to be described and dimensioned.

There are three enhanced receiver options in the LTE standard:

- enhanced type 1 UEs (user equipment) with receive diversity;
- enhanced type 2 with chip-level equalisation;
- enhanced type 3 with adaptive receive diversity and chip level equalisation.

The spectral efficiency evolution for user equipment with two antennas is illustrated in Figure 3.1 showing the relative difference between HSPA Release 6, HSPA+ and LTE.

These examples show the gains that can be achieved from increasing the granularity of the scheduling.

The scheduling algorithms in an e node B will be different to the scheduling algorithms used in a microcell or macrocell. In larger cells round-trip signalling delay means that the shorter scheduling options (for example half millisecond) cannot be used. Also, in the lower

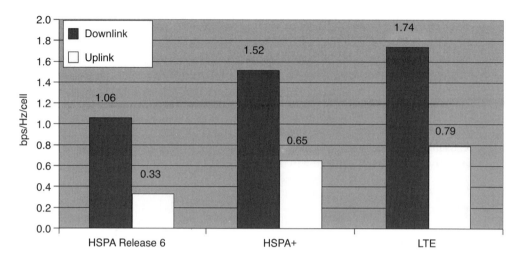

Figure 3.1 Evolution of spectral efficiency. Reproduced with permission of Nokia Siemens Networks.

bands it is unlikely that the 20-MHz channel bandwidths used to get high peak data rates in WiFi or higher-band LTE networks could be deployed. This is not a problem but a function of fundamental channel properties at different frequencies.

Note that channel-sensitive scheduling does not deliver gain in larger cells or for users moving at faster than walking pace. Fast-moving users become part of the multichannel averaging process as they move through the continuously changing multipath channel.

The point about LTE is that it has been designed to be flexible enough at the physical layer to scale across a wide range of channel bandwidths and respond to a wide range of channel conditions. The ability to support 1.4 MHz, 3 MHz and 5 MHz, 10, 15 and 20 MHz channel spacing is one example.

This should deliver substantial throughput gains. However, we have said that LTE must also deliver more efficient throughput in terms of joules per bit and/or megabytes per watt hour both at the e node B and the user device (to extend the user data duty cycle) and in small cells and large cells and for stationary users and highly mobile users (potentially up to 500 kph and certainly 350 kph).

We have stated that scheduling at the physical layer in small cells can be in one-millisecond increments and/or across different OFDMA subcarriers in the frequency domain and/or across spatial channels. Scheduler signalling load is localised to the e node B.

Physical layer scheduling can be either max CQI,[6] proportional fair or round robin. Max CQI scheduling allocates LTE resource blocks on the basis of measured channel quality. This is the most efficient scheduler from an operator/network perspective in terms of throughout against available bandwidth and power. However, edge of cell users would suffer very poor service. This would result in inconsistent user experience metrics that could trigger high product return and churn.

---

[6] Channel quality indicators.

Round-robin services all users and devices in turn. It delivers the best user experience and hence lowest product return and churn,[7] however, it is theoretically the least efficient option from an operator/network perspective (throughput against bandwidth and power). Proportional fair scheduling is a compromise point between these two extremes and can be set to meet specific user experience expectations.

Irrespective of where the scheduling compromise point is set, there is a directly beneficial relationship between user equipment performance and scheduler efficiency gain. Baseband interference cancellation and advanced receiver techniques will be particularly effective in interference-limited conditions; improved RF selectivity will also help. Improved RF sensitivity will be particularly helpful in noise-limited conditions.

Scheduling is also performed at packet level in order to differentiate best-effort traffic from streamed or interactive or conversational traffic and to respond to resource availability that in turn is a function of physical layer scheduling efficiency. Traditional handover to other channels and/or other bands is also supported providing opportunities to deliver macroscheduling gain.

This difference between micro- and macroscheduling gain is often underappreciated – both mechanisms work together to deliver efficiency gains, although as we are arguing these gains may be realised as spectral efficiency gain or network efficiency gain rather than per user power-efficiency gain.

LTE microscheduling and macroscheduling both rely on channel-feedback mechanisms (channel-quality indication and channel state indication) but are implemented at different time resolution, as small as a millisecond for microscheduling and typically a few seconds or several minutes for handover, depending on how fast the user is moving and the cell geometry and cell density. Packet scheduling (microscheduling) is realised in both the frequency and time domain. The CQI is computed from frequency-domain and time-domain reference symbols that are seeded into the channel data stream. The CQI measurements then determine whether QPSK, 16 QAM or 64 QAM modulation is used, the level of error coding, whether MIMO will be used and whether and how frequency selective scheduling is used. All of these will be influenced by how fast the user is moving. Frequency selective scheduling would typically be choosing say 25% of the subcarriers with the highest CQI.

In terms of e node B energy consumption, increased sensitivity in a user device operating in a noise-limited environment will result in a higher reported downlink CQI that will translate into a higher allocation of downlink time and frequency subcarrier resource blocks. A better link budget also translates into being able to support higher-order modulation and lighter error protection/correction overheads, which means that the ratio of information bits to error protection bits improves. Thus, any improvement in sensitivity will result in greater information throughput for a given amount of input energy. Note that channel coding and other adaptive physical layer mechanisms mitigate the impact of poor user equipment sensitivity or selectivity, but the bandwidth and power cost is effectively transferred to other users. The relationship of user equipment performance to overall network QOS and QOE is therefore direct and inescapable.

Given that the base station is sharing a finite amount of transmit power, for example 20 watts in a macro- or large microsite, then this means that more power is available to serve other user

---

[7] Churn is an industry term used to describe the migration of users from one service provider or network operator to another usually for reasons of poor service. If handset subsidies are available this can be a particularly expensive problem. Churn rates in some markets at some periods have been as high as 25% per year.

downlinks or put another way, the same amount of power is being used more efficiently across all users being serviced from the site.

A similar argument applies for devices with increased receive selectivity operating in an interference-limited environment.

On the uplink, user device TX efficiency will influence the user data duty cycle but TX linearity is equally important. The base station has to demodulate a signal modulated with noise- and phase-sensitive higher-order modulation. Much effort is expended in mitigating the impact of channel distortions on this complex signal but if the signal is distorted before it leaves the antenna of the user device there is not much that the base station can do with the received signal energy. This is the reason for the nearly $5\times$ more strict EVM LTE specifications compared to WCDMA.

As with RX sensitivity, error vector magnitude is a specified performance metric described precisely in the conformance standard and user devices would generally be designed and cost engineered to exceed this standard by an acceptable but not unnecessarily large margin. A lower EVM with additional margin over and above the conformance requirement would, however, translate into higher uplink data rates and more uplink capacity at the base station.

Improving the RF performance in the front end of the user's device reduces baseband coding overheads, which reduces baseband DC power drain. Applications will also run faster, which further reduces power drain and adds to user experience value. The benefit will accrue irrespective of where a user is in the cell though the maximum opportunity cost is likely to be incurred in edge of cell conditions. This can be said to generally apply in all macro-, micro-, pico- and femtocell topologies and also applies irrespective of whether the user is stationary or highly mobile.

3.7.1   LTE QOE and Compression

Some significant improvements are being made in compression techniques, particularly in digital TV where the DVB T2 standard is now complete and close to implementation.

However, digital TV is not power constrained and is essentially a one-way delivery medium. It also has a relatively generous first-order and second-order latency budget (first order is the amount of latency, second order is the variation between worst and best latency).

In a mobile broadband network, indeed in any two-way radio network, an increase in compression ratio means that more information can be loaded onto each bit sent. This means that the energy per bit relative to the noise floor improves. However, each bit is now carrying more information so if the bit is received in error proportionately more information is lost.

Even with all of the adaptive process in an LTE radio network, the radio channel is still an error-prone channel when compared to broadcast networks and the errors have a habit of occurring in bursts that can translate at higher layers of the protocol stack into packet errors and retries that are disruptive to compression techniques that rely on memory – the result is error extension.

Additionally, higher-order compression is computationally complex, power hungry and time hungry – the extra clock cycles needed to compress a signal in digital TV can be easily accommodated in the power and time budget both at the transmitter and the receiver (which is normally connected to the mains).

This is not the case in a mobile broadband network.

Doing higher-order compression on the user equipment TX path for instance will introduce compression delay, absorb energy and require performance parameters such as error vector magnitude to be more closely controlled (which in turn will have an impact on the user's energy budget).

Compression techniques have generally been developed either to maximise broadcast transmission efficiency (DVB T as an example) or memory bandwidth efficiency (MP3 and MP4). These efficiency gains do not translate comfortably into a mobile broadband channel. Improvements in compression techniques may have some beneficial impact on future spectral efficiency but these benefits are likely to be marginal rather than fundamental and less useful than RF performance and baseband performance gain.

3.7.2   MOS and QOS

Measuring the quality of voice codecs is done on a well-established mean opinion score (MOS) basis and similar processes are being agreed for video perceived quality benchmarking. Note that if actual video quality stays the same but audio quality improves the perceived video quality will improve.

Similarly, user experience opinion scores will generally be higher when the delivered quality is relatively constant and will be higher than a service that is sometimes good and sometimes bad. This has an impact on scheduling implementation but it can be generally stated that any improvements in user equipment RF and baseband performance will translate directly into a better and more consistent user experience MOS.

The three companies referenced[8] have web sites with information on voice and video MOS comparisons. Some additional standards work on application performance benchmarking may be useful in the longer term.

3.7.3   The Cost of Store and Forward (SMS QOS) and Buffering

Present and future mobile broadband networks have a mix of store and forward processes whose purpose is to reduce the cost of delivery both in terms of bandwidth and energy used. An example is SMS where in extreme circumstances several hours might elapse between a message being sent and received.

At the physical layer best-effort data will be buffered in the e node B or user's device to smooth out traffic loading (which reduces the offered traffic noise) and/or to take advantage of CQI-based scheduling gain. There is an alternative argument that says short term (buffering) or longer term (store and forward) has an associated memory cost that needs to be factored in to the efficiency equation – fast memory in particular incurs a capital and operational (energy) cost.

There is also a plausible and probably understated relationship between receive sensitivity and data throughput in which it could be shown that an extra 3 dB of sensitivity translates into a doubling of the downlink data rate in a noise-limited environment. This could also be equated to a doubling of the user data duty cycle or a halving of the user power budget for the

---

[8] http://www.opticom.de/.
http://psytechnics.net/.
http://www.radvision.com/.

same amount of data. Similarly, it can be argued that a 3-dB improvement in selectivity would translate into a doubling of the downlink data rate in an interference-limited environment.

Some RX RF performance improvements can deliver a double benefit. Achieving better/lower phase noise in a front-end PLL and LNA improves sensitivity, but also improves the accuracy with which antenna adaptive matching can be realised.

The impact of a link budget gain in a mobile broadband network is similar to but different from the impact of a link budget gain in a legacy voice and text network. Improving the link budget in a voice and text network improves capacity and coverage, depending on whether any particular link at any particular time is noise or interference limited. The user experiences better voice quality (though the operator can trade this against capacity gain by using a lower rate codec), improved geographic coverage and in building penetration better, fewer blocked calls, fewer dropped calls and a longer duty cycle (minutes of use between battery recharge).

Improving the link budget in a mobile broadband network increases coverage and capacity and average per user data throughout rates and improves end-to-end latency. This translates into reduced application latency for the user, lower blocked session rates, lower dropped session rates and a longer data duty cycle (megabytes of use between battery recharge).

Assuming a mobile broadband network will also be supporting voice and assuming voice traffic is on an IP virtual circuit then the link budget will need to be sufficient to ensure users have the same voice quality, coverage and talk time that they have on present devices.

The link budget can be improved by increasing network density, improving base-station efficiency up to an interference threshold, or by improving user-equipment efficiency or a mix of all three.

Increasing network density has a capital and operational cost implication in terms of site acquisition, site rental, site energy costs and backhaul. The increased burstiness of offered traffic in a mobile broadband network means that backhaul needs to be overprovisioned to cope with higher peaks versus the traffic average. Backhaul costs are therefore becoming more important over time.

Improving the link budget by improving user equipment performance does not have these associated capital or operational costs. Additionally, improving TX and RX efficiency in user equipment can be shown to deliver nonlinear gains in scheduler efficiency (microscheduling gain).

If RF and baseband performance improvement in user equipment can be combined with extended band flexibility then additional gain can be achieved by implementing more comprehensive interband interoperator handover. However, every additional band increases cost and decreases sensitivity and selectivity. Additionally, interband measurements are slow and absorb network signalling bandwidth and power.

Ideally, user equipment would be capable of accessing all available bandwidth in all countries with a parallel capability to scan the available bandwidth for the best-available connectivity. To be efficient this would need a separate receive function in the user's device (effectively a dual receiver). Such a device does not presently exist.

3.8 Chapter Summary

Using a combination of averaging, adaptation and opportunistic scheduling including in certain channel conditions spatial multiplexing and multiuser diversity, LTE delivers spectral efficiency benefits over and above HSPA Release 6 by roughly an order of three.

Power-efficiency gains are, however, presently proving to be more elusive both in WiFi and LTE devices and unless this is resolved user experience will be constrained.

OFDM in mobile broadband is allowing for high per user data rates.

Traffic-scheduling and queuing algorithms translate this into gains that can be enjoyed by multiple users cosharing the access resource, in this example a specific LTE channel resolved in both the time and frequency domain.

This is effectively algorithmic innovation at Layer 1 and Layer 2 and provides the basis for a plausible anticipation that mobile broadband will capture more access traffic from other guided and unguided access options than might have been originally expected with the only real caveat being power efficiency.

To an extent scheduling increases user device power efficiency but adequate gains are going to require a combination of algorithmic and materials innovation, a topic to which we return in later chapters.

4

Telecommunications Economies of Scale

4.1 Market Size and Projections

The starting point of this chapter is the remarkable rise of mobile connectivity relative to fixed connectivity as a percentage of the world market expressed in terms of global connections.

Figure 4.1 is sourced from our colleagues at The Mobile World, as are all the market and business statistics used in this book, and shows a crossover point in 2002 when mobile matched fixed for the first time and then a year-on-year transition to where we are today with an almost five-to-one relationship.

This of course has had a profound and largely positive impact on the part of the industry supply chain that services the mobile market, though we argued in our introduction that overspending on spectral investment has constrained R and D spending in some critical areas including some areas of RF component innovation that in turn has made it more difficult to get a return from that investment.

Table 4.1 provides a simplified overview of this supply chain.

The arrows denote risk and value distribution that can either be horizontal or vertical. Risk is a poison chalice that tends to get moved around. Sometimes some parts of the supply chain take the risk and others realise the value. The risk–reward equation, as we shall see, is influenced significantly by market volume. Figure 4.2 shows actual sales to 2010 and forecast sales through to 2013.

Specifically in this chapter and the next we set out to track the supply-chain changes that have occurred between 2007 and 2012, a period in which sales will have grown from 1.1 billion units per year to 1.7 billion units per year and project that forward to take a view on how the supply chain will look in the future.

All those unit sales suggest a lot of people have mobile phones and that a lot of people have multiple mobile phones which explains why many countries, Finland being one example, have penetration rates significantly higher than 100%.

Figures 4.3 and 4.4 below also show the mix of devices split as ordinary phones, iPhones and other smart phones, iPAD and similar tablet form factor devices and lap top devices.

Making Telecoms Work: From Technical Innovation to Commercial Success, First Edition. Geoff Varrall.
© 2012 John Wiley & Sons, Ltd. Published 2012 by John Wiley & Sons, Ltd.

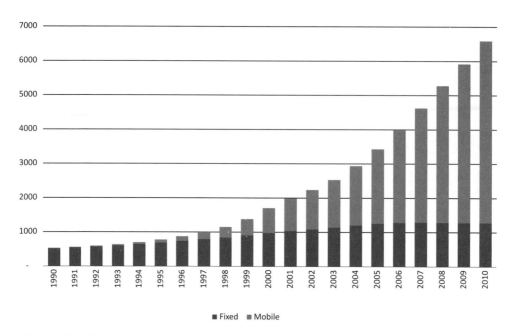

Figure 4.1 Global connections, 1990 to 2010. Reproduced with permission of The Mobile World.

The iPhone launched in 2007 still has a relatively small global market share but spawned a new product sector of enhanced functionality smart phones. This has had a major impact on service revenues and service and delivery costs. The launch of the iPad in 2010 appears to be creating another new sector that will likely have a similarly transformative impact that should be positive, provided delivery and support cost can be rigorously controlled. A topic to which we return on several occasions.

The trend clearly shows ordinary phones becoming a smaller part of the mix. Given that much of the volume growth is from developing economies with low GDP relative to developed economies this might seem surprising. It probably shouldn't be. People in

Table 4.1 Industry supply chain

Materials innovation	⇕	Baseband chip vendors				
Process innovation		RF IC vendors				
Packaging Innovation		RF component vendors	Device manufacturers	Network Operators	Service providers	End users
Manufacturing innovation		Display and peripheral component vendors				
Algorithmic innovation		Software vendors				

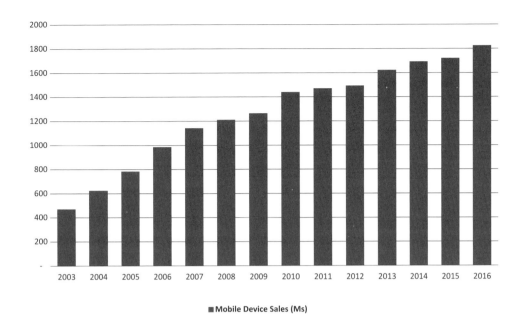

Figure 4.2 Mobile device sales, 2003–2016. Reproduced with permission of The Mobile World.

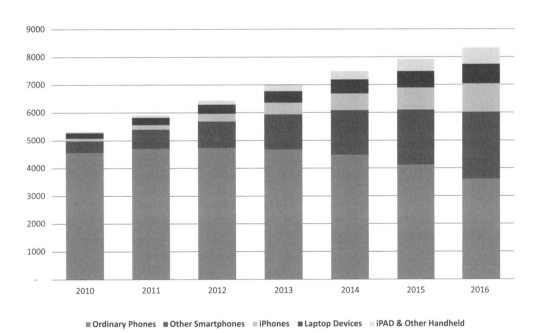

Figure 4.3 Global mobile device population 2010 to 2016. Reproduced with permission of The Mobile World.

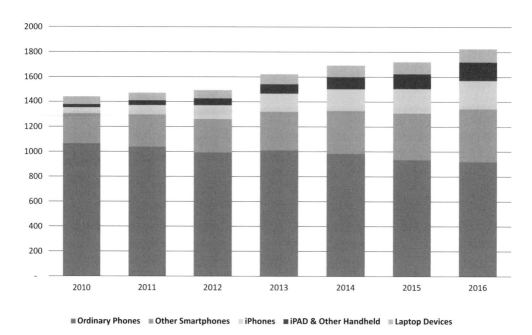

Figure 4.4 Mobile device sales by type 2010 to 2016. Reproduced with permission of The Mobile World.

fast-growing economies are willing to spend a larger percentage of their monthly income on mobile connectivity both for practical reasons and as a demonstrable token of ambition and/or success. Some developing countries also have high multiple-occupancy ratios, the number of people living in one household, which means that the disposable income available for this type of discretionary spending is higher than might be expected.

This shift in product mix is also having a significant impact on offered traffic. An ordinary phone supporting predominantly voice and text will typically generate an average of 30 megabytes or so of offered traffic per month. A smart phone will generate between 300 and 500 megabytes and a lap top with mobile broadband connectivity about 2 gigabytes. The result in many markets is that data volumes are presently quadrupling on an annual basis. Most forecasts, including our own, have been underestimating the rate of growth and present forecasts may also prove overconservative.

Figures 4.5 and 4.6 takes the unit volumes and offered traffic data points and calculates the offered traffic on a per device basis. Note that data volume does not necessarily translate into data value particularly in markets where 'all you can eat' tariffs are supported. This is because a high data user can inflict significant collateral damage to other users in the cell. This in turn creates an opportunity cost that in a percentage of cases will be higher than the income generated from the device. Traditionally users/subscribers have been analysed by network operators in terms of average revenues per user (ARPU) but increasingly now should be measured either in terms of AMPU (average margin per user) or more accurately AMPD (average margin per device).

The scale shift is, however, quite remarkable. An exabyte by the way is a million terabytes.

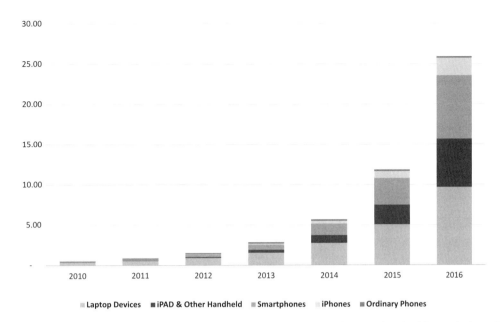

Figure 4.5 Mobile data volume per month (Exabytes) 2010 to 2016. Reproduced with permission of The Mobile World.

The impact on the relative revenue mix between voice and data is shown in Figure 4.7 with projections forward to 2016.

However, revenue is not the same as profitability. If a network is lightly loaded, incremental loading, as long as it has some revenue attached to it, will improve network investment return rates. If the network is heavily loaded and additional load requires additional capacity to be added, then return on investment is dependent on the investment per subscriber

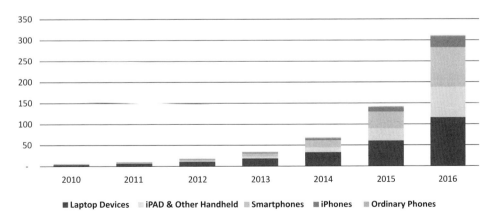

Figure 4.6 Annual mobile data traffic (Exabytes) 2010 to 2016. Reproduced with permission of The Mobile World.

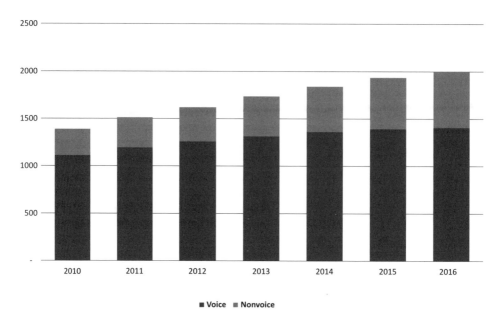

Figure 4.7 Mobile service revenue – 2010 to 2016. Reproduced with permission of The Mobile World.

per megabyte. This is not a metric that operators have previously needed to be concerned about.

Adding capacity may also imply an investment in new spectrum. The economics of this are complex as a return on investment is dependent on ensuring an adequate supply of performance competitive market competitive user devices are available. For reasons that we will analyse this is often much harder to achieve than one might expect and is a function of RF cost economics. RF cost economics are a function of the band plan but also the technology deployed in each supported band.

Figure 4.8 shows the absolute dominance of GSM on a global basis. This is effectively a technopoly – a market dominated by a legacy technology. This has advantages and disadvantages. If you don't happen to have GSM deployed in your network it is definitely a disadvantage as you will be exposed to an alternative technology that is subscale.

A design and manufacturing team will typically get a better return from servicing a large established mainstream market and certainly would be wary of the opportunity cost implicit in diverting resources away from what should be a dominant income stream.

But also there is a question as to how fast newer technologies are adopted. The adoption rate is driven by the scale difference in user experience between the old and new technology and the relative cost of delivery. To date neither the scale difference between 3 G and 2 G systems or the relative delivery economics or the user experience difference have been sufficient to drive a radically fast transition. As Figure 4.8 shows even taking WCDMA and CDMA together they still only represent 25% of the market ten years after market introduction.

We also need to take into account relative shifts in terms of regional importance and how these might affect future value ownership in the industry. In our introduction we highlighted the rapid rise of China and Asia Pacific in general as the globally dominant market for

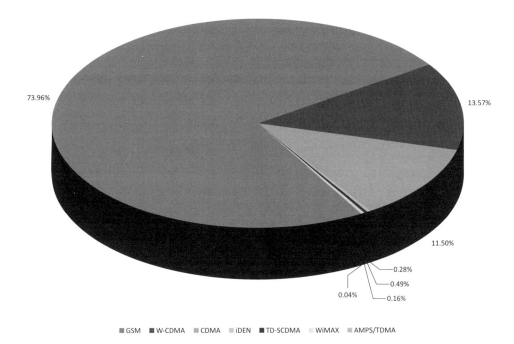

Figure 4.8 Mobile connections by technology Q1 2011. Reproduced with permission of The Mobile World.

telecommunication products and increasingly as a manufacturing and development centre of telecommunication products, both for its internal market and rest of the world markets.

Figure 4.9 tracks this trend over the five years from 2005 to Q1 2011 and shows the relative decline of the US as a market measured in terms of mobile connections – a remarkable change considering that 15 years earlier the US represented 70% of the global market by volume and value.

Figure 4.10 shows the relative market presence of the top twenty operators with China Mobile as the dominant player.

Figure 4.11 proportionate connections taking into account crossownership where operators have shareholdings in other operators.

Figure 4.12 shows China Mobile's dominance relative to other carriers in the China market. China Mobile acquired China Telecom's mobile businesses in four phases. China Unicom was formed by the merger of China United and Great Wall. China Telecom re-entered the mobile business in 2008 by acquiring Unicom's CDMA business. All three use different 3G technologies, a triumph of politics over technical common sense.

4.2 Market Dynamics

In this chapter we study the impact of these market dynamics on the supply chain, how economies of scale determine the allocation of R and D resource, related gravitational investment effects, the influence of standards making on this process, the influence of the

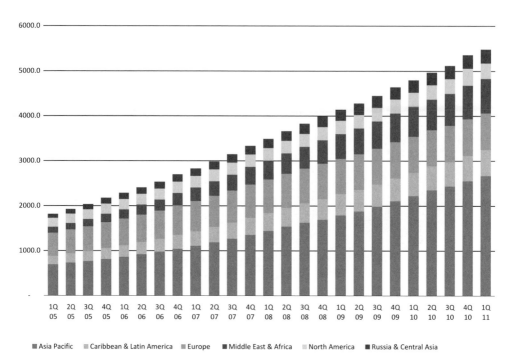

Figure 4.9 Mobile connections by region 2005 to Q1 2011. Reproduced with permission of The Mobile World.

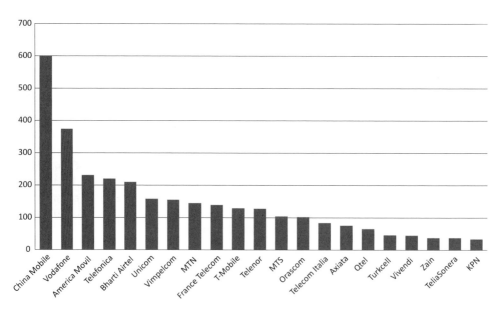

Figure 4.10 Leading mobile operators controlled connections, Q1 2011. Reproduced with permission of The Mobile World.

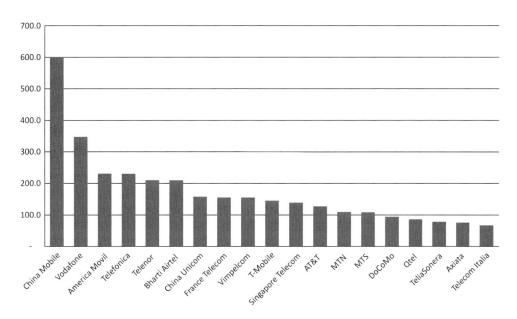

Figure 4.11   Leading multinational mobile operators proportionate connections, Q1 2011. Reproduced with permission of The Mobile World.

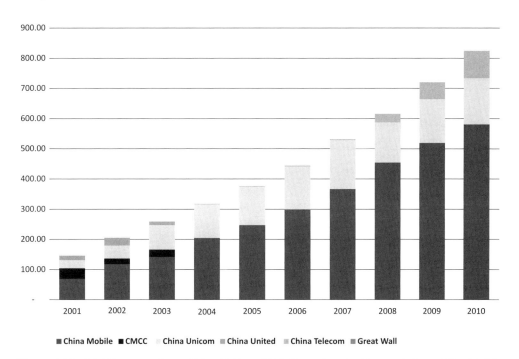

■ China Mobile ■ CMCC   China Unicom ■ China United   ■ China Telecom ■ Great Wall

Figure 4.12   Mobile connections, China 2001 to 2010. Reproduced with permission of The Mobile World.

spectral auction and allocation process, the role of enabling technologies at component level and the role of component technologies at system level.

We find that even in China and even for China Mobile, the economics of deploying a mobile broadband network into nonstandard spectrum are substantially subscale. It might be expected that China Mobile's apparent market leverage could be translated into aggressive component price reduction and aggressive innovation. Actually, this proves not to be the case. Attempting to also deploy nonstandard technologies into these bands introduces additional risks and costs. The associated costs are incurred partly on the network side (base station costs) but predominantly get reflected in handset and user equipment costs both in terms of direct component cost, design risk, time to market risk and performance risk, all of which have additional cost and revenue-loss implications.

An analysis of this can be found in the RTT Study RF Cost Economics for Handsets undertaken for the GSM Association in September 2007 and available as a free download,[1] but the following is a synopsis to avoid you having to read a 92 page document.

The opening assumption is that there is an underlying and ongoing need to lower user device costs and provide a continuously improving user experience in terms of data rates *and* duty cycles and a parallel need to support multiple simultaneous data streams. The ongoing need for lower user device costs is debatable but let's go with it for the time being.

Cost reduction in all consumer electronics products is typically achieved by realising higher levels of integration. However, RF functions have been traditionally difficult to integrate. For example, the higher power levels used in wide-area cellular systems make it problematic to place devices such as RF power amplifiers in close proximity to other RF and non-RF functions. RF devices do not scale in the same way as other devices and obey Ohms law rather than Moore's law.

In terms of spectral policy making, it is assumed that future cellular phones will become increasingly frequency transparent, able to access multiple frequencies across multiple bands. However, this requirement is at odds with the parallel need to increase integration levels to achieve ever lower component and production cost targets. A more highly integrated phone does not necessarily make it easier to add new bands for example, and it can be easier to add bands with more discrete solutions. It is certainly easier and cheaper to fix mistakes in less highly integrated designs.

As integration levels increase, the number of RF components reduces, the RF bill of materials (RF BOM) goes down and production costs go down. However, moving to a higher level of integration implies an increase in nonrecurring engineering costs. The cost of design mistakes becomes higher.

An increase in nonrecurring RF engineering and design costs also implies that higher market volumes are needed to achieve RF economies of scale. NRE costs can be reduced by adopting lower levels of device integration but the RF BOM will increase and the form factor of the phone will increase. RF performance may or may not decrease but will be more variable from phone to phone (handset to handset, batch to batch variations from the production line).

Although handsets are becoming more integrated over time, the practical implementation of a single-chip software-definable phone remains elusive. It is relatively easy to count at least 100 separate components in a phone, including modules that themselves contain multiple

---

[1] http://www.rttonline.com/resources.html.

functions on separate parts of a common substrate. About 75% of these components in present cellular phones are passive, inductors, capacitors and filters. These devices are frequency specific.

Supporting additional frequencies in a handset implies an increase in the number of passive components. This implies higher material and manufacturing costs. Active devices such as the power amplifier can be designed to cover relatively wide frequency bands but become harder to match and lose overall efficiency. Nonstandard frequency allocations therefore have an impact on passive and active device requirements.

New MEMS (microelectrical mechanical system) -based technologies offer the potential opportunity to integrate many of these functions onto an RFIC, including for example, switch and tuneable filter functions. Tuneable structures integrated with other active components can be used to implement wideband power amplifiers, broadband tuneable matching networks and adaptive antenna matching.

Similarly, MEMS devices may be used to vary the load impedance of power amplifiers so that they will work efficiently at varying power levels over a relatively wide range of frequencies. MEMS also potentially address the problem of duplexing, particularly in 3G phones. In GSM, duplexing, (the separation of transmit and receive channels within a specific frequency band) can be achieved with a front-end switch as the phones are not transmitting and receiving at the same time.

In WCDMA, HSPA and LTE, transmission takes place at the same time as reception. Adding a band means another duplex filter needs to be added that has an associated direct cost and an associated indirect cost (takes up additional board space and needs matching components).

There are presently proposals for MEMS-based active digitally tuneable duplexers, also known as digital duplexers, which may potentially resolve these band-specific 3G specific duplex cost overheads. These techniques together will enable a transition towards single-chip software-defined radios that will help eliminate many present spectrally specific device and design issues.

Five years ago when this study was written these devices were claimed to be ready for market. However, five years later their application in terms of functional replacement of traditional RF components is limited for reasons that we explore in Chapter 5.

This highlights a pervasive disconnect in our industry. Spectral allocation and auction policy and technology policy, or rather the absence of technology policy known as 'technology neutrality' is generally based on an assumption that technology is actually capable of solving all problems within defined time scales. There are several problems with this problem. Some performance metrics are defined by the basic rules of physics and cannot be resolved by material, component level or algorithmic innovation. Additionally, problems that can be resolved by innovation take longer to resolve than expected. In wireless we find that RF component innovation in particular moves slowly and introduces additional unexpected costs throughout the supply chain, especially for the operator and service provider community. Also, sufficient incentives need to be in place to ensure adequate innovation occurs at the right time and sufficient engineering effort is applied to translate that innovation into practical fiscal gain. This sets the narrative for both the remainder of this chapter and Chapter 5.

Figure 4.13 shows the main functional components in a modern multimedia handset. The 'front-end' filters and diplexers deliver a signal path to and from the baseband signal

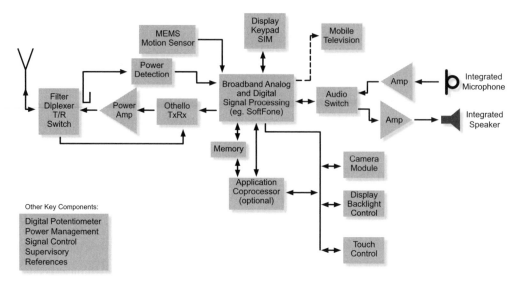

Figure 4.13 Functional block diagram of a modern multimedia handset. Reproduced with permission of MediaTek (previously Analog Devices).

processor. On the receive path, a power-detection function measures the received signal strength that determines the amount of transmit power being used on the transmit path. The transmit path includes the power amplifier. The baseband signal processor filters the received signal and provides an interface to all other devices in the phone including the display, audio paths, voice paths and camera module. This example also includes a MEMS motion sensor.

From the above, it might be assumed that the cost of the RF functions in the phone are reducing as an overall percentage of the total costs as additional non-RF functions are added.

However, the RF bill of materials has stayed more or less constant as a percentage of the overall bill of materials This is due to the addition of additional frequency bands and additional functionality, for instance in the above example, the addition of a mobile TV receiver.

Actually, as most of us will have noticed very few phones ever ended up with mobile TV receive functionality, for reasons we explore in Chapter 15. However, additional bands have been added that have introduced at least equivalent amounts of complexity. Additionally, the broadcast community is promoting the development and deployment of ATSC and DVB T services to portable devices in spectrum that is immediately adjacent to the US 700-MHz and European 800-MHz mobile broadband spectrum. This is unsurprising as these bands have been created from the switchover from analogue to digital TV.

One issue therefore is to consider the behaviour of user equipment front ends when handling a broadcast receive path and mobile broadband transceiver functionality. The broadcast receive path is relatively wideband and therefore vulnerable to interference from any transmit energy produced in the user's device or by other proximate users. The characterisation of this transmit energy is in turn a function of the band plan and spectral masks deployed in the mobile broadband user device.

4.3   Impact of Band Allocation on Scale Economics

In 2007 there were nine RF duplex spaced cellular frequency bands between 800 MHz and 2.6 GHz specified by 3GPP (the 3G Partnership Project) either presently used by GSM or UMTS and/or are considered suitable for longer-term UMTS implementation.

The nine bands are shown in Table 4.2.

Five years later in 2012 this had expanded to 25 bands including eight additional TDD bands. These are shown in Table 4.3.

Release 9 of the 3GPP specifications define four further bands, three being extensions or variations of the 850 MHz band and the other being around 1500 MHz. Bands 18 and 19 extend the existing 850 band adding 10 MHz at the lower end of the present Band V allocation to provide a 35-MHz rather than a 25-MHz pass band. Band performance relative to Band V would be degraded by about 1 dB.

The bands made available by the digital switch over at 700 MHz in the USA and 800 MHz in Europe have also introduced particular challenges, which we review in Chapter 5.

On the 28 June 2010, the US Obama administration issued a memorandum stating an intention to auction a further 500 MHz of spectrum over and above present allocations. Some but not all of this will line up with rest of the world band plans and may or may not include spectrum in L band at 1.4, 1.5 and 1.6 GHz and S band allocations presently ring fenced for hybrid satellite/terrestrial networks.

Additionally in some form factors, for example tablets, slates and lap tops, there may be a future need to integrate ATSC TV receive and/or DVB T2 receiver functionality in the 700-MHz band for portable rather than mobile reception. This receive capability has to coexist with locally generated frequency duplexed (FDD) and time-domain duplexed (TDD) transmit energy.

4.3.1   FDD versus TDD

At this point a short digression into the relative merits of FDD and TDD is needed.

A traditional FDD duplex band plan is shown in Table 4.4. The duplex gap in a cellular/mobile broadband handset keeps the TX power from one set of users getting in to the

Table 4.2   Band allocations and duplex spacing

Band	3GPP	Allocation (MHz)	Uplink	Duplex spacing (MHz)	Downlink	Region
I	2100	2 × 60	1920–1980	190	2110–2170	Present UMTS
II	1900	2 × 60	1850–1910	80	1930–1990	US PCS
III	1800	2 × 75	1710–1785	95	1805–1880	GSM Europe, Asia, Brazil
IV	1700/2100	2 × 45	1710–1755	400	2110–2155	New US
V	850	2 × 25	824–849	45	869–894	US and Asia
VI	800	2 × 10	830–840	45	875–885	Japan
VII	2600	2 × 70	2500–2570	120	2620–2690	New
VIII	900	2 × 35	880–915	45	925–960	Europe and Asia
IX	1700	2 × 35	1750–1785	95	1845–1880	Japan

Table 4.3 Twenty-five band allocations

Band	Uplink		Downlink		Duplex	Mode
1	1920	1980	2110	2170	130	FDD
2	1850	1910	1930	1990	20	FDD
3	1710	1785	1805	1880	20	FDD
4	1710	1755	2110	2155	355	FDD
5	824	849	869	894	20	FDD
6	830	840	875	885	35	FDD
7	2500	2570	2620	2690	50	FDD
8	880	915	925	960	10	FDD
9	1749.9	1784.9	1844.9	1879.9	60	FDD
10	1710	1770	2110	2170	340	FDD
11	1427.9	1452.9	1475.9	1500.9	23	FDD
12	698	716	728	746	12	FDD
13	777	787	746	756	21	FDD
14	788	798	758	768	20	FDD
15	1900	1920	2600	2620	700	FDD
16	2010	2025	2585	2600	575	FDD
17	704	716	734	746	30	FDD
33	1900	1920	1900	1920	0	TDD
34	2010	2025	2010	2025	0	TDD
35	1850	1910	1850	1910	0	TDD
36	1930	1990	1930	1990	0	TDD
37	1910	1930	1910	1930	0	TDD
38	2570	2620	2570	2620	0	TDD
39	1880	1920	1880	1920	0	TDD
40	2300	2400	2300	2400	0	TDD

receive band of another set of users. The duplex spacing keeps locally generated transmit power out of the receive path in the user's device.

In a TDD system the function of duplex spacing is performed in the time domain by ensuring that transmit time slots do not overlap receive time slots. The function of the duplex gap can only be achieved in the time domain if all networks are synchronised together and if all base stations are co sited. Specifically, the frame structure configuration needs to be coordinated between adjacent cells to avoid simultaneous TX and RX on the same frequency at the same time.

Table 4.4 Traditional FDD duplex band plan

Other users	Guard band	Operational bandwidth lower duplex	Duplex gap	Operational bandwidth upper duplex	Guard band	Other users
			Duplex spacing			

An argument can be made that a time division duplexed (TDD) radio layer is more efficient at handling mobile broadband data traffic than present FDD networks.

In terms of traffic handling a TDD radio layer can be made to be asymmetric as required, so it is potentially equivalent to an ADSL rather than VDSL connection.

As the radio channel is reciprocal (the same frequency is used on the uplink and downlink) it is easier to characterise the channel. This in turn makes it easier to extract some peak data rate gain from multiple antenna (MIMO) systems (covered in more detail in Chapter 5). It is also easier to adaptively noise match the receive chain and power match the transmit chain.

At the radio planning level there is no need for a duplex gap. This releases additional spectrum. However, there are some practical problems with TDD. Even if internetwork synchronisation is achieved, users in different networks may have varying asymmetry, which means that one set of users TX slots overlap another set of users receive slots. The assumption here is that there is generally enough physical distance between users to mitigate any interference effects. If interference does occur it is managed by handing over to other time slots or channels.

Both WiFi and WiMax use TDD and substantial licensed TDD spectrum has been allocated in the bands listed in Table 4.5.

China has pursued its own TDD standards with TD SCDMA and band allocation policy with Band 34 at 2010 to 2025 MHz, Band 39 at 1880 to 1920 MHz and Band 40 at 2300 to 2400 MHz. The assumption is that TDD LTE would also be supported as an option in Bands 38, 39 and 40.

China is one of the few countries with sufficient market scale to support a nationally specific standard and nationally specific band allocations. However, this does not necessarily mean it's a good idea.

Generally, it can be stated that nonstandard standards and nonstandard spectral allocations have hampered the pace of past progress. The decision by Japanese regulators in the late 1980s/early 1990s to introduce PHS (an alternative to DECT and the UKs ill fated CT2 cordless

Table 4.5 TDD band plan including China allocations

Band	Frequencies (MHz)	Total Bandwidth (MHz)	Deployed in
33	1900–1920	20	
34	2010–2025	15	China
35	1850–1910	60	
36	1930–1990	60	
37	1910–1930	20	
38	2570–2620	50	
39	1880–1920	40	China
40	2300–2400	100	China
WiFi (unlicensed)	2400–2480		
Total bandwidth		365 MHz	
		305 MHz excluding overlaps	
		150 MHz excluding China	
		230 MHz including WiFi	

standard) and PDC, a nonstandard implementation of GSM into nonstandard spectrum at 800 and 1500 MHz was designed to create a protected internal market that could be used by local vendors to amortise R and D and provide the basis for innovation incubation.

In practice, it proved hard to translate this innovation into global markets and the R and D opportunity cost made Japanese handset vendors and their supply chain less rather than more competitive internationally. Korean vendors have faced similar challenges from nationally specific mobile broadband and broadcast standardisation. This has introduced unnecessary opportunity cost without a proportionate market gain.

The decision might alternatively be justified on the assumption that TDD will become more dominant as a mobile broadband bearer but several caveats apply. TDD does not work particularly well in large cells as additional time-domain guard band needs to be introduced to avoid intersymbol interference between transmit and receive time slots. TDD does not work particularly well in small cells as base stations and users within a cell radius are likely to be closer together and therefore more likely to create mutual interference. This will be particularly noticeable with more extreme duty cycles, for example when some users require uplink rather than downlink asymmetry.

As a prior example of this, the three PHS networks deployed in Japan in the mid-1990s were not intersynchronised. PHS networks were also deployed in China, Taiwan and Thailand but never gained an international market footprint. This was partly due to the intersymbol interference issue but also due to the fundamental fact that TDD devices have poor sensitivity. Power in transmit devices does not disappear instantaneously and substantial residual power can still be visible within the RX time slots. This does not matter when the duty cycle is relatively light, for example in a basic GSM voice path where only one slot in eight is transmitting. It certainly matters if the TX duty cycle is high, which in many mobile broadband cases it will be and/or if high peak data rates are supported. If the asymmetry is changing rapidly over time the signalling bandwidth will also become disproportionate.

So in practice, any theoretical gains available from TDD will disappear in implementation loss. TDD may provide higher headline peak data rates but the average data throughput will be lower. TDD next to a LTE FDD receive channel will also be particularly bad news both at the network level, base station to base station, and in the user equipment receive path. Given that it is unlikely that FDD will disappear for the foreseeable future, most user equipment will need to be dual mode, so any potential component savings, eliminating duplex filters for example, would not be possible. Although the RF specifications are similar they are not the same and will incur additional conformance test time and cost. As always, the impact on user equipment cost and performance tends to get overlooked or ignored.

Even at the system level it is arguable whether there is any efficiency gain. The argument is that if the uplink is lightly loaded in an FDD system then valuable bandwidth is being wasted. However, all that happens is that the noise floor at the e node B reduces. This improves the performance of all other FDD uplinks in the cell, reducing power drain for all users served by the cell site.

So you might ask why TDD appears to work in WiFi. The answer is that WiFi is only spectrally efficient and energy efficient because it is low power, 10 milliwatts of transmit power from the user's device rather than 125 or 250 milliwatts in TDD LTE. The occupied channel bandwidth of WiFi at 2.4 GHz is 22 MHz with a channel spacing of 25 MHz within 80 MHz of operational bandwidth so that's three channels within 80 MHz. This is not in itself efficient. The efficiency comes from the channel reuse that is in turn a function of the low transmit power. There is some trunking gain achievable from a 20-MHz channel but much of this disappears if bidirectional differentiated quality of service needs to be supported.

It is hard to avoid the conclusion that TDD at least for general use within licensed spectrum is one of those technology cul de sacs that the industry has managed to drive into with no reverse gear engaged. The combination of reasons outlined above explain why wide-area TDD WiMax networks never performed as well as expected as offered traffic loading increased over time.

Anyway, back to FDD. As we have said, in an FDD band plan the bands are subdivided into transmit and receive duplex pairs. The duplex separation varies between 45 MHz (800/850/900 MHz bands) and 400 MHz (Band IV US AWS). The lower duplex in 2007 was always mobile transmit as the propagation conditions are more favourable. However, five years later the allocations are a mix of standard and reverse duplex for reasons that we discuss in Chapter 5.

4.3.2   Band Allocations, Guard Bands and Duplex Separation. The Impact on Transceiver Design, Cost and Performance

The choice of frequency, the guard bands between band allocations and the duplex separation of the uplink and downlink within each individualband all have a profound influence on the architecture of the phone and the active and passive devices used in the phone. Additionally, legacy spectral allocations may need to be supported in some handset frequency plans and future repurposed UHF TV allocations other than the 700 and 800 MHz bands may need to be accommodated.

These 'wide-area' cellular radio transceiver functions may also need to physically coexist with local area (WiFi) and personal area transceivers and with easily desensitised receive only functions such as GPS or ATSC/ DVB TV. Digital filtering techniques and architectural innovations, for example direct conversion receivers, translational (GSM) and polar loop (EDGE and WCDMA) transmit architectures, have been developed that minimise the present RF component count and RF component cost implications of multiband and multimode handsets. As a result, it would be reasonable to assume that RF component costs represent a declining percentage of the BOM of a modern cellular handset.

However, RF costs have remained relatively stable as a percentage of the total BOM over time and are likely to remain so. This is due to the increase in RF complexity. It might have also been expected that operator and user expectations of RF performance and functionality, for example higher average data rates would also have encouraged more highly specified RF front ends, but this has not been the case.

Figure 4.14 gives an indication of the typical value split between functions in 2007, in this case in an ultralow-cost handset.

We have said that the RF BOM is staying relatively constant over time as a percentage (between 7 and 10%) of the overall BOM of the phone and that this is true irrespective of whether the phone is an entry level, mid-tier or high-end device.

There are exceptions to this. A mid-tier camera phone in 2007 for example had a value split of about 5% for the RF (including Bluetooth). The logic and digital circuits accounted for about 30%, memory at 12%, the LCD at 10%, the camera at 11%, PCB and electromechanical components at 13%, mechanical components at 14% and 'other bits' at 6%. It could be argued of course that the imaging bandwidth of this device would have deserved a more

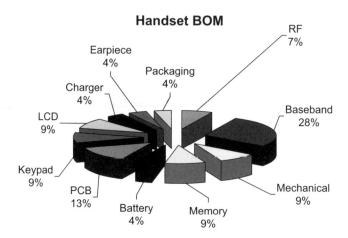

Figure 4.14   The RF BOM compared to other components 2010. Reproduced from RF Cost Economics for Handsets May 2007[2]

highly specified RF functionality. It is, however, true that a need to support additional access technologies will introduce additional costs.

These costs include nonrecurring expenditure (NRE) and component cost. This implies that volume thresholds are needed to support a supply of handsets that can be considered to be 'cost economic' in terms of RF-related NRE and RF-related component cost.

In 2007 it was clear that this volume threshold was higher than generally acknowledged by many in the industry and was increasing over time. In addition, it was hard to achieve competitive RF performance unless certain volume thresholds were achieved. This implied that 'performance volume thresholds' and related 'performance scaling effects' needed to be factored in to spectral valuation and spectral allocation policy.

It also highlighted the escalating cost and risk factors associated with geographically specific nonstandard spectral allocations, particularly in countries with relatively small addressable markets. These cost and risk factors were compounded by the implementation of nonstandard technologies. This explains why five years later, despite a large expansion in band allocations globally, most phones only support at most four or five bands and at most two wide-area radio access standards, typically GSM and WCDMA.

Any discussion of the economics of adding an additional frequency band to a handset therefore has to comprehend the technology or technologies used to access that band. The technology used (GSM and/or GSM/UMTS in 1997 or GSM and UMTS/LTE in 2012) influences the RF architecture of the phone, the RF component cost of the phone and the nonrecurring engineering cost of getting that phone to market.

It was of course possible to produce UMTS only devices and such devices were developed and sold into some markets, for example Japan. Similarly today, there are iPhone products for the US market that are dualband and some LTE-only dongles. There is little point in adding additional bands if additional cost is incurred.

---

[2] http://www.rttonline.com/documents/rfcosteconhsetss.pdf.

Table 4.6   Assumed future spectrum allocations by technology in 2007

Single-mode GSM	Dual-Mode GSM/UMTS	Refarming Handsets	Frequent Traveller	Heavy User	Rural coverage	Handsets supporting other technologies
Quadband 850/900/ 1800/1900	Quadband and UMTS/LTE 1900/2100	Handsets with UMTS/LTE 900/1800	Handsets with UMTS at 850/1900	Handsets with UMTS/LTE at 2.6 GHz	Handsets with UMTS at UHF 470–862 MHz	For example, 2.6 or 3.5 GHz Wi Max
Quadband	Quinband dual mode	Quinband dual mode	Quinband dual mode	Sextuplet band dual mode	Septuplet band	Octo- or nonoband

Additionally, each extra band incurs additional insertion loss and loss of band-to-band isolation. The same applies to multiple technologies. On the other hand, a dualband or single-band single-technology phone can only be shipped to a small local market and will have limited global roaming capability though this only matters to people that regularly travel (less than 10% of the global population).

The 2007 study assumed seven handset categories each supported by a specific band plan. This is shown in Table 4.6.

This excluded the probable need to support Band IV in the US (the AWS band at 1700/2100 MHz where T Mobile was the dominant spectral investor) and possible need to support Band VI (800 MHz) and Band IX (1700 MHz) for Japan but there was a general assumption that somehow a world phone would be produced that would cover ten bands or so and multiple access technologies. Such a phone could be shipped to multiple markets and thereby achieve global-scale economy. The devices were commonly and erroneously described as software-defined radio.

4.3.3   Software-Defined Radio, WiMax and the 2.6-GHz Extension Bands

Five years later WiMax has failed to achieve global scale and is therefore no longer on most user equipment vendor development road maps and software-defined radios still do not exist or more specifically the RF functions are still implemented with discrete components.

The 2.6- and 3.5-GHZ bands remain unallocated and unauctioned, though this may have changed by the time most of you read this. What has happened instead is that the top of the UHF TV bands have been allocated and auctioned in the US at 700 MHz and in Europe at 800 MHz. In addition, AT and T has purchased T Mobile in the US to consolidate the AWS band with AT and T 700 and 850 MHz spectral investments, and most phones are still four or five band at most.

It was actually clear to quite a few people in 2007 that software-defined radios would only become available at mass-market prices and mass-market volumes as and when specific device and integration issues were resolved and as and when there were clear financial incentives to realise such a device.

In practice, each additional band whether 'standard' (as defined by 3GPP) or 'nonstandard' (a country-specific allocation not referenced by 3GPP) incurs substantial nonrecurring investment cost and (related) component cost multipliers.

The market volumes needed to economically cost justify additional band support and/or technology support are substantial and are increasing over time. In 2007 we stated that sufficient market volumes were unlikely to be achieved in any countries or regions other than China, India, Europe and possibly the USA/Latin America. Even in these 'large local markets' the entry costs and risks of nonstandard bands were in practice much higher than generally acknowledged. Take intermodulation as an example.

4.3.4   Intermodulation

Whenever two or more frequencies are mixed intentionally or unintentionally within a handset, they will produce sum and difference frequencies. This is known as intermodulation and can occur at receive frequencies and transmit frequencies or between transmit and receive frequencies. Frequencies are mixed or multiplied together intentionally within the frequency synthesiser to create new wanted frequencies to be modulated. The unwanted sum or difference frequency (the image) is filtered out.

When frequencies mix together unintentionally, sum and difference products may translate unwanted signal energy into other bands within the phone or into other proximate devices. A new frequency band introduced into a handset will create a new set of intermodulation products that will need to be managed in terms of their potential impact on the other bands and radio systems used in the phone. Resolving these issues adds to the nonrecurring engineering cost, may result in time to market delay and may add to component cost if additional filtering or reciprocal mixing has to be introduced.

4.3.5   The Need to Support Higher Frequencies and Linear Power Efficiency

The new extension/expansion band at 2.6 GHz and bands at 3.5 GHz creates new challenges. Many countries have legacy wideband radar systems at 2.6 GHz that are problematic but the 2.6 GHz transceivers also have to work with WiFi transceivers at 2.4 GHz These are low power devices, generating not more than 10 milliwatts of transmit power.

Wide-area cellular systems require handsets to transmit at higher powers, typically 250 milliwatts, and to be able to reduce this output power in defined steps down to a few milliwatts (the dynamic output power range over which the phone has to operate).

Handset vendors have a choice of power-amplifier technologies that are typically either based on CMOS (complementary metal-oxide-semiconductor) or SiGe (silicon germanium) or GaAs (gallium arsenide) semiconductor processes.

Simplifying a rather complex story, CMOS is lower cost and supports more aggressive integration but does not perform as well as GaAs particularly at higher frequencies. SiGe combines some of the advantages of CMOS and GaAs. GaAs also has some advantages in terms of delivering a better linearity/amplifier efficiency trade off, an important metric for UMTS and LTE and related technologies using a combination of phase and amplitude modulation. The requirement specifically is to deliver good linearity and efficiency at maximum power and good efficiency at minimum power.

Table 4.7 Linearity requirements by technology

Generation	Technology	Peak to average ratio (dB)	Power control dynamic range
1G	AMPS	0	25 dB
	ETACS	0	25 dB
	JTACS	0	25 dB
2G	GSM	0	30 dB
	PDC	3–5	30 dB
	TDMA/EDGE	3–5	35 dB
3G	UMTS rel 99	5	80 dB
	UMTS rel 6/7	5–8	35 dB
	UTRAN/LTE Release 8,9,10,11,12	8–17 (includes multicarrier bonded channel options)	TBD

An optimum PA (power-amplifier) technology choice for 3.5 GHz is unlikely to be the same as an optimum PA technology choice for 700 MHz. As a general rule, it gets harder to deliver gain without introducing excessive noise as frequency increases. An optimum PA technology choice for GSM is unlikely to be the same as an optimum PA technology choice for UMTS that requires more linearity to preserve the AM (amplitude-modulation) characteristics in the modulated signal envelope.

Table 4.7 shows the overall trends over the past 20 years (first- and second-generation cellular) and likely trends over the next five to ten years in terms of the peak to average ratio of the modulated signal envelope (which determines the amount of linearity needed in the amplifier) and the power control dynamic range (which determines the upper and lower power outputs required from the device).

GSMK was chosen for GSM because the modulated signal envelope has no intentional amplitude modulation and could/can therefore use Class C amplifiers (as used in FM analogue systems) albeit with an increased dynamic range (lowest to highest output power). These power amplifiers could be/can be up to 55% efficient. All other evolving technology options including evolved variants of GSM (EDGE) have used/use a combination of phase and amplitude modulation to modulate the signal envelope.

This requirement combined with a wide dynamic power control range has created a number of optimisation challenges for GSM EDGE and UMTS/LTE handsets in terms of RF power efficiency and linearity. Various techniques have been developed that take out the envelope modulation and reintroduce it after signal amplification has been achieved by the PA. These are known variously as polar modulation and/or translational loop architectures and are part of a family of post distortion and predistortion feedback and feedforward techniques that correct for amplifier nonlinearity. These techniques work well but require careful calibration or tracking of RF PA behaviour and the adaptive circuits under varying load characteristics and over temperature and time. In parallel, some of the dynamic range requirements have been reduced by implementing adaptive coding schemes and adaptive modulation schemes that will ease some of these RF PA characterisation issues.

It is therefore important to consider what technology or mix of technologies will be used in the allocated spectral band. Choosing a new nonstandard band for network deployment or failing to mandate a technology for a specific band can have major implications on both the design and function of the RFPA including cost and RF performance (efficiency and linearity). Efficiency loss translates directly into a decrease in talk time. Insufficient linearity translates into a loss of modulation accuracy at high power that will cause a loss of uplink throughput and potential interference to other users.

Power amplifiers can be designed to work over large frequency ranges, for example from 150 to 2500 MHz, but this does not mean that they are necessarily the best choice of technology across the band or have the capability of being acceptably efficient across the band. It is not just the availability of the PA that is important but the filter and matching components needed to make it work efficiently both in the chosen band and across the other bands also supported by the handset.

Power-amplifier pricing in the public domain is often based on relatively small minimum-order quantities, for example 10 000 units. However, these volumes assume multiple customers are likely to be available that will meet and exceed this MOQ criteria. If this looks at all doubtful, or if better returns look achievable from other applications, then the products just will not appear.

Similar design issues need to be considered when validating device performance in multi-mode handsets where more than one RF PA may be operating simultaneously, for example a (relatively high power) LTE PA generating signal energy in parallel and proximate to a (relatively low power) Bluetooth and/or WiFi transmitter.

This is directly relevant to handsets using a mix of 'other technologies'. The transmitted signals need to be kept apart from each other and frequency plans need to be carefully validated to prevent intermodulation/mixing of these multiple transmit frequencies both into other transmit bands and into the receive bands supported in the handset. The receive bands could easily desensitise receive functions such as GPS or a DVB T or ATSC receiver. The resolution of these issues can incur substantial nonrecurring engineering cost that need to be recovered in the RF BOM and/or absorbed over substantial market volumes.

If RF functions in the phone are used in an either/or mode rather than simultaneously, there will be a need to 'mode switch' in the front end of the phone to provide a dedicated signal path for a particular service. Thus, the choice of a nonstandard band may have an impact on the performance of other radio transceiver functions in the phone but will also require additional components.

There can be several switching functions in the front end of the phone.

4.3.6   The TX/RX Switch for GSM

There may (probably will) be a TX/RX switch that provides a time-duplexed separation between the GSM transmit burst and the receive burst received after a 'two-slot' delay (just over a millisecond).These devices will be switching at the frame rate (217 frames per second) and are designed to be reliable over 100 billion cycles or more. The switch speed and duty cycle of these functions makes them presently unsuitable for other technologies, for example MEMS-based switching solutions.

4.3.7 Band Switching

This switch function routes the signal to the appropriate SAW diplex filter that will band pass the wanted signal energy from that band and band stop unwanted signal energy.

4.3.8 Mode Switching

This switch function routes the signal depending on the modulation and air interface standard being used within the band of interest, for example GSM or UMTS. These band-switching and mode-switching devices need to be efficient (offer low insertion loss). They also need good linearity to preserve the integrity of the amplitude-modulated waveforms used in UMTS/LTE and other third-generation air interfaces and to avoid intermodulation and unwanted harmonics. An increased requirement for linearity implies a larger die size (increased cost) and an increase in insertion loss for these devices. There is therefore both a dollar cost and a performance cost to be considered. These devices are typically GaAs devices, though hybrid CMOS/silicon on sapphire processes are also now available. RF MEMS devices may also provide an alternative option for this function.

4.4   The Impact of Increased RF Integration on Volume Thresholds

The power amplifier is presently a separate device not integrated into the RF IC. This is because it is high power (250 milliwatts is equivalent to 24 dBm into 50 ohms). It generates heat. It has to coexist with received radio signals that can be as low as -120 dBm (0.001 of a picowatt). It has to be isolated from the other mixed and digital baseband signals on the chip.

Single-chip phones may be available within the next two to three years, although some vendors suggest this is significantly overoptimistic. However, the availability of these devices whether sooner or later will increase rather than decrease the volume threshold at which nonstandard RF handset designs become economic.

In 2007 most mobile phone designs had the PA, SAW filters and antenna switch (RF front end components) off chip. The VCO (voltage-controlled oscillator) and synthesiser (quite a noisy device) would have been off chip in earlier designs, but by 2007 were generally integrated. As we shall see in Chapter 5 not much has changed in the succeeding five-year period. Partly this is due to the NRE costs and risks of aggressive integration.

The development time and development cost for a retuned integrated receiver rises nonlinearly with integration level. The mask costs are higher; typically about 1 million dollars for a 0.13-micrometre process and increased with integration level as the industry transitioned to 90 and 65 nm.

'Single-chip' phones do not make it easier, but rather make it harder to support nonstandard frequency allocations. To realise a truly frequency agile single-chip device requires the integration of diplexing and duplexing onto the die. MEMS-based tuneable filters provide an opportunity to integrate these remaining front-end components on to the RFIC. This provides the basis for a software-defined radio but such products are not presently available at mass-market volumes or mass-market prices.

Integration of these functions onto a device with significant temperature variations will be a particular challenge and it is likely that most if not all vendor solutions will continue to support off-chip RF power amplification in order to ensure device reliability and performance consistency. Even as and when these RF PA and RF MEMS integration challenges are resolved, there will still be frequency-specific components that have to be added to the device, for example the antenna and passive components to match the antenna to the RFIC.

Although MEMS-based functions integrated onto/within an IC potentially offer an ability to have tuneable functionality across a wide range of frequency and band allocations, there will be optimisation limitations. For example, a highly integrated RFIC would be optimised to tune across specific frequency bands with specific channel spacing with specific RF signal characteristics. Additional frequency bands may require hardware optimisation of the IC. At this point, an approximately $6 million 'entry cost' is incurred. Hence, our contention in 2007 was that the volume threshold for nonstandard band support would increase rather than decrease as integration levels increased.

This holds true for the transition to 90 nm, the transition to 65 nm and for sub-50-nm processes but also this is an RFIC perspective of the world and we have assiduously noted that a lot of RF components are still not included on the RFIC.

In general, it can be said that it is good practice to study the practical present handset cost multipliers and performance issues implicit in nonstandard band allocations and bear in mind, when developing economic models, that present (NRE) entry costs may increase rather than decrease over time.

Other considerations may be significant, for example the mechanical form factor of the device. The present trend towards superslim phones (a height of less than 7 mm) is dependent on the availability of low form factor passive devices (capacitors, inductors, resistors and other resonant components including the antenna) that are specific to the supported frequency bands.

Small volumes (in terms of space) make it proportionately harder to realise antennas that resonate efficiently across widely spaced frequency bands. 'Small' market volumes imply a risk that these 'difficult to design' components will not be readily available to the handset vendors. In the intervening period between 2007 and 2012 there has been a general market assumption that high peak data rates are needed and that multiple antennas are needed in mobile devices to achieve this. This has compounded the space/volume problem for antenna designers. As antenna volume reduces, operational bandwidth reduces. This decreases total isotropic sensitivity on the receive path and total radiated power on the transmit path and makes the device more susceptible to hand- and head-proximity detuning effects.

But performance is not just a function of physical volume but also of market volume. High market volumes mean that the performance spread of RF components can be more carefully controlled. This can be best described as 'performance scaling', the specific market volumes needed to achieve consistent and acceptable RF performance in practical handset designs.

4.4.1 Differentiating RF Technology and RF Engineering Costs

Technology costs are the recurring costs in the device and a composite of the component technologies needed to support the chosen air interface, or interfaces in single-mode, dual-mode and multimode devices. SAW filters for example are one of the enabling technologies used in the RF section of a modern cellular handset. They have a defined function (to achieve selectivity) and an associated cost that may or may not decrease over time and over volume.

Engineering costs are more typically (though not always) nonrecurring in that they are a composite of the engineering time and effort needed to achieve a certain desired result using a mix of available technologies. Nonrecurring engineering costs have to be amortised over a certain production volume within a certain time.

Cost implies risk and risk implies a business need to achieve a certain return on investment (ROI). Thus, the price charged for components and for the engineering effort needed to turn those components into finished product will directly reflect the return on investment criteria.

This return on investment criteria is not static and may change over time. More significantly, the return on investment will be determined by the number of vendors competing to supply components and finished product to a defined market.

If a market is too small in terms of either volume or value then the likely outcome is that the market will be undersupplied both in terms of the number of component vendors and the amount of engineering effort needed to turn those components into cost and performance competitive product. This will inflate realised prices, limit choice and compromise operator time to market.

Additionally, the handsets that are available will probably perform poorly in terms of their RF functionality. This in turn will limit achievable user data rates (capacity) and the data/voice geographic footprint of the network (coverage). As a rule of thumb, every dB of sensitivity or selectivity lost in a handset translates into a required 10% increase in network density to maintain equivalent coverage/capacity. Handset sensitivity and selectivity is therefore directly related to the overall investment and running cost of the network.

The impact of production volume on RF performance therefore needs to be carefully quantified. The metric is not simply volume but volume over time, effectively a 'maturity threshold' that has to be reached in order to support an adequate supply of performance competitive price competitive handsets.

So we need to define the 'volume thresholds' and 'maturity thresholds' needed to achieve a supply of 'economically efficient' handsets. Economically efficient handsets are handsets that have reached a volume threshold at which their component costs do not significantly decrease with additional volume. This implies that an acceptable return of investment has been achieved both in terms of component development investment and the engineering effort needed to turn those components into finished competitive product.

However, we are also saying that economically efficient handsets must also have reached a volume and maturity threshold at which handset RF performance is effectively as good as it can be, given the capabilities of the technology used, in other words a maturity performance threshold.

4.4.2   The Impact of Volume Thresholds and Maturity Performance Thresholds on RF Performance – A GSM Example

In 1992, when GSM single-band 900-MHz phones first became available, it was a major design and production challenge to make phones that would meet the basic conformance sensitivity specification of -102 dBm. There were similar problems meeting other RF performance parameters, for example adjacent channel selectivity and phase errors on the transmit path.

Only just achieving the conformance requirement significantly increases production costs. This is because the variation in component tolerances from phone to phone and batch to batch (a function of component volume) will mean that a significant number of phones will fail to pass basic RF performance production tests. This metric is known as 'RF yield'. There may be limited opportunities to rework and retest devices but essentially a 'low' RF yield will translate directly into an increase in the RF bill of materials for those phones that did actually make it through the production test process.

By 1997 (5 years on), two things had happened. First, most vendors had been through two or three or more design iterations. This had delivered more safety margin in terms of designed performance over and above the conformance specification. Secondly, the major vendors had sufficient volume to negotiate with their RF component vendors to tighten component tolerances to reduce the handset to handset and batch to batch differences that had previously compromised RF yield. So in practice, a significant number of handsets were being shipped to market with a sensitivity of around −107 dBm, 5 dB better than the conformance specification. Note that this did not apply to all handsets from all manufacturers and the spread between best and worst handsets was between 3 and 5 dB.

At this point, GSM phones achieved a lower cost point, provided better and more consistent voice quality, longer talk and standby times, additional functionality and a smaller form factor than analogue cellular phones.

Another ten years (2007) and the best handsets could be measured at −110 dBm (static sensitivity). There was still a 'best to worst' spread of between 3 and 5 dB between manufacturers and sometimes also between different handsets from the same manufacturer, but essentially GSM handset performance from an RF perspective was as good as it was ever going to get. The 'best to worst' spread still existed partly because of device and design differences but also because not all handsets have sufficient production volume to realise a gain in performance.

Note that over this period, design engineers also had to deliver additional band support. Phones were initially single band (900 MHz), then dualband 900/1800 MHz (from about 1995), then triband 900/1800/1900 (from about year 2000), then triband, and then (2005 to 2007), quadband (850/900/1800/1900).

4.4.3   The Impact of Volume Thresholds and Maturity Performance Thresholds – A UMTS Example

The fact that GSM has more or less reached its development limit in terms of RF performance provided one of the motivations for moving to UMTS. UMTS traded additional processing overhead to achieve what can be rather oversimplistically described as 'bandwidth gain (analogous to the benefits that broad band FM delivered over narrowband AM systems in the 50 years between 1940 and 1990). Bandwidth gain can be translated into more capacity (more users per MHz of allocated spectrum and/or higher data rates per user) and/or coverage.

There are other RF potential costs benefits, for example the wider channel spacing (5 MHz rather than 200 kHz) relaxes the need for channel to channel RF frequency stability. However, similar rules on volume and maturity performance thresholds apply.

In 2002, the first UMTS phones barely managed to meet the conformance specification of −117 dBm.

Figure 4.15 shows the results of reference sensitivity test on four presently available phones (2006/2007). The best device is over 5 dB better than the conformance specification. Note the

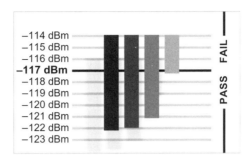

Figure 4.15 Sensitivity measurements on four UMTS phones. Reproduced with permission of Spirent Communications – originally published in RF Cost Economics for Handsets Study May 2007.[3]

difference between the best and worst handsets was about 5 dB (the worst handset only just passed).

So five years after market introduction, the best handsets available were 5 dB better than specification. In other words both GSM and UMTS handsets improved their sensitivity by 1 dB per year over the first five years of their initial market launch. This improvement has now levelled off to the point where we are as close to the ultimate sensitivity of the device as the existing technology will allow (about another 3 dB) providing justification for the transition to the next generation of technology (LTE).

Note that these performance curves are volume specific, technology specific and frequency specific. The performance gains are achieved by a mix of engineering effort (amortised over significant product volumes) and improved RF component tolerance control. Similar gains over time and over volume are realised in terms of adjacent channel selectivity in the receiver. Similar gains over time and over volume are also realised in terms of transmitter performance, particularly in terms of error vector magnitude, specifically the phase and amplitude accuracy of the modulated signal waveform and the close-in and wideband noise characteristics of the device. An example would be the performance variation from unit to unit of SAW filters over temperature. High market volumes over time gradually erode these device-to-device variations.

Closer tolerancing of active and passive components with volume therefore translates directly into uplink and downlink performance gain. These uplink and downlink gains translate directly into an improved link budget that in turn translates into either higher data rates per user and/or more users per MHz of spectrum and/or improved coverage (a decrease in network density for a given user and data density).

Additionally, as the performance margin improves, RF yield improves, typically from 90% to close to 100%. Note that low RF yield will also choke handset availability, which in turn can lead to significant losses in terms of market opportunity.

These performance benchmarks need to be achieved across multiple bands. Initially most UMTS handsets were designed for 1900/2100 MHz, but as with GSM (the single-band to dualband to triband to quadband transition) UMTS /LTE handsets will need to work equally effectively at 850 MHz, 900 and 1800 MHz, 1700/2100 MHz, at 800 and 1700 MHz in Japan, at 700 MHz at 800 MHz and at 2.6 GHz.

---

[3] http://www.rttonline.com/documents/rfcosteconhsetss.pdf.

However, as we shall see in Chapter 5, there is a significant risk that LTE devices and parallel HSPA + Release 8, 9, 10 and 11 devices will not follow this positive improvement curve and may well exhibit a far wider best to worst spread than UMTS or GSM. This should be of major concern to all operators or analysts with responsibility for assessing mobile broadband access economics.

As with GSM and UMTS, the addition of each incremental LTE band implies substantial nonrecurring engineering expenditure and small but significant additional component costs in terms of diplex and duplex filter functions. Adding nonstandard bands over and above these presently allocated bands will be problematic in terms of engineering resource allocation (not enough engineers available to do the required design and optimisation work). As we shall see, this explains why vendors work on high 'opportunity cost' multipliers when asked to produce handsets for nonstandard bands and/or nonstandard technologies. Additionally, as we have stated, each additional band introduces additional performance loss. It is quite possible that this incremental performance loss will be more significant for LTE devices than for UMTS or GSM devices.

4.5 The RF Functions in a Phone

RF functions in a cellular phone include the selective RF front end, the receiver low-noise amplifier (LNA), RF to IF mixing, the frequency synthesiser and the transmitter power amplifier (PA). The function of the front end is to capture signals of interest on the receive path and to propagate a signal on the transmit path. The receiver LNA amplifies the signal of interest on the receive path.

The mixing process takes the incoming signal and mixes it with a second frequency to create an intermediate frequency (IF) at which the signal will be processed. In direct conversion receivers, the second frequency is identical to the receive frequency but with a 90 degree phase offset.

The frequency synthesiser takes the stability of a frequency reference such as a quartz crystal and translates that reference to the required frequency to be demodulated (receive path) or modulated (transmit path). The transmitter power amplifier amplifies the signal to be transmitted.

4.5.1   RF Device Functionality and Useful Inventions Over the Past 100 Years

For the past 100 years radio devices have been required to oscillate, resonate, filter, switch and amplify. The efficiency with which these tasks are performed defines the overall efficiency of the radio system.

Fleming's thermionic valve in 1904 and Lee de Forest's triode valve in 1907 were major moments in radio device development. These devices, combined with resistors, inductors, diodes and capacitors provided the basis for Marconi's development of tuned circuits during the First World War.

In retrospect, the discovery of the piezoelectric effect by Pierre and Jacques Curie in 1880 was probably at least as significant. The Curie brothers discovered that when pressure was applied to certain crystals, an electrical voltage was generated. Conveniently for the radio industry, this proved to be a bidirectional effect. Applying electrical voltage to certain crystals would cause them to vibrate at a specific frequency.

In 1917, Paul Langevin used quartz crystals in a sonar device for submarine detection and from then on quartz became the basis for detecting and creating specific audio and radio frequencies. In the Second World War, similar research in the US, Japan and the Soviet Union showed that certain classes of ceramics exhibited piezoelectric behaviour. Courtesy of two world wars we were provided with a choice of quartz crystals and/or ceramic-based devices as the basis for providing accurate frequency and time referencing in radio products.

The invention of the transistor in 1947 and the integrated circuit in 1958 used in combination with these devices provided the basis for the power-efficient and spectrally efficient radio transceivers that have powered the wireless industry for the past 50 years and the cellular industry for the past thirty years. However, 50 years on these RF functions are still typically realised as discrete components, existing alongside rather than inside present integrated circuits.

4.5.2   Present-Day Issues of RF Device Integration

Present-day issues of RF device integration are as much mechanical as electrical.

Radio reception starts with an *antenna*.

Antennas in hand-held devices are either electrical dipoles, small loops, helical, meander antennas or patch antennas. Patch antennas, also known as planar internal antennas are increasingly popular as embedded antennas. Typically, these are used with grounding that shifts the antenna resonance to a lower frequency with a slot added to increase the electrical length, a design known as planar inverted F antennas (PIFA).

Antenna size can also be reduced by using dielectrics with a high dielectric constant. Another option is to use fractal-based antenna patterns to use whatever space is available reasonably effectively. However, any antenna, when constrained within a space that is significantly less than a quarter wavelength of its centre frequency will be inherently inefficient and sensitive to de tuning, for example by hand capacitance effects.

The following are significant contributors to this loss of efficiency.

The imperfect impedance match of the antenna – especially at the band edges – gives rise to significant reflection loss particularly at lower frequencies (850/900/800/700 MHz). Ohmic and dielectric losses convert useful RF energy into heat in the antenna and any associated matching circuits. RF currents may be coupled into other components within the handset, dissipating RF energy inside the phone.

Candy bar, clam shell and slider phones all have similar but different challenges in terms of antenna efficiency. Some antenna designs in present products when used close to the head have negative gains of −8dB or more.

4.5.3   The Antenna TX/RX Switch Module for GSM, Duplexers for UMTS

Another option for improving sensitivity is not to transmit and receive at the same time. This is used in GSM-only phones where there is a two-slot offset between transmit and receive frames. Switching is normally implemented with a GaAs device or pin diodes. WCDMA/UMTS and LTE FDD phones, however, send and receive at the same time, and therefore require a duplexer.

GSM/WCDMA/UMTS/LTE phones therefore typically end up with a duplexer and a GSM TX/RX switch in the front end of the phone. Each additional UMTS/LTE band requires an additional duplexer.

4.5.4   Other Front-End Switch Paths

In addition, there is a need to band switch and mode switch. In an ideal world you would not introduce these switch paths. They create loss and distortion and dissipate power.

More bands and additional modes therefore add direct costs in terms of component costs and indirect costs in terms of a loss of sensitivity on the receive path and a loss of transmitted power on the transmit path.

4.5.5   MEMS Devices

One alternative is to use MEMS (microelectrical mechanical system) -based switches.

The idea of building microelectrical mechanical switches has been around for twenty years or so but is now becoming increasingly practical and has the benefit of sharing available semiconductor fabrication techniques. MEMS components are manufactured using microma-chining processes to etch away parts of a silicon wafer or to construct new structural layers that can perform mechanical and electromechanical functions.

A MEMS-based switch would have low insertion loss, good isolation and linearity and would be small and power efficient. In addition, it is essentially a broadband device. It is electrostatically activated so needs a high voltage which is inconvenient, but low current (so practical).

MEMS devices are sensitive to moisture and atmospheric contaminants so have to be hermetically sealed, rather like a quartz crystal. This packaging problem would disappear if the device could be sealed at the wafer level during manufacture with additional overmoulding to provide long-term protection.

Integrated MEMS devices are therefore a plausible candidate for band switching and mode switching within the next three to five years. TX/RX switching (for GSM or other time division multiplexed systems) would be more ambitious due to the duty-cycle requirements but still possible using optimised production techniques. There is also a potential power handling and temperature cycling issue. The high peak voltages implicit in the GSM TX path can lead to the dielectric breakdown of small structures, a problem that occurred with early generations of SAW filters. Because MEMS devices are mechanical, they will be inherently sensitive to temperature changes.

This suggests a potential conflict between present ambitions to integrate the RF PA on to an RFIC and to integrate MEMS devices to reduce front-end component count and deliver a spectrally flexible phone. The balance between these two options will be an important design consideration. The optimal trade off is very likely to be frequency specific.

For example, if the design brief is to produce an ultralow-cost handset, then there are arguments in favour of integrating the RFPA on to the RFIC. However, this will make it difficult to integrate MEMS components on to the same device. You can either have frequency flexibility or low cost, but not both together.

4.5.6   Filtering Using Surface Acoustic Wave (SAW), Bulk Acoustic Wave (BAW) Devices and MEMS Resonators – Implications for Future Radio Systems

These devices are covered in detail in the next chapter but are potentially key enabling technologies.

It seems inevitable that the regulatory environment will require the industry to produce handsets that are capable of working across ever more numerous multiple bands and that the standards-making process will ensure that handsets will also have to support ever more numerous multiple radio standards. This increases RF component cost and makes it harder to deliver consistent RF performance across such a wide range of possible RF operational conditions. This trend also highlights that some of the traditional RF device technologies that have served us faithfully for 50 years or more are nonoptimum for these extended operational conditions.

From a business perspective, there is evidence of a closer coupling between companies with antenna and shielding expertise and silicon vendors. Similar agreements are likely between the MEMS community and silicon vendors to meet the perceived 3- to 5-year need for a closer integration of RF MEMS functionality with next generation silicon. At that stage, but not before, the software-defined radio finally becomes a practical reality. Long-term reliability issues of MEMS devices (given that they depend on mechanical movement) also still need to be resolved.

4.5.7   Handset RF Component Cost Trends

Table 4.8 shows the RF BOM of a Tri Band GSM handset in 2003 (from RTT RF Cost Economics Study for the GSMA September 2007 – original source price points from Plextek[4] for the manufactured product costing for a multimedia smart phone designed for a customer.

Table 4.8   Triband GSM RF BOM Reproduced by permission of Plextek

Component	Quantity	Cost in dollars
RF VCO	1	0.94
TXVCO	1	1.6
RF balun	1	0.09
Dual digital transistors	4	0.18
TCVCXO	1	1.53
Triband transceiver	1	2.32
RF front-end module	1	2.19
Transmit power control IC	1	1.08
Triple-band power amplifier	1	2.48
High-speed LDO (2 V voltage regulator)	1	0.11
Total		12.52

---

[4] http://www.plextek.com/.

The 12.52 dollars equated to 7% of the total BOM cost of $178. The RF PA was the biggest single line item at $2.48. The baseband value was $49.84.

In 2007, a comparative device, admittedly quadband rather than triband would be about 6 dollars costed on a similar volume and the RF BOM would still be about 7% of the total bill of materials. Generally, vendor forecasts in 2007 suggested that the RF BOM cost would halve again to 3 dollars so the RF BOM would be 10% of a 30-dollar handset (for the ultralow-cost handset market) or approximately 7% of a 40-dollar handset. This would only have been achievable if it had been proved feasible to integrate the PA and/or front-end matching in to the RFIC. TI attempted to achieve this, failed, had to write off $250 million of R and D investment and as a result exited the sector that a few years before they had dominated.

Several things happened that in retrospect might seem obvious but were not very evident at the time. The introduction of the iPhone in 2007 raised the realised average price of smart phones, but actually this also meant that the average realised price of phones in general increased rather than decreased on a year-to-year basis.

Inclusion of the iPhone in an operators range plan became and remains a trophy asset and its absence can result in a loss of market value for the operator. Any time-to-market delay therefore assumes a disproportionate importance. This means that the RF design team must choose design and integration options that represent the least risk in terms of time to market delay and the safest way to achieve this is to have relatively unaggressive levels of integration with each separate duplex band having its own discrete RX and TX component and processing path.

This has been great news for the RF component community. A five-band iPhone has (at time of writing at least) five RF power amplifiers and all that implies in terms of component and real estate added value and unit prices are relatively robust due to the high overall value of the handset BOM.

Also, the ultralow-cost handset market never materialised to the extent anticipated. As stated earlier this is partly due to the aspirational effect that smart phones have had on the lower end of the market. Price point expectations in newly dominant markets particularly China are presently aggressive and are linked to a stated need to move rapidly to a 11-band phone. The performance cost of implementing this would be of the order of 7 dB. Even taking low price points for individual components the overall BOM would be substantially higher than most other phones shipped into most other markets.

This makes no commercial sense either for the operator community in China or for the supply chain or for end users who would end up with a phone with poor sensitivity and selectivity that would translate directly into dropped calls and depressingly short voice and data duty cycles.

Poor RF performance would also reduce coverage and absorb operator network capacity and increase customer support costs. However, there is still a conundrum here in that additional bands mean that one phone can be shipped to more markets, thus achieving greater scale gain and savings in inventory management. However, this is more than offset by additional component cost and performance loss.

The Holy Grail is therefore to be able to add bands and/or technologies without additional cost and without performance loss. We examine the practicalities of this in Chapter 5. In the meantime, RF component vendors can breathe at least a temporary sigh of relief as their price point erosion has proved less severe than might have been expected five years ago.

Effectively they have been saved by the smart phone, a device costing $200 rather than $20 dollars selling at $500 rather than $50 dollars with various add on value opportunities of which insurance against damage and loss is bizarrely important.

4.6 Summary

The dynamics of the RF supply chain are complex and occasionally counterintuitive but exhibit the classic gravitational effect typical of large-volume markets with dominant product sectors. In wide-area wireless the dominant legacy technology is GSM. The theoretical spectral efficiency benefits of UMTS/WCDMA might have been expected to encourage a relatively rapid transition from the 'older' to newer' technology, but this has not happened for several reasons. The gravitational effect of the large legacy market has meant that vendors make more money out of innovating for established markets with known volume. The purpose of the innovation is to reduce cost and improve performance. A failure to innovate has an associated opportunity cost. A similar lag effect can now be observed in LTE and will be examined in more detail in the next chapter.

RF innovation has also had to deliver on two fronts – new bands and evolving standards that have become more complex over time. GSM standardisation has moved forward with the introduction of EDGE and higher-order modulation and coding and channel aggregation options. WCDMA has evolved into HSPA and HSPA has evolved into HSPA + and at least partially integrated with the LTE standards road map from Release 8 onwards. Competing standard such as WiMax have further diluted R and D effort.

The impact of this to date is that WCDMA handsets, HSPA handsets and LTE devices have not performed in RF terms as well as they might have done. This has made it harder to realise the theoretical spectral efficiency benefits achievable from the newer technologies. In parallel, there is scant evidence to suggest that energy efficiency on a bit-delivered basis has substantially improved other than as a result of increased network density that incurs additional capital and operational cost.

Once upon a time handset manufacturers and network hardware manufacturers were one and the same. Motorola manufactured handsets and networks, so did Nokia, so did Ericsson, so did Alcatel, even Nortel had a handset division.

One by one these companies either abandoned handset manufacture or decoupled their handset and network development and manufacturing activities. The result is that network equipment and system vendors no longer have control over handset performance. Unfortunately, as we shall see in the next chapter neither do operators. This is an even greater cause for concern.

5

Wireless User Hardware

5.1 Military and Commercial Enabling Technologies

I am not sure where to start in this chapter. The beginning is always a good idea but it is hard to settle on a date when wireless hardware first came into use. Arguably it could be the photophone that Alexander Graham Bell invented (referenced in Chapter 2) or before that early line-of-sight semaphore systems. The ancient Greeks invented the trumpet as a battlefield communications system and the Romans were fond of their beacons.[1]

In December 1907 Theodore Roosevelt dispatched the Great White Fleet, 16 battleships, six torpedo boat destroyers and 14 000 sailors and marines on a 45 000-mile tour of the world, an overt demonstration of US military maritime power and technology. The fleet was equipped with ship-to-shore telegraphy (Marconi equipment) and Lee De Forest's arc transmitter radio telephony equipment. Although unreliable, this was probably the first serious use of radio telephony in maritime communication and was used for speech and music broadcasts from the fleet – an early attempt at propaganda facilitated by radio technology.

In 1908 valves were the new 'enabling technology' ushering in a new 'era of smallness and power efficiency'. Marconi took these devices and produced tuned circuits – the birth of long-distance radio transmission. Contemporary events included the first Model T Ford and the issue of a patent to the Wright brothers for their aircraft design.

Between 1908 and the outbreak of war six years later there was an unprecedented investment in military hardware in Britain, the US and Germany that continued through the war years and translated into the development of radio broadcast and receiver technologies in the 1920s.

In the First World War portable Morse-code transceivers were used in battlefield communications and even in the occasional aircraft. Radio telegraphy was used in maritime communications including U boats in the German navy. These events established a close link between military hardware development and radio system development. This relationship has been ever present over the past 100 years. Examples include the use of FM VHF radio in the US army in the Second World War (more resilient to jamming) and the use of frequency

---

[1] Telecommunications is a combination of the ancient Greek word 'tele' meaning far off and the Latin word 'communicare', to impart.

Making Telecoms Work: From Technical Innovation to Commercial Success, First Edition. Geoff Varrall.
© 2012 John Wiley & Sons, Ltd. Published 2012 by John Wiley & Sons, Ltd.

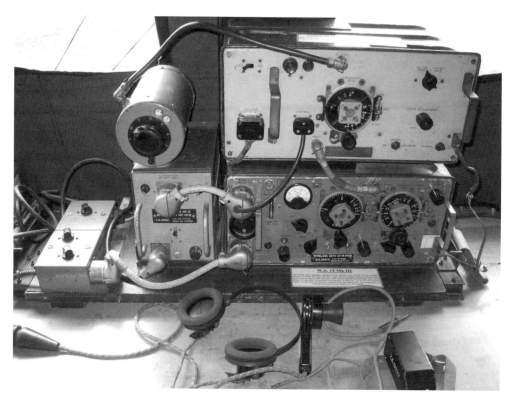

Figure 5.1 Pye WS19 dualband short-wave/VHF radio. Reproduced with permission of the Pye Telecom Historic Wireless Collection.[2]

hopping and spread spectrum techniques used in the military prior to their more widespread use in commercial radio systems.

However, it is not unusual for innovation to flow in the opposite direction. Radio and radar systems in the Second World War borrowed heavily from broadcast technologies developed in the 1930s.

When the British expeditionary force arrived in France in 1939 they discovered that the German infantry had significantly superior radio communications, using short wave for back to base communications and VHF for short-range exchanges.

In the UK a rapid design project was put in place to produce a dualband short-wave/VHF radio that could offer similar capability. This has been described as 'the radio that won the war' and is certainly one of the earliest examples of a dualband dual-mode radio. The device shown in Figure 5.1 is a WS (wireless set) 19 with a separate RF amplifier No 2 producing 25 watts of RF power.

Long procurement cycles and a need to amortise costs over a significant service period can also mean that military systems remain in use for longer than civilian systems, sometimes as long as thirty years. If this coincides with rapid development in the commercial sector then

---

[2] http://www.pyetelecomhistory.org/.

military handsets can seem bulky and poorly specified when compared for example to modern cellular phones.

Additionally, the development of a radio technology for military applications does not necessarily guarantee that the technology will be useful in a commercial context. Military radio equipment does not have the scale economy that cellular phones enjoy. On the other hand, military radios do not have the same cost constraints and may not have the same weight constraints and may be able to use more exotic battery technologies than cellular phones. Components can also be aggressively toleranced to improve performance.

This means that the development of a radio technology for military applications does not necessarily guarantee that the technology will be useful in a commercial context. Two examples are software-defined radio, a radio that can self-redefine its modulation and channel coding and cognitive radio, a radio that can scan and then access available radio resources.

Software-defined radio is used extensively in military radio systems, partly to overcome interoperability issues. Translating software-defined radio techniques to cellular radio has, however, proved problematic partly because of cost and performance issues including battery drain but also because in the commercial sector interoperability issues have been addressed more effectively through harmonisation and standardisation though this is now beginning to change.

Cognitive radio provides another example. Usually combined with software-defined radio, cognitive radio is useful in a military context because radio systems can be made flexible and opportunistic. This makes them less vulnerable to interception and jamming. Cognitive radio is presently being proposed for use in white-space devices in the US UHF TV band and being propositioned for use in other markets including the UK. These devices are designed to sit in the interleaved bands not being used by TV broadcasting (covered in more detail in Chapter 15)

There is, however, an important difference. In a military context it does not necessarily matter if cognitive radios interfere with other radio systems. In a commercial context, a guarantee that radios will not create interference is a precondition of deployment. Interference avoidance in the white-space world is based on a composite of location awareness and channel scanning to detect the presence or absence of TV channels and beacon detection to detect the presence of wireless microphones.

However robust these mechanisms might be, other users of the band, TV broadcasters, wireless microphones and more recently 700-MHz cellular operators at the top of the band and public safety users will need to be convinced that white-space devices do not present a threat either technically or commercially. We return to this topic in Chapter 15.

There are also similarities between military radios and specialist mobile radios including radio equipment used by public protection and disaster-relief agencies. This includes a need to ruggedise equipment, make equipment waterproof rather than splashproof, make equipment dustproof and secure. Encryption and access security are covered in Chapter 10 of 3G Handset and Network Design.

In the US specialist mobile radio user equipment and network equipment is standardised by the Association of Public Communication Officers. Analogue radios were specified as APCO 16 and digital radios are specified as APCO 25. In the UK and parts of Asia these radios comply with the TETRA radio standard though a standard called TETRAPOL with narrower channel spacing is used in some countries including France. The physical layer implementation of these radios is covered in the wireless radio access network hardware of Chapter 12.

Figure 5.2 Inmarsat IsatPhone Pro hand-held terminal. Reproduced with permission of Inmarsat Global Limited.

APCO-compliant user devices have 3-watt power amplifiers to deliver deep rural coverage and are relatively narrowband with narrow channel spacing in order to realise good sensitivity on the receive path. Other specialist low-volume but high-value markets include satellite phones. Inmarsat, for example, have a new portable handset[3] designed by EMS Satcom (hardware) and Sasken (the protocol stack). This is illustrated in Figure 5.2.

Light Squared[4] have plans to use a version of this phone in a dual-mode network in the US combining terrestrial coverage with geostationary satellite coverage covered in more detail in Chapter 16.

---

[3] http://www.isatphonelive.com/.
[4] http://www.lightsquared.com/.

Table 5.1 Degrees of smartness

Smart Phones	Smarter phones	Even smarter phones	Even smarter smarter	Very smart phones	Even smarter very smart phones	The smartest phones
1985					2007–2012	The future?
Voice						
Voice	Text					
Voice	Text	Data	Audio			
Voice	Text	Data	Audio	Imaging		
Voice	Text	Data	Audio	Imaging	Positioning and touch-screen displays	
Voice	Text	Data	Audio	Imaging	Positioning and touch-screen displays	Sensing

Other satellite networks, including the Iridium low-earth orbit satellite network, have radios that are optimised to work on a relatively long distance link – 35 000 kilometres for geostationary or 700 kilometres for low-earth orbit. These networks support voice services, data services and machine to machine applications (covered in more detail in Chapter 16)

Specialist phones for specialist markets include Jitterbug,[5] a phone optimised for older people manufactured by Samsung and a phone being developed for blind women.[6]

5.2 Smart Phones

But this chapter was supposed to be predominantly addressing the evolution of mobile phone and smart phone form factors and our starting point has literally just arrived by group e mail from my friend Chetan Sharma in the US announcing that smart phone sales have just crossed the 50% share mark in the US and that the US now accounts for one third of all smart phone sales in the world (as at lunch time Monday 9 May 2011).[7]

The term smart phone has always been slightly awkward – is the phone smart or is it supposed to make us smart or at least feel smart and all phones are smart so presumably we should have categories of smart, smarter and smartest phones.

Over the last twenty five years the evolution of smartness expressed in terms of user functionality can be described as shown in Table 5.1.

In the beginning, cellular phones were used mainly for business calls; they were expensive, and predominantly used for one-to-one voice communication.

Then text came along, then data, originally a 1200-bps or 2400-bps FFSK data channel in first-generation cellular phones, then audio, imaging and positioning and most recently a range of sensing capabilities. Smart phones were originally intended to help make us more

---

[5] http://www.samsung.com/us/mobile/cell-phones/SPH-A310ZRADYN.
[6] http://www.zone-v.com.
[7] http://www.chetansharma.com/usmarketupdateq12011.htm.

efficient, predominantly in terms of business efficiency. Mobile phones have become tools that we use to improve our social efficiency and access to information and entertainment. Both in business and socially, modern mobile phones allow people to be egocentric. Text has become twitter and cameras and audio inputs enable us to produce rich media user-generated content. Positioning and sensing are facilitating another shift and producing a new generation of devices that allow us to interact with the physical world around us.

Who would have thought, for example, that a whole new generation of users would master SMS texting through a phone keypad, achieving communication speeds that are comparable with QWERTY key stroke entry albeit with bizarre abbreviations.

The question to answer is how fast are these behavioural transforms happening and the answer to that can be found by looking at how quickly or slowly things have changed in the past.

A time line for cellular phone innovations can be found at the referenced web site[8] but this is a summary.[9] Cellular phones did not arrive out of the blue twenty five years ago. The principle of cellular phones and the associated concepts of frequency reuse were first described by DH Ring at the Bell Laboratories in 1947, a year before William Shockley invented the transistor.

In 1969 Bell introduced the first commercial system using cellular principles (handover and frequency reuse) on trains between Washington and New York. In 1971 Intel introduced the first commercial microprocessor. Microprocessors and microcontrollers were to prove to be key enablers in terms of realising low-cost mass market cellular end-user products.

In 1978 Standard Telecommunication Laboratories in Harlow introduced the world's first single-chip radio pager using a direct conversion receiver and the Bahrain Telephone company opened the first cellular telephone system using Panasonic equipment followed by Japan, Australia, Kuwait and Scandinavia, the US and the UK.

In 1991 the first GSM network went live at Telecom Geneva.

In 1993 the first SMS phones became available.

In 1996 triband phone development was started and an initial WCDMA test bed was established

In 1997 Nokia launched the 6110, the first ARM-powered mobile phone but significant because it was very compact, had a long battery life and included games. The Bluetooth Special Interest Group was founded in September.

The world's first dualband GSM phone the Motorola 8800 was launched in the spring.

The first triband GSM phone the Motorola Timeport L7089 was launched at Telecom 99 in Geneva and included WAP, a wireless access profile designed to support web access. WAP was slow and unusable. On 5 March 2002 Release 5 HSDPA was finalised.

Hutchison launched the '3' 3G service on 3 March 2003.

In 2010 100 billion texts[10] were sent from mobile phones.

The particular period we want to look at is from 2002/2003 onwards, comparing state-of-the-art handsets then and now. The objective is to identify the hardware innovations that have

---

[8] http://www.cambridgewireless.co.uk/cellular25/technology/.

[9] A nice pictorial history of mobile phones can also be found at http://www.cntr.salford.ac.uk/comms/mobile.php.

[10] An SMS text is limited to 140 bytes or 1120 bits that is equivalent to 160 seven-bit characters or seventy 16-bit Unicode characters using UCS2. UCS 2 is used for Arabic, Chinese, Japanese and Korean script. Concatenated SMS or long SMS can be used to string multiple SMS texts together but is not supported in all phones. Enhanced SMS supports the inclusion of image and audio and animation but is not supported by all phones.

created significant additional value for the industry value chain and identify how and why they have added value. These include audio capabilities, imaging capabilities, macro- and micropositioning coupled with touch-screen interactive displays and last but not least sensing.

5.2.1   Audio Capabilities

Adding storage for music was an obvious enhancement that came along in parallel with adding games in phones and was the consequence of the rapid increase in storage density and rapidly decreasing cost of solid-state memory.

Rather like ring tones, operators and handset vendors, in particular Nokia despite substantial investment have failed to capitalise on the audio market that has largely been serviced by other third parties for example iTunes and Spotify.

In February 2010 Apple announced that 10 billion songs had been downloaded from the iTunes store[11] with users able to choose from 12 million songs, 55 000 TV episodes and 8500 movies 2500 of which were high-definition video.

Given that it is easy to side load devices from a PC or MAC there seems to be insufficient incentive to download music off air. Even if it's free it flattens the battery in a mobile device. There is even less incentive to download TV or movie clips. This is just as well as the delivery cost to the operator would probably exceed the income from the download which in any case goes to Apple not the operator.

Audio quality from in-built speakers as opposed to head phones has not improved substantially even in larger form factor products – lap tops including high-end lap tops as one example still have grim audio performance unless linked to external audio devices.

Innovations in 2002 included new flat speaker technologies actuated across their whole surface with NXT as the prime mover in the sector. NXT no longer exists as a company though a new Cambridge-based company Hi Wave Technologies PLC is working on similar concepts in which the display can double as a speaker and haptic input device.[12]

So much for audio outputs, how about audio inputs in particular the microphone.

Emile Berliner is credited with having developed the first microphone in 1876 but microphone technology moved forwards in leaps and bounds with the growth of broadcasting in the 1920s.

George V talked to the Empire on medium wave-radio in 1924 from the Empire Exhibition at Wembley using a Round–Sykes Microphone developed by two employees of the BBC, Captain Round and AJ Sykes. Over time microphones came to be optimised for studio or outdoor use, the most significant innovations being the ribbon microphone and lip microphone.

Marconi–Reisz carbon microphones were a BBC standard from the late 1920s to the late 1930s and were then gradually supplanted by the ribbon microphone. Over the following decades an eclectic mix of microphone technologies evolved including condenser microphones, electromagnetic induction microphones and piezoelectric microphones.

Over the past ten years the condenser microphone has started to be replaced by MEMS-based devices sometimes described as solid-state or silicon microphones. The devices are usually integrated onto a CMOS chip with an A to D converter.

---

[11] CUPERTINO, California—February 25, 2010—Apple® today announced that music fans have purchased and downloaded over 10 billion songs from the iTunes® Store (www.itunes.com).

[12] http://www.hi-wave.com/.

The space saved allows multiple microphones to be used in consumer devices and supports new functions such as speech recognition. These functions have existed for years but never previously worked very well. This new generation of microphones also enable directional sensitivity, noise cancellation and high-quality audio recording. Companies such as Wolfson Microelectronics, Analogue Devices, Akustica, Infineon, NXP and Omron are active in this sector.

Audio value is probably more important than we realise. It has powerful emotional value. Audio is the last sense we lose before we die, it is the only sensing system that gives us information on what's happening behind us. Perceptions of video quality improve if audio quality is improved.

Ring tones also sound better. How and why ring tones became a multibillion pound business remains a compete mystery to me but then that's why I don't drive a Porsche and have time to write a book like this.

5.2.2   Positioning Capabilities

Positioning includes micropositioning and macropositioning. Many iPhone applications exploit information from a three-axis gyroscope, an accelerometer and a proximity sensor to perform functions such as screen reformatting. This is micropositioning. Macropositioning typically uses GPS usually combined with a digital compass.

Gyroscopes are MEMS elements that sense change in an object's axis of rotation including yaw, pitch and roll. Traditionally, a separate sensor would be needed for each axis but the example shown from STMicroelectronics in Figure 5.3 uses one MEMS-based module taking up 4 by 4 by 1 mm. An accelerometer takes up about 3 mm square. By the time you read this these two devices will probably be available in a single package. The devices have a direct cost but an indirect real-estate cost and a real-estate value. The value is realised in terms of additional user functionality. In the previous chapter we referenced products in which the display turns off when the phone is placed to the ear.

The gyroscope can be set at one of three sensitivity levels to allow speed to be traded against resolution. For example, for a gaming application it can capture movements at up to 2000 degrees per second but only at a resolution of 70 millidegrees per second per digit. For the point-and-click movement of a user interface – a wand or a wearable mouse – the gyroscope can pick up movements at up to 500 degrees per second at a resolution of up to 18 millidegrees per second per digit.

The 3-axis gyroscope illustrated draws 6 milliamps compared to 5 milliamps for each of the three devices needed two years ago. The device can be put into sleep mode and woken up when needed within 20 milliseconds, for example to support dead reckoning when a GPS signal is lost. MEMS-based devices measure the angular rate of change using principles described by Gaspard G de Coriolis in 1835 and now known as the Coriolis effect.

The diagram in Figure 5.4 from an Analogue Dialogue article[13] shows that as a person or object moves out towards the edge of a rotating platform the speed increases relative to the ground, as shown by the longer blue arrow.

---

[13] http://www.analogue.com/library/analogDialogue/.

Figure 5.3   STMicroelectronics MEMS-based module. Reproduced with permission of STMicro-electronics.

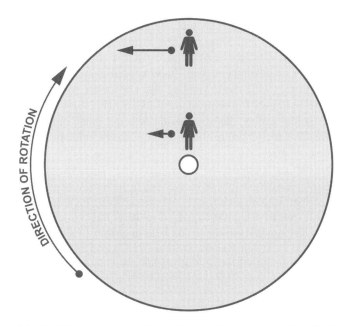

Figure 5.4   Coriolis acceleration. Reproduced with permission of Analog Devices.

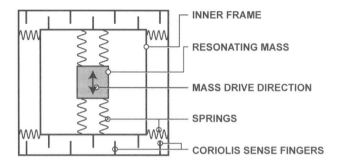

Figure 5.5   Springs and mechanical structure. Reproduced with permission of Analog Devices.

The movement is captured by an arrangement of sensors, as shown in the springs and roundabouts diagram, see Figure 5.5 above.

Integrating microsensing with GPS based macrosensing is covered in more detail in Chapter 16.

5.2.3   Imaging Capabilities

Imaging includes imaging capture through a camera or multiple cameras and image display. Camera functionality is also covered in Chapter 7. For the moment we want to just study the technology and business impact of display technologies. First, one that didn't make it.

In Chapter 5 of 3G Handset and Network Design we featured a miniaturised thin cathode ray tube based on an array of microscopic cathodes from a company called Candescent. The thin CRT had a number of advantages over liquid crystal displays including 24-bit colour resolution, high luminance, wide-angle viewing and a response time of 5 milliseconds that compared with 25 milliseconds for contemporary LCDs. The company announced plans in 1998 to invest $450 million in a US-based manufacturing plant in association with Sony. Candescent filed for Chapter 12 protection in 2004. Candescent.com is now a web site for Japanese teeth whitening products.

The moral of this particular tale is that it is hazardous to compete with established legacy products (liquid crystal displays) that have significant market scale advantage that in turn translates into research and development and engineering investment to improve performance and lower cost. This is exactly what has happened to the liquid crystal display supply chain.

The LCD business has also effectively scaled both in terms of volume and screen size with screens in excess of 100 inches that as we shall see in Chapter 6 has a specific relevance to future smart phone functionality. The lap top market owes its existence to LCD technology and lap tops as we know now come with embedded WiFi and either a 3G dongle or sometimes embedded 3G connectivity.

The global lap top market is of the order of 100 million units or so per year though the number depends on whether newer smaller form factors such as the iPad are counted as portable computers. A similar number of television sets are sold each year. Even combined, these numbers are dwarfed by mobile phone shipments some of which are low value, an increasing number of which are high value.

So the question is will the iPad change the lap top market in the same way that the iPhone has changed the smart-phone market.

By March 2011 Apple claimed (3rd March press release) to have shipped a total of 100 million iPhones. iPad shipments since market launch nine months prior totalled 15 million. This is a modest volume compared to Nokia. Nokia have suffered massive share erosion from over 40 to 27% but this is still 27% of 1.4 billion handsets or 378 million handsets per year. A design win into an Apple product is, however, a trophy asset and fiercely fought over by baseband and RF and display component vendors.

This gives them a design influence that is disproportionate to their market volume, though perhaps not to their market value. Apple sales volume into the sector represents at time of writing about 4% of the global market by volume but captures something of the order of 50% of the gross margin achieved across the sector.

This means that component vendors are in turn under less price pressure. The iPhone has an effective realised price including subsidies and rebates that is higher than many lap tops and although the display is small relative to a lap top its added value is higher and its cost is lower. The display in an iPhone 4 for example is estimated[14] to have a realised value of about ten dollars, rather more than the RF components used in the phone.

So while we are on the topic of imaging let's skip to another value domain – display-based interaction.

5.2.4   *Display-Based Interaction – A Game Changer*

When Apple released its league table of the best-selling iPhone apps from its 'first 10 billion' downloads earlier this year, 8 out of 10 were games. Clearly games on mobile make money but the value realisable from games is at least partially dependent on the screen and screen input capabilities of the host device.

The iPhone and the iPad both use capacitive touch screen displays with a flexible transparent surface and transparent conductive layer. An oleophobic coating is used to stop greasy fingers leaving marks on the screen.

The capacitors are organised in a coordinate system. Changes in capacitance across the coordinate points are input into gesture interpretative software that is referenced against knowledge of the application that is running at the time. This makes possible all those thumb and finger pinch and spread features that people now take for granted. Simple really but no one had thought to do this prior to 2007 or rather no one had actually done it. Apple benefited substantially from having first-mover advantage.

Table 5.2 shows some of the added feature sets across the first four generations of iPhone.

The iPhone 4 has a 3.5-inch diagonal 24-bit colour display with 960 by 640 pixel resolution at 326 ppi, an 800 to 1 contrast ratio, 500 candelas per sq metre brightness and a 5.25-watt hour battery. As a comparison, the Nokia 9210, which was a market leading smart phone (shown in Figure 5.6) product in 2002, had a 4000 colour (12 bit) 110 by 35 mm display at 150 dpi with a

---

[14] http://www.isuppli.com/Teardowns/News/Pages/iPhone-4-Carries-Bill-of-Materials-of-187-51-According-to-iSuppli.aspx.

Table 5.2 Four generations of iPhone

January 2007	July 2008	June 2009	June 2010
iPhone	iPhone 3G	iPhone 3Gs	iPhone 4
2-megapixel camera	2-megapixel camera	3.2-megapixel camera	5-megapixel camera plus video camera
	Assisted GPS	Compass Voice control	Ambient light sensor
WiFi and GSM	WiFi and EDGE	3G wide-area connectivity	

refresh driver cycle time of 50 milliseconds (limiting the frame rate to 12 frames per second). The device ran off a 1300-mAh lithium ion battery. A contemporary 2011 N8 Nokia smart phone by comparison has a 16.7 million colour capacitive touch screen display with orientation sensor (accelerometer), compass (magnetometer) and supports high-definition video recording and play back with 16 GB of memory, all running off a 1200 mAh battery (Figure 6.4 in Chapter 6).

The iPad takes the same principles of the iPhone but translates them into a larger tablet form factor with a 9.7-inch diagonal display with 1024 by 768 pixel resolution. The device has back and front cameras and can do high-definition video recording at VGA resolution at up to 30 frames per second output to an HDTV. The device is powered by a 25-watt lithium polymcr battery.

Figure 5.6 Nokia 9210 Introduced in 2002. Reproduced with permission of Nokia Corporation.

5.2.5   Data Capabilities

Data is data but includes text, which includes as we said earlier Twitter and related text-based social-networking entities.

Much of the narrative of Chapter 5 and other chapters in our original 3G Handset and Network Design centred on the transition from delivery bandwidth added value to memory bandwidth added value.

This happened much as predicted but we missed the significance of search algorithms as a key associated enabling technology. This had been spotted in 1998 by a newly incorporated company called Google, now worth upwards of $150 billion dollars and delivering content from over one million servers around the world. A topic we return to in our later chapters on network software.

Text is well documented as being a surprise success story of the mobile phone sector with the highest value per delivered bit of all traffic types and the lowest per bit delivery cost.

However, another chapter (literally) is unfolding in terms of text-based value with telecoms as part of the delivery process – the e book, a product sector created by the electrophoretic display. Electrophoretic displays were described in Chapter 5 of 3G handset and Network Design but at that stage (2002) were not used in mainstream consumer appliance products. Electrophoretic displays do not have a backlight but rely on reflected light. This means they work well in sunlight or strong artificial light. Electrophoretic displays also do not consume power unless something changes. For this reason they are often described as bistable displays.

In an electrophoretic bistable display, titanium oxide particles about a micrometre in diameter are dispersed in a hydrocarbon oil into which a dark-coloured dye is introduced with surfactants and charging agents that cause the particles to take on an electric charge. This mixture is placed between two parallel, conductive plates. When a voltage is applied the particles migrate electrophoretically to the plate bearing the opposite charge. The particles that end up at the back of the display appear black because the incident light is absorbed by the dye; those at the front appear white because the light is reflected back by the titanium oxide particles.

These displays have enabled a new product sector to evolve, the e book, with the Kindle as, arguably the exemplar product in the sector. Table 5.3 shows the typical performance today of one of these displays.

The Kindle 1 was introduced at a Press Conference hosted by Jeff Bozo of Amazon on 19 November 2007 as a product that 'kindles' (sets fire to) the imagination (and sells books via Amazon in the process). The Kindle 2 was launched in February 2009, moving from a 4-level to 16-level grey display and packed in to a substantially slimmer form factor.

Initial market launches were in the USA. The device had a CDMA 1EV DO transceiver supplied by Qualcomm. Later versions for Europe and rest of the world are 3G enabled.

The device can hold up to 3500 books; everything is free except the books of which there 650 000 to choose from, plus autodelivered newspapers, magazines and blogs. Some of the books are free as well. It takes about 60 seconds to download a book and your book and the collection is backed up remotely. Extracts can be shared with friends via an integrated Twitter or Face book function. The QWERTY keyboard can also be used to annotate text and there is an in-built dictionary and access to Wikipedia plus an ability to edit PDF documents.

The differences between the Kindle 2 and the Kindle 1 include a redesigned key board. The ambition of the Kindle and other e books is to reinvent and enhance the reading experience and change the economics of book distribution by reducing the cost of delivery.

Table 5.3 Electrophoretic display

Display Type	Reflective electrophoretic
Colours	Black, white and grey-scale
Contrast ratio	10 : 1 minimum
White state	70 minimum
Dark state	24 maximum
Bit depth	4-bit grey, 16 levels
Reflective	40.7%
Viewing angle	Near 180 degrees
Image update time	120 milliseconds
Size	2 to 12 inches (diagonal)
Thickness	1.2 mm
Resolution	Better than 200 dpi
Aspect ratio	Typically 4 : 3

On 27 January 2011 Amazon.com Inc. announced a 36% rise in revenue for the final quarter of 2010 with revenue of $12.95 billion and a net income of $416 million or 91 cents per share compared to $384 million or 85 cents per share a year earlier.

Analysts were disappointed and suggested the costs associated with the Kindle were having an overall negative impact on the business with each unit being shipped at a loss once product and distribution costs including payments to wireless carriers were factored in. Amazon has never published sales figures for the device though stated in a Press Conference that they had shipped more Kindle 3 units than the final book in the Harry Potter series. Additionally, operators are beginning to sell advertisement-supported Kindles with 3 G connectivity at the same price as WiFi-only models.[15]

Given that this is a business with a market capitalisation of $90 billion it can be safely said that some short-term pressure on margins will probably turn out to be a good longer-term investment and this may well be an example of short-term investor sentiment conflicting with a sensible longer-term strategic plan. Only time will tell.

Time will also tell if other bistable display options prove to be technically and commercially viable. Qualcomm, the company that provides the EVDO and now 3 G connectivity to the Kindle have invested in a MEMS-based bistable display described technically as an interferometric modulator and known commercially by the brand name Mirasol.

The display consists of two conductive plates one of which is a thin-film stack on a glass substrate, the other a reflective membrane suspended over the substrate. There is an air gap between the two. It is in effect an optically resonant cavity. When no voltage is applied, the plates are separated and light hitting the substrate is reflected. When a small voltage is applied, the plates are pulled together by electrostatic attraction and the light is absorbed, turning the element black. When ambient light hits the structure, it is reflected both off the top of the thin-film stack and off the reflective membrane. Depending on the height of the optical cavity, light of certain wavelengths reflecting off the membrane will be slightly out of phase with the light reflecting off the thin-film structure. Based on the phase difference, some wavelengths will

---

[15] AT and T Mobility US press release, Summer 2011.

constructively interfere, while others will destructively interfere. Colour displays are realised by spatially ordering IMOD[16] elements reflecting in the red, green and blue wavelengths. Given that these wavelengths are between 380 and 780 nanometres the membrane only has to move a few hundred nanometres to switch between two colours so the switching can be of the order of tens of microseconds. This supports a video rate capable display without motion-blur effects. As with the electrophoretic display if the image is unchanged the power usage is near zero.

5.2.6 Batteries (Again) – Impact of Nanoscale Structures

One of the constraints in the design of portable systems is the run time between recharging. Battery drain is a function of all or certainly most of the components and functions covered in this chapter.

Batteries are a significant fraction of the size and weight of most portable devices. The improvement in battery technology over the last 30 years has been modest, with the capacity (Watt-hours kg) for the most advanced cells increasing only by a factor of five or six, to the present value of 190 Wh/kg for lithium-ion. Other performance metrics include capacity by volume, the speed of charge and discharge and how many duty cycles a battery can support before failing to hold a charge or failure to recharge or failure to discharge. Self-discharge rates also vary depending on the choice of anode and cathode materials and electrolyte mix. For example, lithium-ion batteries using a liquid electrolyte have a high energy density but a relatively high self-discharge rate, about 8% per month. Lithium polymer self discharge at about 20% per month. As stated in Chapter 3 high-capacity batteries can have high internal resistance, which means they can be reluctant to release energy quickly. Lead acid batteries for example can deliver high current on demand, as you will have noticed if you have ever accidentally earthed a jump lead. Using alternative battery technologies such as lithium (same capacity but smaller volume and weight) requires some form of supercapacitor to provide a quick-start source of instant energy.

The University of Texas[17] for example is presently working on nanoscale carbon structures described generically as graphene that act like a superabsorbent sponge. This means that a supercapacitor could potentially hold an energy capacity similar to lead acid with a lifetime of 10 000 charge–discharge cycles. Although automotive applications are the most likely initial mass market application there may be scaled down applications in mobile devices that require intermittent bursts of power.

Energy density for more mainstream power sources in general is presently improving at about 10% per year. The energy density of high explosive is about 1000 Wh/kg so as energy density increases the stability of the electrochemical process becomes progressively more important.

The lead acid battery was discovered by Gaston Plante in 1869. Batteries store and release energy through an electrochemical oxidation–reduction reaction that causes ions to flow in a particular direction between an anode and cathode with some form of electrolyte in between, for example distilled water in a lead acid battery. The first lithium-ion-based products were

---

[16] Interferometric modulator display.
[17] http://bucky-central.me.utexas.edu/publications/journal_in_group.html.

introduced by Sony in 1990 and have now become the workhorse of the cellular phone and lap top industry.

Lithium is the third element of the periodic table and the lightest element known on earth. Most of the world's lithium supply is under a desert in Bolivia dissolved in slushy brine that is extracted and allowed to dry in the sun. It is a highly reactive substance often combined with a combustible electrolyte constrained in a space. Much of the material cost, as much as 90%, is absorbed into overcharge/overdischarge protection. Even so, from time to time there are incidents when batteries ignite. In April 2006 Sony, Lenovo and Dell encountered battery problems and had to do a major product recall. The problem becomes more acute in applications requiring relatively large amounts of energy to be stored, electric vehicles being one example. Nonflammable electrolytes including polymer electrolytes are proposed as one solution. Petrol can be pretty explosive too, so this is one of those technical problems that can almost certainly be solved once industrial disincentives to innovate (sunk investment in petrol engine development) have evaporated. Similar industrial disincentives exist in the telecommunications industry. RF power-amplifier vendors for example are understandably reluctant to produce wideband amplifiers that mean six amplifiers are replaced with one amplifier, particularly if the unit cost stays the same.

But back to chemistry. Interestingly, electrochemical reactions at the nanoscale are different from electrochemical reactions at the macroscale. This opens up opportunities to use new or different materials. The small particle size of nanoscale materials allows short diffusion distances that means that active materials act faster. A larger surface area allows for faster absorption of lithium ions and increased reactivity without structural deterioration.

Shape changing includes structural change or surface modification, sometimes described as flaky surface morphology. A shape or surface change can increase volumetric capacity and/or increase charge and discharge rates.

For example, carbon nanotubes have been proposed as anodes. Silicon is a possible anode material with high capacity in terms of mAh/g, though the capacity fades quickly during initial cycling and the silicon can exhibit large changes in volume through the charge/discharge cycle. Film-based structures might mitigate these effects. Conventional thinking suggests that electrode materials must be electronically conducting with crystallographic voids through which lithium ions can diffuse. This is known as intercalation chemistry – the inclusion of one molecule between two other molecules. Some 3D metal oxide structures have been shown to have a reversible capacity of more than 1000 mAh/g. Low-cost materials like iron and manganese have also been proposed.

One problem with lithium-ion batteries has been cost and toxicity with early batteries using cobalt as a cathode material. Cobalt has now largely been replaced with lithium manganese. In 2002 other battery technologies being discussed included zinc air and methanol cells, but these have failed to make an impact on the mobile phone market for mainly practical reasons.

Batteries have some similarities with memory – both are storage devices and both achieve performance gain though materials innovation, packaging and structural innovation. Memory technologies are addressed in more detail in Chapter 7.

Batteries and memory are both significant markets in their own right both by volume and value. New and replacement sales of batteries into the mobile-phones sector alone are approaching two billion units per year and account for between 4 and 5% of the individual BOM of each phone. Even assuming say between five and ten dollars per phone this is a $10 to $20 billion market.

Memory represents a higher percentage of the BOM at around 9 to 10% though with less year-on-year replacement sales (apart from additional plug-in memory). We address the technology economics of memory in Chapter 7.

5.3   Smart Phones and the User Experience

Battery constraints are of course one of the two caveats that apply to the potential usability and profitability of smart phones. If a smart-phone battery runs flat by lunchtime then two things happen. The user gets annoyed and the operator loses an afternoon of potential revenue. The smarter smart phones deal with this by automatically switching off functions that are not being used. For example, there is no point in doing automatic searches for WiFi networks if the user never uses WiFi. There is no point in having a display powered up if the user cannot see it.

Smart phones as stated earlier can also hog a disproportionate amount of radio access bandwidth, a composite of the application bandwidth and signalling bandwidth needed to keep the application and multiple applications alive.

This does not matter if there is spare capacity, but if a network is running hot (close to full capacity) then voice calls and other user data sessions will be blocked or dropped. When blocked sessions or dropped sessions reach a level at which they become noticeably irritating to the user then customer support and retention costs will increase and ultimately churn rates will increase, which is very expensive for the industry.

As this paragraph was written (17 May 2011) Vodafone's annual results were announced[18]

> Vodafone's annual profits received a smart phone boost despite an overall fall in net profits of 7.8 per cent. Despite a £6 bn+ impairment charge on operations in its European "PIIGS" markets (Portugal, Ireland, Italy, Greece, Spain), a 26.4 per cent boost in mobile data revenue made a significant contribution to its overall profit of £9.5 bn for the year.

This suggests that smart phones are unalloyed good news for operators. However, this is only true if the operator can build additional capacity cost effectively. This may of course include network sharing or acquisition. However this is achieved it is wise to factor opportunity cost into any analysis of device economics. A device is only profitable if the income it generates is greater than the cost of delivery.

5.3.1   Standardised Chargers for Mobile Phones

Anyway, if a user's battery goes flat at lunchtime, at least he/she will soon be able to borrow a charger to substitute for the one they left at home. 2011 was the year in which a common charger standard was agreed using the micro-USB connector already used for the data connection on many smart phones and finally overcoming the frustration and waste of multiple connection techniques. Hurrah we say.

---

[18] http://www.vodafone.com/content/annualreport/annual_report11.

5.4 Summary So Far

The thesis so far in this chapter is that innovations in display technologies have created two new market sectors.

Electrophoretic displays have created an e book sector that did not previously exist.

Displays with capacitive sensing and touch-screen 'pinch and spread' functionality have reinvented and repurposed the smart phone.

The Nokia 9210 featured by us in 2002 was essentially a business tool. Its purpose was to make us more efficient.

The iPhone's use of capacitive sensing has transformed a business tool into a mass-market product that changes the way that people behave, interact with each other and manage and exploit information. The addition of proximity sensing, ambient light and temperature sensing, movement and direction sensing both at the micro- and macrolevel enable users to interact with the physical environment around them and if used intelligently can help battery life by turning off functions that cannot or are not being used.

Fortuitously, wide-area connectivity is an essential part of the added value proposition of all of these newly emerged or emergent product sectors.

The iPhone has also changed the anatomy of the smart phone. Figure 5.7 shows the main components including RF functions.

The added value of these components is, however, realised by different entities.

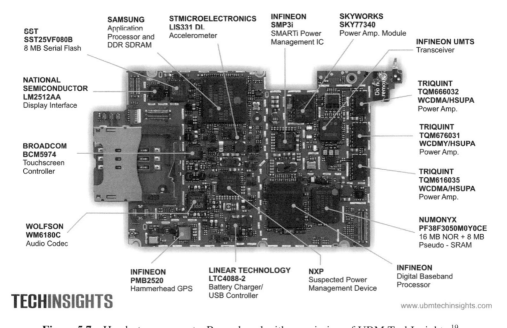

Figure 5.7   Handset components. Reproduced with permission of UBM TechInsights.[19]

---

[19] http://www.ubmtechinsights.com.

Some of the components support functionalities that help sell the phone. In the iPhone the accelerometer would be one example, the GPS receiver another. These components also realise value to companies selling applications from an APP store, for example a gaming application or navigation-based or presence-based application.

The memory functions are either user facing, supporting user applications or device facing, supporting protocol stack or physical layer processing in the phone (covered in more detail in Chapter 7).

The application processor, display interface, touch-screen controller, audio codec, battery charger and USB controller and power-management chip and digital baseband processor all perform functions that are necessary for and/or enhance the user experience and in some cases help sell the phone in the first place, for instance the display and audio functions.

The network operator is, however, predominantly interested in how at least some of this added value can be translated into network value. This brings us to connectivity that is the domain of the transceiver and power-amplifier modules with some help from the baseband processor and all the other components that are not identified but in practice determine the RF connectivity efficiency of the phone.

As with displays and other user-facing functions, RF efficiency and enhanced functionality is achieved through a combination of materials innovation, manufacturing innovation and algorithmic innovation. Manufacturing innovation includes silicon scaling but also provides opportunities to exploit changes in material behaviour when implemented at the nanoscale level.

We have noted the increasing use of MEMS in for example 3-axis gyroscopes. Gyroscopes, accelerometers and ink-jet printer heads all constitute a significant part of the established $45 billion market for MEMS devices.

MEMS (microelectromechanical systems) already determine the performance of some parts of the front end of a mobile phone. The SAW and/or FBAR filters used in the transmit and receive path are an example. NEMS (nanoelectromechanical systems) are a logical next step.

Nanoscale devices are generally referred to as structures sized between 1 to 100 nanometres in at least one dimension. The devices require processes that can be implemented at atomic or molecular scale.[20] The deposition of a single layer of atoms on a substrate is one example. Structures exhibit quantum-mechanical effects at this scale. Materials behave differently. As a reminder a nanometre is one billionth of a metre. A microscale device is generally described as a structure sized between 1 and 1000 millimetres in one dimension. A millimetre is one thousandth of a metre.

One of the traditional problems with many RF components is that they do not scale in the same way that semiconductor devices scale. Applying NEMS techniques to RF components may change this. But first, let's remind ourselves of the problem that needs to be solved.

Mobile broadband data consumption is presently quadrupling annually and operators need to ensure that this data demand can be served cost efficiently. Part of the cost-efficiency equation is determined by the RF efficiency of the front end of the phone.

The increase in data consumption is motivating operators to acquire additional spectrum. However, this means that new bands have to be added to the phone. Each additional band supported in a traditional RF front end introduces potential added insertion loss and results

---

[20] A molecule consists of two or more atoms bonded together.

in decreased isolation between multiple signal paths within the user's device. Additionally, a market assumption that users require high peak data rates has resulted in changes to the physical layer that reduce rather than increase RF efficiency. Two examples are multiple antennas and channel bonding.

Economists remain convinced that scale economy in the mobile-phone business is not an issue, despite overwhelming industry evidence to the contrary. This has resulted in regulatory and standards making that have resulted in a multiplicity of band plans and standards that are hard to support in the front end of a user's device. The 'front end' is an industry term to describe the in and out signal paths including the antenna, filters, oscillators, mixers and amplifiers, a complex mix of active and passive components that are not always easily integrated. The physicality of these devices means they do not scale as efficiently as baseband devices.

These inefficiencies need to be resolved to ensure the long-term economic success of the mobile broadband transition. The band allocations are not easily changed so technology solutions are needed.

In Chapter 4 we listed the eleven 'legacy bands' and 25 bands now specified by 3GPP. The band combination most commonly supported in a 3 G phone is the original 3G band at 1900/2100 MHz, described as Band 1 in the 3GPP numbering scheme, and the four GSM bands, 850, 900, 1800 and 1900 MHz (US PCS).

We also highlighted changes to existing allocations, for example the proposed extension of the US 850 band and the related one-dB performance cost incurred. This directly illustrated the relationship between operational bandwidth, the pass band expressed as a ratio of the centre frequency of the band, and RF efficiency.

Additionally in some form factors, for example tablets, slates and lap tops, we suggested a future need to integrate ATSC TV receive and/or DVB T or T2 receiver functionality in the 700-MHz band.

Within existing 3GPP band plan boundaries there are already significant differences in handset performance that are dependent on metrics such as the operational bandwidth as a percentage of centre frequency, the duplex gap and duplex spacing in FDD deployments and individual channel characteristics determined, for example, by band-edge proximity.

It is not therefore just the number of band plans that need to be taken into account but also the required operational bandwidths, duplex spacing and duplex gap of each option expressed as a percentage of the centre frequency combined with the amount of unwanted spectral energy likely to be introduced from proximate spectrum, a particular issue for the 700- and 800-MHz band plans or L band at 1.5 GHz (adjacent to the GPS receive band).

The performance of a phone from one end of a band to the other can also vary by several dBs. Adaptive-matching techniques can help reduce but not necessarily eliminate this difference. At present, a user device parked on a mismatched channel on the receive path will be, or at least should be, rescued by the handover process but the associated cost will be a loss of network capacity and absorption of network power (a cost to others). So there are two issues there, a performance loss that translates into an increase in per bit delivery cost for the operator, and commercial risk introduced by either late availability or nonavailability of market competitive performance products capable of accessing these new bands.

The market risk is a function of the divide down effect that multiple standards and multiple bands have on RF design resource. The industry's historic rate of progress with new band implementation has been slow. GSM phones were single band in 1991, dualband by 1993, triband a few years later, quadruple band a few years later and quintuple band a few years

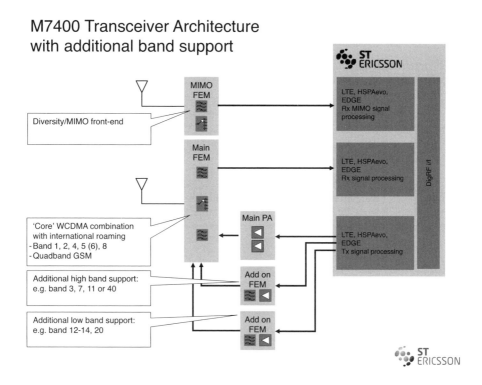

Figure 5.8 Block diagram of multiband multimode RF front end. Reproduced with permission of ST Ericsson.[21]

later. A five-band phone remains a common denominator product. Self-evidently if a band per year had been added we would have twenty-band user devices by now, the average has been a new band every three years or so yet here we have the international regulatory community assuming a transition from a five-band phone to a phone supporting ten bands or more within two to three years at the outside. This could only be done if a significant performance loss and significant increase in product cost was considered acceptable. Note that one operator's ten band phone will not be the same as another operator, further diluting available RF design and development capability.

The US 700-MHz band plan is particularly challenging for RF design engineers both in terms of user-device design and base-station design. We revisit this topic in Chapter 12, Staying with user-device design considerations, Figure 5.8 illustrates the complexity introduced even when a relatively small subset of 3GPP bands is supported.

Table 5.4 shows the core bands supported and the additional bands and band (either or) combination options. Note the separate MIMO front-end module (FEM) to support the diversity receive path.

---

[21] http://www.stericsson.com/home/home.jsp.

Table 5.4   Core band combination with international roaming support

Core band combination band 1 plus other bands for international roaming support

	Band 1	3G WCDMA/LTE	1900/2100 MHz
	Band 2	US PCS 1900	1800/1900 MHz
	Band 4	US AWS 1700	1700/2100 MHz
	Band 5 and 6	US 850 band (plus Asia and Japan)	824/840 869/894
	Band 8	Europe and Asia 900 MHz	880/915 925/960
Additional bands			
	Band 3	1800 MHz Europe and Asia	1710/1785 1805/1880
	Band 7	2.6 GHz band	
	Band 11	1.5 GHz for Japan	
	Bands 12 and 14	US 700 MHz	
	Band 20 or	European 800 MHz	
	Band 40	2.3 GHz for China	

To support progress to or beyond requires substantial RF component innovation. Although most of these components have already been introduced in Chapter 3 we now need to revisit present and future performance benchmarks and how these devices may potentially resolve the multiple band allocation problem described above and related cost and implementation issues.

5.5   RF Component Innovation

5.5.1   RF MEMS

The application of radio-frequency microelectromechanical system microscale devices is being expanded to include MEMS switches, resonators, oscillators and tuneable capacitors and (in the longer term) inductors that potentially provide the basis for tuneable filters.

RF MEMS are not a technical panacea and as with all physical devices have performance boundaries, particularly when used as filters. If any device is designed to tune over a large frequency range it will generally deliver a lower Q (and larger losses) than a device optimised for a narrow frequency band.

An alternative approach is to exploit the inherent smallness of RF MEMS devices to package together multiple components, for example packaging multiple digital capacitors on one die. The purpose of these devices to date has been to produce adaptive matching to changing capacitance conditions within a band rather than to tune over extended multiband frequency ranges.

In an ideal world additional switch paths would be avoided. They create loss and distortion and dissipate power. More bands and additional modes therefore add direct costs in terms of component costs and indirect costs in terms of a loss of sensitivity on the receive path and a loss of transmitted power on the transmit path. MEMS (microelectrical mechanical system) -based switches help to mitigate these effects.

There is also a potential power handling and temperature cycling issue. The high peak voltages implicit in the GSM TX path can lead to the dielectric breakdown of small structures, a problem that occurred with early generations of SAW filters. Because MEMS devices are mechanical, they will be inherently sensitive to temperature changes. This suggests a potential conflict between present ambitions to integrate the RF PA on to an RFIC and to integrate MEMS devices to reduce front-end component count and deliver a spectrally flexible phone.

The combination of RF power amplification and an RFIC is also problematic with GSM as the excess heat generated by the PA can cause excessive drift in the small signal stages of the receiver, including baseband gain stages and filtering. At the lower levels of UMTS and LTE this becomes less of a problem but still requires careful implementation. The balance between these two options will be an important design consideration. The optimal trade off is very likely to be frequency specific.

For example, if the design brief is to produce an ultralow-cost handset, then there are arguments in favour of integrating the RFPA on to the RFIC. However, this will make it difficult to integrate MEMS components onto the same device

MEMS are also being suggested as potential replacements for present quartz crystal-based subsystems. The potential to use microelectrical mechanical resonators has been the subject of academic discussion for almost 40 years and the subject of practical research for almost as long.

The problem with realising a practical resonator in a MEMS device is the large frequency coefficient of silicon, ageing, material fatigue and contamination. A single atomic layer of contaminant will shift the resonant frequency of the device. As with MEMS switches and filters, the challenge is to achieve hermetically robust packaging that is at least as effective as the metal or ceramic enclosures used for quartz crystals, but without the size or weight constraint. There are products now available that use standard CMOS foundry processes and plastic moulded packaging.

While some 2G handsets make do with a lower-cost uncompensated crystal oscillator, 3G and LTE devices need the stability over temperature and time, accuracy and low phase noise that can only presently be achieved using a voltage-controlled temperature-compensated device. Temperature-compensated MEMS oscillators (TCMO), particularly solutions using analogue rather than digital compensation, are now becoming a credible alternative and potentially offer significant space and performance benefits. A MEMS resonator is a few tenths of a millimetre across. A quartz crystal is a few millimetres across, one hundred times the surface area.

MEMS resonator performance is a function of device geometry. As CMOS geometries reduce, the electrode gap reduces and the sense signal and signal-to-noise ratio will improve, giving the oscillators a better phase noise and jitter specification. As MEMS resonators get smaller they get less expensive. As quartz crystals get smaller they get more expensive. MEMS resonators therefore become increasingly attractive over time.[22]

SAW filters use semiconductor processes to produce combed electrodes that are a metallic deposit on a piezoelectric substrate. SAW devices are used as filters, resonators and oscillators and appear both in the RF and IF (intermediate frequency) stages of present cellular handset designs. SAW devices are now being joined by a newer generation of devices known as BAW (bulk acoustic wave) devices.

In a SAW device, the surface acoustic wave propagates, as the name suggests, over the surface of the device. In a BAW device, a thin film of piezoelectric material is sandwiched between two metal electrodes. When an electric field is created between these electrodes,

---

[22] A white paper on MEMS resonators is available via this URL http://www.sand9.com/white-papers/.

an acoustic wave is launched into the structure. The vibrating part is either suspended over a substrate and manufactured on top of a sacrificial layer or supported around its perimeter as a stretched membrane, with the substrate etched away.

The devices are often referred to as film bulk acoustic resonators (FBAR). The piezoelectric film is made of aluminium nitride deposited to a thickness of a few tens of micrometres. The thinner the film, the higher the resonant frequency. BAW devices are useful in that they can be used to replace SAW or microwave ceramic filters and duplexers in a single component. BAW filters are smaller than microwave ceramic filters and have a lower height profile. They have better power-handling capability than SAW filters and achieve steeper roll-off characteristics. This results in better 'edge of band' performance.

The benefit apart from the roll-off characteristic and height profile is that BAR devices are inherently more temperature resilient than SAW devices and are therefore more tolerant of modules with densely populated heat sources (transceivers and power amplifiers). However, this does not mean they are temperature insensitive. BAR filters and SAW filters all drift with temperature and depending on operational requirements may require the application of temperature-compensation techniques. However, present compensation techniques compromise Q.

Twelve months ago a typical BAW duplexer took up a footprint of about 5 by 5 mm^2 with an insertion height of 1.35 mm. In the US PCS band or 1900/2100 band these devices had an insertion loss of about 3.6 dB on the receive path and 2.7 dB on the transmit path and delivered RX/TX isolation of 57 dB in the TX band and 44 dB in the RX band.

A typical latest generation FBAR duplexer takes up a footprint of about 2 by 2.5mm^2 and has an insertion height of 0.95 mm max. In the US PCS band or UMTS1900 band these devices have a typical insertion loss of about 1.4 dB on the receive path and 1.4 dB on the transmit path and deliver RX/TX isolation of 61 dB in the TX band and 66 dB in the RX band. More miniaturised versions (smaller than 2 by 1.6 mm) are under development.

SAW filters also continue to improve and new packaging techniques (two filters in one package) offer space and cost savings.

FBAR filters and SAW filters could now be regarded as mature technologies but as such are both still capable of substantial further development. Developments include new packaging techniques. For example, FBAR devices are encapsulated using wafer-to-wafer bonding, a technique presently being extending to miniaturised RF point filters and RF amplifiers within a hermetically sealed 0402 package.

SAW filters are, however, also continuously improving and many handsets and present designs assume SAW filters rather than FBAR in all bands up to 2.1 GHz though not presently for the 2.6-GHz extension band.

Some RF PA vendors are integrating filters with the PA. This allows for optimised (lower than 50 ohm) power matching, though does imply an RF PA and filter module for each band. There are similar arguments for integrating the LNA and filter module as a way of improving noise matching and reducing component count and cost. A present example is the integration of a GPS LNA with FBAR GPS filters into a GPS front-end module.

RF PA vendors often ship more than just the power amplifier and it is always an objective to try and win additional value from adjacent or related functions, for example supplying GaAs-based power amplifiers with switches and SAW and BAW filter/duplexers.

In comparison to SAW, BAW and FBAR filters, RF MEMS realistically still have to be characterised as a new technology in terms of real-life applications in cellular phones and mobile broadband user equipment. RF MEMS hold out the promise of higher-Q filters with good temperature stability and low insertion loss and could potentially meet aggressive cost expectations and quality and reliability requirements.

However, the NRE costs and risks associated with RF MEMS are substantial and need to be amortised over significant market volumes. Recent commercial announcements suggest that there is a growing recognition in the RF component industry that RF MEMS are both a necessary next step and a potential profit opportunity. The issue is the rate at which it is technically and commercially viable to introduce these innovations.

5.5.2   Tuneable Capacitors and Adaptive TX and RX Matching Techniques

The rationale for adaptive matching is that it can deliver TX efficiency improvements by cancelling out the mismatch caused by physical changes in the way the phone is being used. A hand around a phone or the proximity of a head or the way a user holds a tablet or slate will change the antenna matching, particularly with internal and/or electrically undersized antennas. A gloved hand will have a different but similar effect.

Mismatches also happen as phones move from band to band and from one end to the other end of a band. TX matching is done by measuring the TX signal at the antenna. In practice, the antenna impedance match (VSWR) can be anything between 3.1 and 5.1 (and can approach 10 to 1). This reduces to between 1.1 and 1.5 to 1 when optimally matched.

Adaptive matching of the TX path is claimed to realise a 25% reduction in DC power drain in conditions where a severe mismatch has occurred and can be corrected. However, in a duplex spaced band matching the antenna on the TX path to 50 ohms could produce a significant mismatch.

Ideally therefore, the RX path needs to be separately matched (noise matched rather than power matched). This is dependent on having a low-noise oscillator in the handset such that it can be assumed that any noise measured is from the front end of the phone rather than the oscillator. Matching can then be adjusted dynamically. Note that low phase noise is also needed to avoid reciprocal mixing to avoid desensitisation.

Adaptive matching is by present indications one of the more promising new techniques for use in cellular phones. The technique is essentially agnostic to the duplexing method used. It does, however, offer the promise of being able to support wider operational band-widths though this in turn depends on the tuning range of the matching circuitry, the loss involved in achieving the tuning range and the bandwidth over which the tuning works effectively.

Note that most of these devices and system solutions depend on a mix of digital and analogue functions. A strict definition of tunability would imply an analogue function, whereas many present device solutions move in discrete steps – eighty switched capacitors on a single die being one example. Of course if the switch steps are of sufficiently fine resolution the end result can be very much the same.

There are some informed industry observers that consider a tuneable front end defined strictly as a well-behaved device that can access any part of any band anywhere between 700 MHz and 4.2 GHz could be twenty years away. This begs the question as to what the difference would be between say a twenty-five-band device in twenty years time (assuming one band is added each year) and a device with a genuinely flexible 700 MHz to 4.2 GHz front end. From a user-experience perspective the answer is probably not a lot, but the inventory management savings of the infinitely flexible device would be substantial, suggesting that this is an end point still worth pursuing.

5.5.3   Power-Amplifier Material Options

RF efficiency is determined by the difficulty of matching the power amplifier to the antenna or multiple antennas and to the switch path or multiple switch paths across broad operational bandwidths. However, transmission efficiency is also influenced by the material and process used for the amplifier itself.

Commercialisation of processes such as gallium arsenide in the early 1990s provided gains in power efficiency to offset some of the efficiency loss implicit in working at higher frequencies, 1800 MHz and above, and needing to preserve wanted AM components in the modulated waveform.

In some markets, GSM for example, CMOS-based amplifiers helped to drive down costs and provided a good trade off between cost and performance but for most other applications GaAs provided a more optimum cost/performance compromise.

This is still true today. WCDMA or LTE is a significantly more onerous design challenge than GSM but CMOS would be an attractive option in terms of power consumption, cost and integration capability if noise and power consumption could be reduced sufficiently.

Noise can be reduced by increasing operating voltage. This also helps increase the passive output matching bandwidth of the device. However, good high-frequency performance requires current to flow rapidly through the construction of the base area of the transistor. In CMOS devices this is achieved by reducing the thickness of the base to micrometres or submicrometres but this reduces the breakdown voltage of the device.

The two design objectives, higher voltage for lower noise and a thinner base for higher electron mobility are therefore directly opposed. At least two vendors are actively pursuing the use of CMOS for LTE user devices but whether a cost performance crossover point has been reached is still open to debate. Low-voltage PA operation also is a direct contributor to PA bandwidth restriction. Because it is *output power* that is specified, low supply voltages force the use of higher currents in the PA transistors. This only happens by designing at low impedances, significantly below the nominal 50 ohms of radio interfaces.

Physics requires that impedance value shifts happen with bandwidth restrictions: larger impedance shifts involve increasingly restricted bandwidths. At a fixed power, broadband operation and low-voltage operation are in direct opposition.

Gallium arsenide in comparison allows electrons to move faster and can therefore generally deliver a better compromise between efficiency, noise and stability, a function of input impedance. However, the material has a lower thermal conductivity than CMOS and is more brittle. This means that it can only be manufactured in smaller wafers, which increases cost. The material is less common than silicon, the world's most common chemical element, and demands careful and potentially costly environmental management. As at today, gallium arsenide remains dominant in 3G user devices for the immediately foreseeable future and is still the preferred choice for tier 1 PA vendors.

Gallium nitride, already used in base station amplifiers is another alternative with electron mobility equivalent to GaAs but with higher power density, allowing smaller devices to handle more power. Gallium nitride is currently used in a number of military applications where efficient high-power RF is needed. However, gallium nitride devices exhibit memory effects that have to be understood and managed.

Silicon germanium is another option, inherently low noise but with significantly lower leakage current than GaAs and silicon. However, an additional thin base deposition of germanium (a higher electron mobility material) is required, which adds cost. Breakdown voltage is also generally low in these devices.

5.5.4   Broadband Power Amplifiers

Irrespective of the process used it is plausible that a greater degree of PA reuse may be possible. For LTE 800 to use the same power amplifier as LTE900 would imply covering an operational bandwidth of over 100 MHz (810 to 915 MHz) at a centre frequency of 862 MHz.

This is 11.6% of the centre frequency. PA operational bandwidths of up to 20% of centre frequency are now possible, though with some performance compromise.[23] As a comparison the 1800 MHz band is 4.3%. Any efficiency loss for a UMTS/LTE 900 handset incurred as a consequence of needing to also support LTE 800 would likely to be unacceptable. This implies incremental R and D spending in order to deliver an acceptable technical solution. The general consensus is that 15% is OK, 30% is stretching things too far, which would mean a power amplifier covering say 698 to 915 MHz would be unlikely to meet efficiency and/or EVM and/or spectral mask requirements. The PA will also need to be characterised differently for GSM (higher power) and for TDD LTE.

The power in full duplex has to be delivered through a highly specified (linear) RX/TX duplexer with a TDD switch with low harmonics. The design issues of the PA and switch paths are well understood but if both FDD and TDD paths need to be supported then the design and implementation becomes problematic, particularly when all the other existing switch paths are taken into consideration.

Wideband amplifiers are also being promoted where one PA (with suitable linearisation and adaptive matching) can replace two or three existing power amplifiers. The approach promises overall reductions in DC power drain and heat dissipation and cost and board real estate but will need to be optimised for any or all of the process choices.

However, even if this broadband PA problem can be solved, within the foreseeable future each full-duplex band requires an individual duplexer. Connecting one broadband PA to multiple duplexers is a lossy proposition with complex circuitry. To match a good broadband PA, a switched (or tuneable) duplexer is required.

5.5.5   Path Switching and Multiplexing

Unlike power amplifiers that drive current into a low impedance, switches must tolerate high power in a 50-ohm system. If connected to an antenna with impedance other than 50 ohms the voltage will be even higher, commonly above 20 Vpk. The switch is in the area of highest

---

[23] This is an area where recent progress has been made. Multiband power amplifiers are now available that cover 1710 to 1980 MHz, a bandwidth of 13%.

Table 5.5 Power-handling requirements. Reproduced with permission of Peregrine Semiconductor[24]

System Standard	Average Power[a]	Peak Power[a]	PAR	V_{PK}	Operating VSWRs	V_{PK} @ VSWR[b]
Two Tone	30	33	3	14.1	n/a	n/a
GSM	33	33	0	14.1	4 : 1	22.6
EDGE	27	30.2	3.2	10.2	4 : 1	16.4
CDMA2000	23	28.1	5.1	8.0	4 : 1	12.9
WCDMA	24	27	3	7.1	4 : 1	11.3
LTE	19.4	27	7.6	7.1	4 : 1	11.3
802.11g	15	22	6.6[c]	4.0	3 : 1	6.0
VHF 2-way	40	40	0	31.6	8 : 1	56.2

[a]Maximum average power at antenna in a 50-ohm system per the system specification.
[b]At angle that produces the highest voltage.
[c]Assumes crest factor reduction.

voltage and has the most severe power-handling and linearity requirements with a need to have an IP3 of >70 dBm.

Table 5.5 gives typical power-handling requirements.

Low-voltage FETS can be 'stacked' in series for high-voltage handling in the OFF condition, but this requires a low loss insulating substrate, either GaAs or silicon on sapphire. Apart from power handling, switches must be highly linear and have low insertion loss.

Switch losses have more than halved over the past five years that has allowed for higher throw counts and cascaded switches but substantial performance improvements are still needed due to the steadily increasing throw count of the antenna side switch, the addition of a PA side switch and the need to support bands at higher frequencies.

Table 5.6 shows the performance of a nine throw switch implemented using a silicon on sapphire process. This provides good isolation and the use of CMOS means that the off-chip logic can be integrated which is generally harder to achieve in GaAs switches. The switch also integrates the low pass filters.

CMOS scaling combined with finer metal pitch technology and improved transistor performance should, however, realise future benefits both in terms of size and insertion loss, a 0.18-μm process would halve the insertion loss and core area and double the bandwidth. The insertion loss and size benefits are shown below. Note that on the basis of these measurements the insertion loss of silicon on sapphire has now achieved parity with GaAs.

Similar flexibility options and cost and performance advantage can be achieved using their proprietary thin-film ceramic material, a doped version of barium strontium titanate (BST). The dielectric constant of the material varies with the application of a DC voltage. The device is claimed to have a Q of 100 at 1 GHz and better than 80 at 2 GHz, and is claimed to be highly linear and provide good harmonic performance with low power consumption, high capacitance density, an IP3 of greater than 70 dBm, fast switching speed and low current leakage.

---

[24] http://www.psemi.com

Table 5.6 Silicon on sapphire process versus GaAs insertion loss comparison. Reproduced with permission of Peregrine Semiconductor

	0.36 μm	0.35+	0.35++	0.18 μm	GaAs
Production Launch	2008	2010	2010	2011	2010
Symmetric Throws	No	No	Yes	Yes	No
TRX Paths	3	3	—	—	3
TX Palhs	2	2	9	9	2
RX Paths	4	4	—	—	4
XtorRon Coff (fs)	587.5	475	360	225	224
TX/TRX IL@ 2 GHz (dB)	0.75	0.6	0.50	0.40	0.45
RX IL @ 2 GHz (dB)	1.0	0.85	0.50	0.40	0.9
Min Iso 1 GHz (dB)	26	30	35	38	25
IMD (Band 1) (dBm)	−111	−115	−118	<−120	−105
Normalised Solution Area	1.0X	0.8X	0.72X	0.59X	5.3X

5.6 Antenna Innovations[25]

5.6.1 Impact of Antenna Constraints on LTE in Sub-1-GHz Bands

The design of the antenna system for LTE mobile devices involves several challenges; in particular if the US 700 MHz (Bands 12, 13, 14, 17) bands have to be supported alongside the legacy cellular bands frequencies in handset-size user equipment. In general, the upper frequency bands (>=1710 MHz) don't create particular difficulties for the antenna design.

The two main challenges are:

- **Bandwidth:** The antenna must have enough bandwidth to cover the additional part of the spectrum, and this is difficult to achieve in the 700-MHz bands as the size of the device is small compared to the wavelength
- **MIMO (multiple-input and multiple-output):** Multiple antennas, at least 2, must be integrated in the device to support the MIMO functionality. Below 1 GHz isolation becomes an issue and the correlation between the radiation pattern is high, limiting the MIMO performance.

The maximum theoretical bandwidth available for a small antenna is proportional to the volume, measured in wavelengths, of the sphere enclosing the antenna. A typical antenna for mobile devices either PIFA or monopole, excites RF currents on its counterpoise, typically the main PCB of the device or a metal frame. The PCB/frame of the device therefore becomes part of the 'antenna', and the maximum linear dimension of the device should be used as the diameter of the sphere.

As a consequence, in the low-frequency bands the bandwidth of the antenna is proportional to the third power of the size of the device using the antenna, and as the frequency is reduced

---

[25] Our thanks to Antenova www.antenova.com for help and advice in providing inputs and background to this section. Any inaccuracies or mistakes will almost certainly be our fault not theirs.

the bandwidth reduces quickly. More bandwidth can be traded off at the expense of antenna efficiency.

For a typical handset with length around 100 mm, it can be already difficult to cover the legacy 3GPP bands 5+ 8 (824–894 MHz and 880–960 MHz), so adding support for an additional 100 MHz (698–798 MHz) at an even lower frequency is a major challenge for small devices; for larger devices, for example tablets, the problem is reduced.

Although a tuneable/switchable antenna can greatly help in meeting the LTE requirements, it comes with its own set of challenges. In particular, the antenna has to satisfy stringent linearity requirements and introduces added cost.

Different choices of active elements are possible. Tuneable digital capacitor arrays can be used to realise 'variable capacitors' with ranges between fractions of a picofarad to tens of picofarads. Alternatively, conventional high-linearity RF switches can be used to create switchable antenna functionality. The active elements can be used in the matching circuit, in which case the function partly overlaps with that of the adaptive matching circuits for the PA discussed below or can be part of the antenna, in which case the active element can be used to change the effective electrical length of the antenna.

In any case, the physical dimension of the antenna cannot be reduced below a certain point even using a tuneable antenna:

- The instantaneous bandwidth of the antenna must be well above that of the channel (20 MHz for bands above 1 GHz).
- If the bandwidth is narrow, a sophisticated feedback loop must be implemented to make sure that the antenna remains tuned correctly in all possible conditions where an external object (head, hand …) could shift the resonant frequency. The feedback loop is typically required if capacitor arrays are used, as the fabrication tolerance can be quite high.
- The above feedback loop is easier to implement in the uplink (UE TX) than the downlink (UE RX), however, the antenna must have enough instantaneous bandwidth so that if it is optimally tuned in the uplink it is automatically also tuned in the paired downlink frequency.

As the bandwidth of the antenna is mostly dictated by the size of the device through some fundamental physical relation, it is unlikely that innovations in materials can bring groundbreaking advancements in miniaturisation.

5.6.2   MIMO

The baseline UE capability as specified in present 3GPP standards is two receive antennas and one transmit antenna. In the downlink, the UE must support receive diversity and dual-stream MIMO. Initial network deployments will be 2×2 or 4×2 in the downlink. In order to work effectively, the two receive antennas must have similar gain and high isolation, preferably better that 10 dB, and their far-field radiation pattern must be orthogonal. If these requirements are not satisfied, the signal received from antenna 2 will be very similar to that from antenna 1, and so the receiver will not be able to see two independent channels created by the multipath environment, collapsing into a single stream MISO (multiple-input and single-output) transmit diversity system.

For a typical handset-size device, designing a MIMO antenna system is not particularly challenging in the upper frequency bands. However, in the lower frequency bands (<1 GHz), the requirements are very difficult to achieve due to the small size of the device compared to the wavelength. The root of the problem is the same as the one causing the limited bandwidth.

In general, a separation of about a quarter wavelength between two antennas is required to achieve a good isolation and low pattern correlation. Although apparently this requirement can be met even at 700 MHz (quarter-wavelength = 107 mm) in a 100-mm long device, for instance placing the antennas at opposite ends of the device, this result is just apparent. In fact, as described above, both antennas will excite the same or a very similar radiation mode on the main PCB of the device, being the counterpoise for both antennas, resulting in very similar (dipole-like) radiation patterns with high correlation (close to 1).[26] In general, the best that can be achieved is to excite two radiation modes aligned with each diagonal of the main PCB: the two resulting dipole patterns will be skewed by an angle about 40–50 degrees, resulting in an envelope correlation coefficient about 0.3, which might be sufficient. The isolation of the antenna will depend on the details of the antenna designs, and methods are available to increase the isolation, for instance artificially introducing a coupling line between the two antennas that partially cancel in phase the natural coupling of the two antennas.

As at least two antennas have to be embedded, the MIMO requirement has a significant impact on the dimensions and mechanical design of the user terminals. We have seen above that one cannot expect a reduction in the size of the antennas even if they are made tuneable or switchable: so, the mechanical designers have to allocate enough space for two antennas of reasonable size (for instance 40 mm (W) × 8 mm (L) × 6 mm (H) if a full ground plane is placed under the antennas), while taking into account all the constraints required to make the MIMO antenna system work properly. In particular, the placement must ensure that the isolation between antennas is sufficient and pattern correlation is low; moreover, the antennas must not be covered or otherwise affected by the user hand and head in typical use conditions, and comply with specific absorption rate (SAR), and hearing-aid compatibility (HAC) and other regulations. It is worth noting that performance in the presence of the hand(s) but not the head will be of greatest importance in high-speed data mode (e.g. streaming videos); it is not sufficient if just one antenna is not affected by the hand(s), as the 2 × 2 MIMO system will effectively collapse into a lower-rate 2 × 1 MISO channel. All this means that it will be even more important than before that the antenna designers are involved from the concept stage of any new terminal design.

To summarise, the decision to deploy LTE in the lower frequency (<1 GHz) bands has created enormous pressure on the antenna design for the user terminals, and set off a race between designers to produce effective solutions, although most of these solutions have yet to be proved in real devices and real networks. Hopefully, the high data rate will demand large displays, and so the typical device size is bound to grow, mitigating some of the problems and challenges.[27]

---

[26] At higher frequency it is possible to excite very different radiation modes, resulting in low correlation, however, although it is possible to excite higher-order modes even at frequencies <1 GHz, with fields confined near the antenna and with a radiation pattern different from that of the fundamental mode, these modes have an extremely narrowband and are not suitable for this application.

[27] The counterargument to this is that 700 and 800 MHz networks will probably need to support smaller form factor devices including dongles, smart phones and standard phones in order to amortise spectral and network investment over a broad user-product mix. Not to do so would place these networks at a commercial disadvantage.

5.6.3   Antennas in Smaller Form Factor Devices Including Smart Phones

A traditional PIFA antenna uses the ground plane as a significant part of the antenna. PIFA antennas tend to have a natural broadband response due to their effective large size due to the interaction with the ground plane but this increases interaction between other antennas and increases susceptibility to hand capacitance effects. Unwanted coupling becomes a particular issue at lower frequencies.

Antennas in space-constrained designs are a fraction of the resonance length, and radiation efficiency and bandwidth are a function of volume. Space needs to be left around the antenna to avoid power being shorted out of the device. The antenna is shielded by the user's hand or head and the antenna needs enough bandwidth (a function of volume) to resist detuning. Antenna designs also need to comply with regulatory requirements for SAR and HAC.

5.6.4   SISO Testing[28]

Antennas in user equipment have morphed from a folding or whip antenna in first-generation devices to smaller helical designs to the first integral patch antenna in 1994, the first of many patch antenna implementations. From an aesthetic perspective patch antennas were a great leap forward, from a performance perspective a potential significant step back. Antenna-performance concerns prompted a set of single input single output test standards which were ratified by CTIA in 2001. These define methods for testing total radiated power (TRP) and total isotropic sensitivity (TIS).

No similar antenna test standards presently exist for MIMO. In terms of functional design the volume required for one good antenna provides sufficient space for two poor antennas and the performance has to be verified over multiple cellular and noncellular frequency bands.

Most user equipment testing is performed by connecting a cable to a 'temporary antenna connector' that bypasses the antenna. The assumption used to be that the device under test antenna could be fairly represented by an isotropic antenna with 0 dBi gain. This was relevant when an antenna was a dipole tuned to a single band of interest but is no longer appropriate with most present antenna implementations.

This means that conducted tests for reference sensitivity and maximum output power do not reflect real-life radiated performance. This can be seen by comparing the requirements for conducted tests with those for radiated tests. The 2001 CTIA tests were followed in 2008 by the 3GPP test specification TS34.114 that defines requirements for TRP and total reference sensitivity (TRS), which is the same as TIS in the CTIA standard.

There is a significant amount of measurement uncertainty in these tests of the order of plus or minus 1.9 dB for TRP and plus or minus 2.3 dB for TRS, the consequence of an error model with over twenty terms. A large measurement error would allow some poor devices to pass the test and some good devices to fail. The solution to this for 3GPP was to relax the minimum requirements set out in the test specification by about half of the acknowledged test uncertainty, and in parallel introduce a stricter nonmandatory recommended performance target.

---

[28] Our thanks to Moray Rumney of Agilent Technologies for these insights into SISO and MIMO performance verification.

The tests are quite thorough, covering azimuth and elevation angles, polarisation, device mechanical and application modes (open, shut, vertical and horizontal, voice and data) and include the low, middle and high channels of each frequency band.

Phantom heads and phantom hands are used to replicate typical usage conditions. Phantom hands are currently right hand but could be extended to the left hand as the interaction between the device and hand can be highly asymmetric. Some device form factors will have contact from two hands at once and possibly more in future.[29] Characterising one device over one band requires thousands of measurements and can take up to two weeks in an expensive anechoic facility.[30]

Tests undertaken by Orange[31] showed that free-space measurements of radiated performance across a range of devices varied by 7 dB for TRS and 4.5 dB for TRP and head and hand loading will increase this spread further. All the devices tested had passed narrower conducted tests suggesting that the larger spread was due to the physical antenna implementation.

The positive point to make is that operators now have access to objective and standardised over the air (OTA) radiated domain performance measurements that provide comparative measurements that are closer to real-world performance than conducted domain measurements (made via a temporary antenna connector).

The introduction of minimum performance requirements has, however, been difficult. Operators wanted to set high targets but user equipment vendors wanted to protect existing designs and small form factor devices that could be termed noncompliant. The result was a compromise that allowed the bulk of legacy devices to remain compliant but with a nonmandatory recommended average performance typically 3 dB better. The recommended TRP performance is still about 6 dB below the nominal power for the conducted test, which suggests substantial room for improvement were it not for shrinking device sizes.

If average data throughput, for example in micro- and macrocells is as important or more important than peak data rate for example in pico- or femtocells in terms of impact on application performance and user experience value then it would seem that optimising SISO antenna performance should be a priority over and above MIMO optimisation. Presently this would not appear to be the case. It is particularly important to ensure that an overaggressive pursuit of peak data rates in small cells does not compromise average data throughput in larger cells. A MIMO antenna design ideally needs to deliver a net gain in a wide range of operational conditions. This is often not the case.

5.6.5   MIMO Testing

MIMO OTA standardisation started in the CTIA and later the 3GPP standards process.

The rather complex relationship between MIMO antenna performance, spatial channel gain and scheduling efficiency makes it difficult to agree on an objective figure of merit that captures real-life conditions and real-life throughput performance.

---

[29] Future multiplayer gaming devices being one example.

[30] See also CTIA/PTRCB test procedures for the US market http://www.cetecom.com/us/en/certification/north-america/ctia/over-the-air-performance.html. The test procedures state a requirement of a minimum of 50% antenna efficiency. Optimum antenna designs can approach 70% efficiency (measured in an anechoic chamber). Some handsets shipped in to some markets have been measured at 10% efficiency, which would appear to validate the need for a minimum acceptable performance threshold to be applied in all markets.

[31] Orange, 'OTA TRP and TRS requirements for GSM 900 and 1800', 3GPP R4-091762.

One approach is to undertake SISO measurements and then emulate arbitrarily complex spatial channel conditions and frequency selective scheduling gain. MIMO gain relative to SISO is dependent on low antenna correlation and high signal-to-noise ratios. Low antenna correlation is more likely to be achieved at higher frequencies.

Such gains are only likely to be significant in cells where the Doppler speed and channel conditions change slowly enough to allow adaptive modulation and coding to be used. In a loaded network, the median SNR will be of the order of 5 dB. This suggests real-life MIMO versus SISO throughput gains will be nearer to 20% than the 100% often claimed. Ideally, a test regime has to be developed that emulates practical rather than theoretical conditions. If SISO OTA radio measurements can be controlled within plus or minus 2 dB then MIMO measurements are unlikely to be more accurate.

A large uncertainty could be argued to invalidate the value of the test. Relaxing the MIMO OTA minimum requirement by some proportion of the test uncertainty would really not help very much.

What all of the above highlights is that a measurement approach needs to be agreed for MIMO devices that results in a methodology that can distinguish good and bad performance. Such an approach has now been agreed after considerable effort in SISO measurements but has yet to be agreed for MIMO measurements.

As of today, it is possible to do comparative measurements of SISO user equipment that are meaningful and provide a benchmark against which subsequent year-on-year performance improvements can be measured. This is not presently possible in MIMO user equipment.

The industry might pause to consider whether the expectations for spatial multiplexing gains in ideal conditions might possibly be disproportionate and unduly influenced by a marketing fascination with headline peak data rates in small-diameter cells. Improvement in robustness from RX and TX diversity may well be of more significance. Some aspects of MIMO, for example RX and TX diversity and beam steering, may deliver enough gain over a sufficiently broad spread of operational conditions to be worthwhile but generally spatial multiplexing gains in the lower bands and in larger-diameter cells will be marginal and it would seem that effort is being expended in areas that will have less impact on the user experience in wide-area access conditions than commonly assumed.

Comparing LTE and WiFi connectivity throughput speed and efficiency suggests that LTE MIMO in local-area applications may also be of marginal value particularly if it compromises wide-area performance. WiFi might be a better (more cost- and energy-efficient) option for load shedding.

5.6.6   SIMO Gains

SIMO refers to the user/device path from the base station from where a modulated waveform is transmitted using specific resource blocks delineated in the time domain and frequency domain (frequency subcarriers within a wideband channel). This 'single input' to the radio channel is then transmitted and is received by multiple antennas at the receiver each with its own version of the signal based on the multipath channel. The receiver then combines these signals to create a better signal than is possible from one receiver. Generally, in almost all small- and large-cell propagation conditions, there will be strong multipath components (with a longer time delay in larger cells).

In WCDMA (and GSM) multipath signals are equalised out at baseband. In HSPA and LTE the additional option of diversity gain is supported (multiple outputs from the channel). If implemented with a separate receive path then additional gain can be achieved in most channel conditions provided antenna efficiencies have not been compromised by the SIMO implementation.

Diversity also mitigates fast fading, though there are other processes, for example channel sensitive scheduling, that have the same or similar effect at least in small-cell environments. Quite often one technique is credited with delivering more gain than it really achieves in practice, it is the combination of adaptive averaging mechanisms that is important. Note that none of these techniques are a substitute for good RF performance but should be seen as techniques for getting higher throughput though a complex rapidly changing multipath channel where fast and slow fading and Doppler effects need to be managed.

As stated above, good RF performance in larger cells may be as important as high peak data rates in small cells and an overemphasis on small-cell performance may compromise large-cell performance metrics.

Although the generalised benefits from spatial multiplexing might be questionable there is strong evidence from operator tests that receiver diversity, which can be described as SIMO (single-input multiple-output) can result in a useful gain in downlink capacity, of the order of 50 to 100% assuming all terminals are diversity enabled. This would suggest that theoretically at least a second antenna and second receiver are well worthwhile. A full bidirectional MIMO implantation of course also requires a second transmit path in the user's equipment that is altogether more challenging.

The second receiver in a SIMO-enabled UE could also be used to do interband CQI measurements to support extended multiband handover that could deliver substantial macroscheduling gain.[32]

5.6.7 Adaptive Matching Power Savings

We have stated earlier that a cross section of RF component vendors are presently working on the implementation of adaptive techniques, in particular adaptive matching on the TX and RX paths and adaptive linearisation on the TX path. This includes SAW filter vendors such as Epcos, RF MEMS vendors and RF PA vendors, usually in partnership with one or more of the RF MEMS suppliers.

Adaptive matching can either be used to improve RX and TX performance within a specific band or be used in multiband RF architectures to minimise performance loss from wider-band TX/RX devices.

Modules presently under development use an electrostatically variable capacitive RF MEMS switch with a Q of 250 at 1 GHz. Typically five switches are used to provide 32 capacitance values to give a 10:1 tuning ratio . Power consumption of the adaptive tuner is presently of the order of 4.4 milliwatts reducing 1 milliwatt in the future, insertion loss is about 0.5 dB, harmonics H2 H3 are <91 dBc and intermods IM3 are <117 dBm. A typical device will need to handle 3 billion cold switching cycles for GSM.

---

[32] There is no present standards work ongoing on this but this would seem to be a useful future work item.

Table 5.7   PA options

Option 1	Separate power amplifiers and individually band optimised TX/RX paths
Option 2	Three separate modules, one for core bands from Band V 850 MHz up to 2.1 GHz, a second for 2.3 and 2.6 GHz, and a third for US and European DDR at 700/800 MHz
Option 3	As option 2 but with the core band module extended down to 700 MHz
Option 4	As option 3 but with the core band module extended up to 2.6 GHz
Option 5	Three separate modules, one for sub-one-GHz, one for existing bands between 1 GHz and 2.1 GHz, one for 2.3/2.6 GHz
Option 6	One broadband PA for all frequencies

Every 1 dB of mismatch loss equates to 1 dB less output power and 1 dB loss of sensitivity. For a WCDMA phone at 1900/2100 MHz, a 1-dB loss will increase current drain by 70 mA. In low band GSM a 1-dB loss will increase current drain by 250 mA due to its higher average power (two watts rather than one watt).[33]

5.6.8   RF PA and Front-End Switch Paths

There are a range of views from RF PA component vendors as to how a multiband phone supporting 700, 800 and 2600 MHz will be implemented: The options are shown in Table 5.7.

Option 1 – a single RF PA and band optimised matching network for each band.

Option 2 – one PA module for Band V (US LTE 850 or E850), Band VIII (LTE 900), Band III (LTE1800), Band II (US LTE1900) and Band I (UMTS/LTE1.9/2.1).

A second module would cover Band 40 (China LTE TDD at 2.3 GHz), Band VII (UMTS/LTE 2.6 GHZ extension band) and Band 38 (LTE TDD in the Band VII duplex gap).
A third module would cover the US and European DDR bands at 700 and 800 MHz.

Option 3 would be as Option 2 but extended down to cover the 700- and 800-MHz DDR bands.

Option 4 would be as option 3 but extended up to cover the 2.6-GHz band.

Option 5[34] would be to have a PA module for bands below 1 GHz, a second module for bands up to and including Band I and a third module for frequencies above 2.1 GHz. However, this option would be significantly different from existing solutions available in the market.

Option 6 would be one wideband power amplifier to cover all bands.

---

[33] Source Peregrine July 2010.
[34] Based on conversations with Nujira 18 Feb 2009, though see earlier comments on LTE energy drain.

Option 6 would presently be ambitious both from a PA design perspective and would be dependent on implementing wideband phase-lock loop architectures with acceptable noise performance, availability of sufficiently efficient wideband adaptive matching and availability of sufficiently efficient wideband linearisation techniques.

Note that PA bandwidth is limited not by the active device but by the bandwidth of the output matching circuit. An output matching circuit is necessary because of the inherent impedance difference between the antenna and the PA device.

Five volts (or less) causes the RF output device to have an output impedance of well below 50 ohms. Devices following the PA for example directional couplers, filters, antennas and isolators have 50 ohm impedance and so an impedance matching circuit (passive) is required. The realisation of this network will limit the bandwidth.

Using a supply above five volts will enable the PA output transistor to be operated nearer to a 50-ohm output impedance thus relaxing but not entirely removing the requirements on the bandwidth-limiting impedance transformation network. By this means, PA bandwidths of an octave become possible.

In order to linearise PA operation, some vendors use envelope tracking. This is achieved by generating a 'tracking control voltage' to modify the supply voltage to the PA transistor. A DC-to-DC buck/boost converter is used both as a control function of the supply and to raise the voltage to greater than 5 V.

Additionally, the PA drive is produced at very low output power levels in order to assist the tracking process through the area where the PA transfer function is extremely nonlinear. Using this technique ideally results in a relatively constant PAE over a widely varying PAPR.

The control voltage function is reconstructed from digital baseband information and due to the PA device variation will require calibration at handset assembly stage. The assumption is that the improvement in efficiency will outweigh the inconvenience. Note that as stated earlier higher data rates require better control of transmit and receive linearity in order to avoid unacceptable EVM on the transmitted signal and/or demodulation errors in the receiver.

Linearisation techniques have to work harder as operational bandwidth increases (see matching comments above), as channel bandwidth increases (from 3, 5, 10 to 20 MHz), as data rates increase and as distance from the cell site increases.

After taking these factors into account the consensus is that Options 1 through 5 should all be realisable for market introduction by 2012/2014. Options 2 or 3 are the favoured options from two/three of the major RFPA vendors as this leverages existing R and D investment.

The generally accepted principle is that component-count reduction, one of the objectives of options 2 through 6, is only commercially worthwhile if performance is at least as good as individually band optimised solutions (option 1).

To be financially and commercially viable from an operator perspective, options 2 through 6 would have to have a similar or better RF performance than option 1 and would have to deliver additional differentiation on the basis of cost and/or component count reduction, real-estate occupancy and power drain.

Note that apparent efficiency gain in RF power amplifiers can be achieved by trading efficiency against modulation accuracy (error vector magnitude). This is not a real gain as the link budget will be the same or possibly worse. From both an individual and group perspective it is better to talk quietly and clearly rather than shout loudly and indistinctly.

5.7 Other Costs

Improved performance in one functional area often introduces cost and complexity in another. These costs can include standardisation costs and the costs of testing component or product compliance with those standards, for example:

5.7.1 RF Baseband Interface Complexity

Mobile RF devices have an RFIC front end with a digital bus on the back end that connects to a baseband IC.

There has been substantial discussion as to how, whether or when, to standardise these control and communication interfaces between the RF and transceiver and baseband functions in the handset. One present initiative is the Dig RF work group that is supported within the MIPI alliance.[35] Version 4 of this standard increases bus speed from 300 Mbps to 1.5 Gbps so that it can be compliant in terms of throughput, timing and latency control with LTE protocols.

The functioning of these protocols requires robust testing. Testing involves the use of active probes with ultralow-capacitive loading (less than 0.15 pF) and high sensitivity (needed to work at 1.5 Gbps and 200 mVp_p). Although the required test investment is of the order of $50 000 dollars or more, the net benefits for handset manufacturers in terms of subsystem interchangeability seem to be substantial.

Handset vendor support of Dig RF is consistent with initiatives in other areas (for example, the standardisation of baseband RF and smart antenna interfaces in base stations) as a mechanism for achieving a broader base of multisource interchangeable components. Many of the new adaptive techniques (adaptive matching and adaptive linearisation) require broader visibility to the physical layer than is presently available. This could include, for example, knowing whether peripheral devices such as Bluetooth are active.

The lack of a standardised approach to what information is needed, what metrics should be measured, how they should be measured and how the measurements should be used is a significant barrier to the general deployment of adaptive techniques. A parallel and related issue is the multiplication of the control lines needed to support these adaptive modules both on the transmit and receive paths.

To address this there is a new working group within MIPI called RF-FE, chaired by Nokia, working to replace the multitude of control lines in modern front-ends with a single serial bus. The first version of this specification was recently approved and released.

5.7.2 Conformance Test Costs

There are a number of features included in the LTE standards that are targeted at reducing conformance test costs both in terms of hours and dollar costs, automatic reset between test runs being one example.

Despite these efforts, the hour and dollar cost conformance costs for additional band costs are substantial even before development test and manufacturing test costs are included. An indicated 100 hours per band for RF conformance and protocol test implies a six-month test

---

[35] http://www.mipi.org/.

programme. This does not include interoperability and network testing or MIMO testing. Aside from the direct costs, a six-month cycle introduces substantial time to market delay and reduced range availability, an indirect cost.

This is an area that definitely demands collaborative attention from handset vendors and the operator and test community.

5.7.3 Conformance and Performance Test Standards and Opportunities

Within the HSPA standards process, 3GPP and interested vendors have identified and promoted opportunities for realising throughput gains from enhanced levels of baseband processing. This has been captured in the 3GPP standards for advanced receivers.

For LTE there is arguably a more compelling case to be made for enhanced RF performance to be captured as a design goal and for the benefits of performance gain over conformance compliance to be more explicitly stated, in effect a standard for an advanced front end. Note there is no reason why this should not be complementary to present and future baseband algorithmic innovation including for example interference cancellation.

Conformance standards are set to be achievable with acceptable manufacturing yield at the time when user products first become available. In practice, it should be possible to realise year-on-year user equipment performance gains from year one of a technology being introduced for at least five years, though these gains may be, and often are, traded against cost-saving opportunities. Operators and standard bodies working together have the opportunity to influence these trade off decisions.

5.7.4 Software-Defined Radio and Baseband Algorithmic Innovation

The baseband processing of Release 99 3G technologies meant that the standard DSP architectures that had begun to dominate GSM and EDGE could no longer be used. A Release 99 handset needed about 1.5 million gates of dedicated hardware logic and associated memory to perform rake receiver, combining and channel decoding. The introduction of HSPA and new equaliser designs added another million gates.

LTE introduces further complexity. The layer 1 multiple high throughput FFT/iFFTF[36] operations, and FIRF[37] and MIMO detection needed to decorrelate the OFDMA subchannels requires a hardware implementation of the LTE physical layer of the order of 3 to 4 million gates coupled to a megabyte of fast memory of a 50 Mbps class device. This has to be coupled with L2/L3 CPU requirements of the order of 150 to 200 MHz to support 100 Mbps operation.

To be viable in terms of cost and power consumption, LTE and LTE-A devices will therefore need to use 32-nm silicon geometries. A 65-nm implementation costs over 40 million dollars, may not work first time and will be obsolete within less than eighteen months. A 32-nm device will cost more than twice as much, be harder to implement and be even more likely to not work first time round introducing additional cost and time to market delay.

---

[36] Fast Fourier and inverse fast Fourier transforms.
[37] Finite impulse response filtering.

Table 5.8 RF component and baseband vendor financial performance comparisons. Reproduced with permission of The Mobile World

	Revenue	Gross Margin	Operating Margin	R&D/Sales	CapEx	Free Cash Flow	Balance Sheet (Equity + Liabilities)	Net Equity	Return on Equity	Net Debt	Debt to Equity
	US$000						US$000	US$000		US$000	
Anadigics	216 714	35.0%	−0.2%	23.1%	—	—	233 812	202 964	0.6%	−97 129	−47.9%
Avago	796 000	49.0%	22.3%	13.4%	—	—	2 157 000	1 505 000	27.6%	−561 000	−37.3%
Broadcom	6 818 319	51.8%	15.9%	25.8%	73 924	1 296 902	7 944 310	5 826 089	18.6%	−1 960 697	−33.7%
Intel	43 623 000	66.1%	36.5%	15.1%	—	—	63 138 000	49 638 000	23.5%	−14 677 000	−29.6%
Mediatek	3 916 048	46.4%	23.7%	17.8%	—	—	4 771 905	3 860 952	27.3%	−2 887 313	−74.8%
Qualcomm	11 669 000	4.8%	30.1%	22.5%	397	3017	31 291	22 736	15.7%	−18 018	−79.2%
RF Micro Devices	1 070 732	37.4%	14.3%	13.1%	—	—	1 053 352	658 364	17.4%	−92 167	−14.0%
Skyworks	1 161 831	43.3%	20.2%	12.1%	—	—	1 614 076	1 411 407	12.1%	−425 645	−30.2%
ST Ericsson	3 012 825	0.0%	−126.3%	0.0% —	—	—	—	—	0.0%	108 667	0.0%
Triquint	878 703	39.9%	13.2%	14.7% —	—	—	978 102	834 019	22.7%	−223 656	−26.8%
Total/ Weighted Average	73 163 172	62.5%	26.0%	17.5%							

The use of dedicated ASICS therefore may be technically efficient but increasingly economically risky and dependent on achieving a stable standards environment, unlikely unless LTE advanced standards become finalised rather faster than present progress would suggest.

As a result, there are opportunities to optimise DSP architectures to support the specific requirements of LTE when implemented as a multimode device with backwards compatibility to present 3G and 2G layer one handsets.

The multimode 'problem' is therefore essentially solved or at least almost solved.

This highlights the need for parallel progress with a solution for the multiband 'problem' or more positively, multiband market opportunity.

5.7.5   RF Design and Development Dilution

Ultimately, progress depends on how motivated the RF design community are to implement change. Partly this is a function of the legacy profitability or lack of profitability relative to other parts of the phone. Table 5.8 shows the large differences in gross margin, operating margin and R and D expenditure for a cross section of RF PA vendors, Anadigics, RF Micro Devices, Skyworks and Triquint and filter vendors (Avago) and compares the margins with baseband vendors. Intel's margins are historically from PC rather than mobile component sales but demonstrate the scaling benefits that can be realised. Conversely not achieving scale can be painful (ST Ericsson as an example).

RF component vendors do not have these same scaling opportunities.

There may also be market disincentives. For example why develop a broadband amplifier that replaces five narrowband amplifiers if the realised price per unit stays the same.

The RF design community is risk averse partly because radical change can introduce either time to market delay or low RF yield on the production line, either outcome being unhelpful in terms of future career progression.

As an example, Nokia recently transferred their RF design assets and RF IC assets to Renesas. Moving the design risk to a third party means Nokia can seek redress when things go wrong.

The highly variable rates of return from this part of the supply chain explain at least to some extent why RF engineering constraints often seem to frustrate fast market adoption.

5.8   Summary

Each individual component in a mobile phone costs money and earns money. The cost includes component cost and real-estate cost balanced against realisable value. The RF front end of the phone is the part that largely determines the network operator's delivery cost and the value realisable from local- and wide-area connectivity.

This is the one area where probably the least rapid progress is being made in terms of practical innovation, which is odd considering the implied opportunity cost that is incurred when spectrum is purchased, networks are built and phones are either not available or do not work very well.

We return to this fundamental disconnect in Chapter 21.

6

Cable, Copper, Wireless and Fibre and the World of the Big TV

6.1 Big TV

How did televisions in our living rooms get to be so HUGE?

The answer is of course that plasma screen technologies and LCD technologies and glass-production technologies have combined together to produce a new generation of monstrously large televisions that do much more than they used to. Specifically these are devices that not only take information from a terrestrial or satellite broadcast but are also connected to the internet and more often than not also include one or more forms of wireless connectivity.

Televisions, set-top boxes and home hubs have therefore become an integral part of the telecommunications value proposition though who owns what parts of that added value remains open to debate. A brief review of technology progress provides some useful context.

Eighty years ago mechanical television systems designed by John Logie Baird in the UK and Charles Francis Jenkins in the US typically had screens of an inch or so wide with a 30-line picture. US TV broadcasting was initially in the AM radio band between 550 and 1500 kHz. In 1930 the Federal Radio Commission allocated channels in the 2 MHz band to support higher resolution (45 and 60 line) pictures.

In the UK 30-line television programmes were broadcast by the BBC from September 1929.

Most mechanical TV broadcasts in the US had stopped by 1933 but carried on in the UK until 1935 and until 1937 in the Soviet Union. Follow the link for a description of how mechanical TV systems worked.[1]

Electronic television required a number of enabling technologies including the valve but also a cathode ray tube to display the image and a camera tube to capture the image. There is a substantial variance of opinion as to who was first with what, basically the French, Russians, US and British all worked on various vital bits.

The BBC set up a committee in 1935 to help decide between two electronic TV systems, a Baird 240-line system and an EMI system with 405 lines. The 405 line system was chosen as it worked better. Content for the network was captured using Marconi's Emitron camera

---

[1] http://www.earlytelevision.org/mechanical_tv.html.

Making Telecoms Work: From Technical Innovation to Commercial Success, First Edition. Geoff Varrall.
© 2012 John Wiley & Sons, Ltd. Published 2012 by John Wiley & Sons, Ltd.

Figure 6.1 Bush TV 22 in production between 1950 and 1952. Reproduced with the permission of the Science Museum/SSPL.

system.[2] Three of these cameras were installed at Hyde Park corner to record the coronation of George V1 in May 1937. The cameras were connected via a lead-sheathed copper cable to Alexandra Palace, a massive undertaking for an audience estimated as at most 10 000 people.

In the US the RCA Victor Company introduced electronic television at the 1939 World Fair though even by 1940 there were only 2000 or so sets in use.

During the war the UK stopped TV broadcasting as it was considered that the VHF transmission could be used by the German air force for direction finding. Postwar 405-line broadcasts started again and carried on until 3 January 1985 in the VHF band (Bands 1, 11 and 111 between 30 and 300 MHz). Band 111 was then reallocated to private mobile radio and later to digital audio broadcasting.

After the Second World War the BBC built a number of regional transmitters each on a different 8 MHz channel. The Bush Radio Company (named after the location of the original factory in Shepherd's Bush) produced the model in Figure 6.1 between 1950 and 1952. It was one of the first televisions capable of being retuned if the owner moved house. It featured a nine-inch screen and an aluminised cathode ray tube that meant that more of the light from the picture came into the room, producing a brighter image. 625-line/50-Hz TV was introduced into the UK in 1964 with colour transmission starting in 1967. The system known as PAL (phase alternating line) was developed by Telefunken in Germany.

In the US a committee known as the National Television System Committee produced a standard in 1941 (the NTSC standard) that came to be used across most of Latin America and Canada with colour being introduced in 1953. This survived until the introduction of digital TV in 2009 standardised by the Advanced Television Systems Committee (ATSC). Standard definition digital TV still retains the 525 line/60 Hz resolution used by the NTSC system.

---

[2] http://www.bbc.co.uk/historyofthebbc/collections/buildings/alexandra_palace.shtml.

Actually, to be pedantic, the visible scan lines for standard definition ATSC are 480, either interlaced or progressively scanned. ATSC high definition increases this to either 720 or 1080 with a 16 : 9 aspect ratio, closer to the natural field of view than a standard TV four by three aspect ratio. DVBT and DVB C and DVB S high definition is similar with a choice of 720, 1035, 1080 or 1152 scan lines.

Given that the human eye only has a certain amount of resolution there would be little point in having an HDTV picture on a small screen. Resolution, however, becomes progressively more useful as screen sizes increase.

Present large-screen displays are either plasma or LCD. Plasma screen use fluorescent phosphors with each pixel having a red, green or blue light source that can be varied in intensity to produce a full colour range. Fluorescent phosphors go dim over time though the degradation is usually sufficiently slow not to be noticeable and they are brighter and have a wider colour range than LCD but use more energy.

6.2   3DTV

An example 3DTV product is the 'smart TV'[3] introduced by Samsung in early 2011 with a viewable screen size of 59 inches.

Similar LCD screen formats are available. All are remarkable thin, the LCD model having a depth of 0.3 inches.

The products with screen sizes of over 40 inches have smart TV features that include access to a Samsung application store and a touch-screen control that allows users to watch TV while a Blu Ray movie is showing on the TV or show additional information relating to a linear TV programme. Accessories include 3D glasses and a Skype-certified high-definition camera. A separate HD Blue Ray player has a 3D sound effect that moves the sound to follow the 3D picture.

Blue lasers operate at a wavelength of 505 nanometres (>500 THz) rather than the 630 nanometres (400 THz) of the red lasers used in standard DVDs. The shorter wavelength means that up to 27 GB of high definition or 3D video can be stored on a single disk, enough for a HDTV feature film or two hours of recorded high-definition content.

This begs the question as to whether it is going to be economic to stream files of this size over the internet. The BBC iPlayer for example is better than standard definition but falls short of high definition, let alone being 3D capable.

Blu-ray 3D uses a sequential system in which the video is produced at 1080p resolution at 24 frames per second, per eye; or 48 frames per second.

Streamed 3D at present uses a side by side system in which a 1080p frame holding both the right and left eye images is sent at 24 frames per second. The TV splits each single frame into two frames and then displays them sequentially with a consequent loss of quality.

Even so, the bandwidth requirement is substantial. As at mid-2011 there are about 200 000 homes in the UK with 3D capable televisions of which about a third are signed up to 3DTV services from Sky or Virgin Media.

The set-top box/personal video recorder market is presently moving in parallel with the integrated smart TV market with features such as triple tuning (the ability to record two

---

[3] http://www.samsung.com/uk/consumer/tv-audio-video/television/.

channels while watching stored content) or quadruple tuning (dual satellite and terrestrial tuning). Typical storage is about 250 GB though this is steadily increasing over time.

6.3   Portable Entertainment Systems

This brings us to portable entertainment systems, lap tops marketed as entertainment centres rather than business aids.

The Vaio F Series is an example of a high-end product from Sony with 750 GB of hard disk memory or a 256 GB flash drive. It is one of the first lap tops capable of showing 3D graphics and converting HD content to 3D with support for active (powered as opposed to passive) 3D glasses via a Bluetooth connection. The LCD-equipped lenses pass different images to each eye in succession (alternate-frame sequencing). Each lens can be made transparent or completely opaque by varying the amount of electric voltage sent to the glass. The LCDs are synchronised with the screen's 240 Hz refresh rate. When running in 3D mode, the display inserts a black screen before it draws the scene for the subsequent eye. To make sure that this black is truly black, the panel's LED backlight goes dark .The device includes an 802.11b/g/n WiFi transceiver with integrated 3G Mobile Broadband connectivity in some markets (Verizon in the US for example).

It would seem obvious to throw in a DVB T2 and ATSC demodulator for terrestrial broad-casting reception but frustratingly flux densities are not adequate to make this an acceptable technical and commercial proposition, a topic to which we return in Chapter 15. Additionally, the power drain and associated duty cycle of these devices can be problematic with some user forums complaining of duty cycles of less than an hour. Just make sure you are never too far away from a mains socket.

This caveat aside, these devices and smart TVs as well are natural hosts for user-generated HD content captured from high-end smart phones and transferred via a high-definition

Figure 6.2   Nokia N8. Reproduced with permission of Nokia Corporation.

multimedia interface (HDMI). These interfaces are critical to realising user value and have to evolve continuously to match the bandwidth capabilities of the devices sitting either side of the connector. For example, the pixel clock rate of HDMI has increased from 165 to 340 MHz to support resolutions up to WQGGA (2560 by 1600 pixels).

The Nokia N8 shown in Figure 6.2 is a contemporary example of a high-end phone capable of capturing high-definition video transferable via an HDMI output port.

6.4   Summary of this Chapter and the First Five Chapters – Materials Innovation, Manufacturing Innovation, Market Innovation

In this first section we have set out to explore the relationship between materials innovation, manufacturing innovation, packaging and interconnect innovation, system innovation, mathematical and algorithmic innovation and market innovation.

In Chapter 2 we highlighted how the production scaling of solid-state memory combined with improved compression techniques provided the basis for Apple to develop the iPod[4] introduced in 2001 and creating a new market sector that closely mirrored Sony's invention of the Walkman twenty two years before.

The development of resistive, capacitive and multitouch interactive displays similarly developed new markets that Apple Inc. has been astoundingly successful at exploiting. The Apple touch screen might be considered as the reason for the success of the iPhone and iPad but it is the algorithmic innovation behind the screen that defines the user experience and that gave Apple the first-mover advantage in a newly created and remarkably profitable smart-phone market sector.

We described Thomas Alva Edison's light bulb moment just over 100 years earlier when observing the uneven blackening within an evacuated glass that led to the theory of thermionic emission and more importantly an understanding of the properties of electrons and photons, an understanding that became crucial to the development of the telecommunications industry throughout the twentieth century.

In 1876 the German physicist Ferdinand Braun demonstrated a rectification effect that could be recreated at the point of contact between metals and certain crystal materials, in effect a semiconductor device. Just over 70 years later in 1947 John Bardeen and Walter Brattain built the first transistor, a device with a gate that could be used to control electron flow. Eleven years later in 1958 Jack Kilby invented the integrated circuit and in 1961 Fairchild Semiconductor produced the first planar transistor in which components could be etched directly onto a semiconductor substrate.

This would suggest that every fifty years or so a fundamental discovery is made at materials level that has a truly transformative effect on the telecommunications industry. Rather like a dormant volcano it must be nearly time for the next big discovery or more likely the discovery has already been made but we have not yet recognised its significance.

And actually it may be a process of rediscovery rather than discovery as we begin to realise that the behaviour of common-place materials can change fundamentally when assembled at the molecular level.

---

[4] http://www.ipodhistory.com/.

Graphene was identified as a possible candidate, a form of carbon constructed as a flat layer of carbon atoms packed into a two-dimensional honeycomb arrangement, practically transparent due to its thinness. As a conductor of electricity it performs as well as copper at a fraction of the cost. As a conductor of heat it outperforms all other materials. At the molecular scale graphene could potentially enable a new generation of ultrafast transistors and may enable a new generation of supercapacitors and related energy-storage solutions.

We also highlighted the transformative effect of transforms tracing a historical path from the Babylonians via Pythagoras, Plato and Euclid, Archimedes, Tsu Ch'ung-chih, Al Khumar Rizmi, Da Vinci, Descartes, Fermat, Newton, Euler and Fourier and the parallel impact of matrix maths via the first and second golden age of Chinese mathematics, the seventeenth-century Japanese mathematician Takakazu Seki, Leibniz, Gauss, JJ Sylvester, Jacques Hadamard, Hans Rademacher and Jo Walsh, an international roll call of mathematical innovation, though we did suggest that we may be entering a third golden age of Chinese mathematics that could potentially dictate how are future evolves.

By Chapter 5 we were busy congratulating ourselves at having presciently forecast the shift from delivery bandwidth value to memory bandwidth value but admitted missing the significance of search and algorithms as a key associated enabling technology.

We pointed out that batteries have many similarities with memory devices. Both are storage devices and both achieve performance gain through materials innovation, packaging and structural innovation.

As with material behaviour, electrochemical reactions at the nanoscale are different from electrochemical reactions at the micro- or macroscale. In particular, the small particle size of nanoscale materials allows short diffusion distances that means that active materials act faster. In lithium batteries for example, nanoscale manufacturing techniques potentially increase surface area and allow for the faster absorption of lithium ions, which results in increased reactivity that means that batteries can store more energy, absorb energy faster and release energy faster when required. We gave the example of carbon nanotubes being proposed as anodes and 3D metal oxide structures that have been shown to have a reversible energy capacity of more than 1000 mAh/g, exactly what those hefty lap tops we featured earlier need in order for them to be useful and useable. You would then need those graphene-based nanostructures to dissipate the heat generated from all that extra available power.

Electrophoretic displays were also credited with having invented a new application sector, the e book, and we name checked Mirasol bistable displays as a potential future game changer. In passing, we waxed lyrical about MEMS-based devices including SAW and FBAR filters, gyroscopes and accelerometers and (potentially) as a replacement for quartz crystals in resonant circuit design but suggested that the rate of innovation in the RF component sector remained a problematic constraint particularly in terms of multiband mobile broadband connectivity capability.

In this final chapter of this section we have briefly reviewed the evolution of the HUGE TV and the potential impact these devices will have on the technology economics of delivery bandwidth and memory bandwidth added value.

Which brings us on to the next five chapters and the topic of user software value.

Part Two

User Software

7

Device-Centric Software

7.1 Battery Drain – The Memristor as One Solution

The last six chapters have looked at how user hardware influences network value and user value. The end point in the last chapter was smart televisions and 3D-capable high-end lap tops.

If it was just one application being supported at any one time then life would be reasonably simple but in practice there is a user expectation that the device and the network to which it is connected should be capable of supporting multiple tasks simultaneously.

An example would be a lap top or tablet user texting or sending e mails while watching a video on You Tube with a Skype session going on in the background. Then the phone rings. At this point the weakest point will cause some form of instability that will become noticeable to the user. This might be caused by a processor bandwidth limitation, a memory bandwidth limitation, an interconnect limitation, a network bandwidth limitation or an energy bandwidth limitation (the battery goes flat). For example, the lap-top user forums referenced in the last chapter are mainly focused on how to disable functionality to reduce battery drain, which rather defeats the point of the product.

There are potential solutions, one of which is the memristor.

The theory of resistance with memory was put forward in 1971 in a paper[1] by Professor Leon Chua, from the University of California Berkeley and was premised on the basis that circuit design has three elements, a resistor, capacitor and inductor but four variables with the fourth being a combination of resistance and memory not realisable from any combination of the other three elements.

The device remained as a theoretical concept up until 2008 when HP Labs[2] demonstrated a nanotechnology scaled titanium oxide structure with atoms that moved when a voltage is applied. The device opens up the prospect of a step function increase in the energy efficiency of computing and switching systems and the opportunity to create memories that retain information without the need for power.

---

[1] http://inst.eecs.berkeley.edu/~ee129/sp10/handouts/IEEE_SpectrumMemristor1201.pdf.
[2] http://www.hpl.hp.com/.

It would appear that very small devices constructed as 3D structures at the molecular level may well be the key enablers for large and/or complex products connected to large and/or complex networks and may make the delivery of 3D images and voice and audio and text and positioning information to portable and mobile devices via mobile broadband networks more cost and energy economic than presently expected.

The only problem is that it will probably be at least twenty years before these components are available commercially, so in the meantime we will just need to make do with existing available technologies. So let's look at a present high-end product and see how the hardware form factor determines software form factor and how the combination of hardware and software form factor determines the user experience.

7.2   Plane Switching, Displays and Visual Acuity

Unsurprisingly, Japanese and Korean TV and PC manufacturers have been working to translate the innovations applied in large-screen displays to small-screen displays including inplane switching. Introduced by Hitachi in 1996, inplane switching uses two transistors for each pixel with an electrical field being applied through each end of the crystal that meant that the molecules moved along the horizontal axis of the crystal rather than at an oblique angle. This reduces light scatter and improves picture uniformity and colour fidelity at wide viewing angles. The two-transistor structure also allows for faster response times.

Inplane switching is used now in a wide range of products including the iPad that has a 9.7-inch 1024-by-768 resolution LED-backlit LCD screen with a viewing angle of up to 178 degrees with 8-bit colour depth. The iPhone 4 uses similar technology (on a smaller screen obviously) as does the iMac. Both the iPhone and iPad have ambient light sensing and an 800 to 1 contrast ratio.

Anyway, back to LG. In 1995 LG partnered with Philips to invest in display technologies and have captured significant market share in high-definition and more recent 3D displays. Their first and probably the world's first 3D smart phone, the Optimus 3D was introduced in February 2011. The device has a 4.3-inch WVGA display provides bright images allowing users to view 2D (up to 1080p) and 3D (up to 720p). LG claim the display is 50% brighter than the iPhone4 producing 700 nits of brightness

A nit by the way is not just something your children catch at school but is a US measurement term for luminance from the Latin 'nitere' to shine. A nit quantifies the amount of visible light leaving a point on a surface in a given direction. This can be a physical surface or an imaginary plane and the light leaving the surface can be due to reflection, transmission and/or emission. A nit is equivalent a one candela per square metre and you do wonder what was wrong with candelas as a measurement term even if candles do seem a touch archaic as a base for comparing luminance levels. Other measurement terms used in the past include the apostilb, blondels, skots, stilbs and the footlambert.

Luminous flux is also measured in lumens with one lux equal to one lumen per square metre. There is no direct relationship between lumens and candelas or nits as luminosity is also a function of the composite wavelengths, so a conversion is only possible if the spectral composition of the light is known.

For example, the maximum sensitivity of the human eye is at wavelengths of 555 nanometres, which is green to you and me and presumably designed so that we can spot edible greenery at a distance. The luminosity function falls to zero for wavelengths outside the visible spectrum. As an interesting if only tangentially relevant aside, The Institute of Cognitive and Evolutionary Anthropology,[3] having measured global skulls from the last 100 years collected together in a dusty basement in Oxford, have discovered that people living in high latitudes have bigger eyes and brains. The bigger eyes are so that they can see better in low light conditions. The bigger brains do not make them more intelligent but are needed to provide the additional visual processing power required. Animals that forage at night similarly have bigger eyes and brains than animals that forage during the day.

Anyway, we use nits to describe display brightness and lux to describe the light conditions that dictate whether we can take pictures with our mobile phone. For the Optimus 3D product these pictures are captured through two five-megapixel lenses set apart to allow for stereoscopic processing to produce 3D images/HDTV that can either be displayed on the phone or on an HDTV or uploaded to the You Tube 3D channel rather slowly if you happened to be using a 3G connection.

7.3 Relationship of Display Technologies to Processor Architectures, Software Performance and Power Efficiency

Actually before you get this far the phone has to process all this stuff, which it does via Texas Instruments OMAP4 dual-core dual-channel dual-memory architecture.

The concept of dual core is to have identical processing subsystems including general-purpose and special-purpose processors running the same instruction set with each processor having equal access to memory, input and output (I/O) outputs with a single copy of the operating system controlling all cores. Any of the processors can run any thread. Some specialised processors including graphic processors may be handling several thousand threads in parallel but the tasks are largely repetitive and similar. Video processing, image analysis and signal processing are all tasks that are inherently parallel.

Power consumption is managed by implementing adaptive power-down modes driven by a policy data base. These trade task latency against power drain, taking into consideration that some tasks are latency and delay sensitive and some are not. Note that it is not just delay but the amount by which that delay varies over time that can be problematic, particularly for tasks that are intrinsically deterministic. User experience expectations will include things like boot time, how long applications take to launch and execute and whether the system appears stable and predictable when multitasking.

All of this then has to be shown to work with legacy Symbian systems, Linux systems including Android and LiMO and Microsoft Windows Mobile. Anecdotally, most software engineers would admit though not generally publicly that most smart phones ship with typically several hundred software faults. Solving these faults results in time to market delay and creates a different set of faults that are generally equally problematic.

These applications processors can be drawing the best part of 5 watts at full steam so have to be carefully managed, as these products would ideally have an eight hour to ten hour duty cycle. Memory and interface specifications also have to be very adequately dimensioned to avoid problems with task latency and task interrupts.

---

[3] http://www.ox.ac.uk/media/news_stories/2011/112707.html.

As at 2011 a high-end smart phone such as the LG device mentioned above and the iPad use low-power double data rate 2 (LPDDR$_2$) mobile DRAM. This has a maximum of 8.5 Gbytes/second peak data throughput at a power consumption of 360 mW. Note this is memory power consumption not processor power consumption. Actually this is four devices housed in a package on package. The clock speed is from 100 to 533 MHz, increasing to a proposed 800 MHz for LPDDR$_3$. Clock speeds of this order are of course running at close to the RF frequencies and therefore have to be carefully managed.

There are various contenders for next-generation mobile DRAM, either evolved versions of LPDDR or alternatives promoted by vendors such as Rambus or Micron or Samsung. Samsung incidentally have almost a 50% share of this multibillion dollar market and support an option called SPMT that replaces the traditional 32 data lane technology with a four-channel, 128-lane technology that enables a total of 512 I/Os with a total bandwidth of 12 Gbytes with a target consumption of 500 mW by 2013.

The average DRAM content is a smart phone or tablet is heading towards a gigabyte. A PC hosts about 3.5 Gbytes but a lot more smart phones now get shipped every year than PCs and people don't keep them as long. Also, mobile DRAM costs two to four times more than a PC DRAM because of the more stringent size and power demands.

A high-end applications processor such as Qualcomm's Snapdragon APQ8064 is claimed to realise speeds of up to 2.5 GHz per core and is available in single-, dual- and quadcore versions.

Drifting back through old RTT technology topics unearthed a study of media processor capabilities in 2004. Back then an application processor was considered to be a media processor with a Java hardware accelerator. The processor supported media-related applications like gaming, 3D processing, authentication and the housekeeping needed for J2ME personal profiles and MIDP (mobile information device profiles).

Most of these devices were ARM based (and still are today) with the exception of Renesas (the SH mobile widely used in many Japanese FOMA phones). Essentially, there were two approaches to delivering performance and power efficiency. Vendors such as Intel and Samsung used fast clock speeds to get performance and then (using Intel as an example), implement voltage and frequency scaling to reduce power drain. Usually, the application processors clocked at a multiple of the original GSM clock reference on the basis that the devices would be used in dual-mode GSM/UMTS handsets. The Intel device frequency scaled from 156 to 312 MHz and the Samsung device scaled up to 533 MHz (13 times 13 MHz). Consider that most baseband processors ticked along at 52 MHz (3 times 13 MHz) this represented a significant additional processor load being introduced to support multimedia functionality.

Given the diversity of media processors being sold or sampled to handset manufacturers and the diversity of vendor solutions, there was an obvious need to try and standardise the hardware and software interfaces used in these devices and to provide some form of common benchmarking for processor performance. This resulted in the formation of MIPI (the Mobile Industry Processor Interface)[4] alliance. This was established by TI, ARM, ST and Nokia as an industry standard for power management, memory interfaces, hardware/software partitioning, peripheral devices and more recently RF/baseband interfaces.

---

[4] http://www.mipi.org/.

In parallel, the Embedded Microprocessor Benchmark Consortium[5] was established to work on a J2ME benchmark suite with a set of standardised processor tasks (image decoding, chess, cryptography) so that processors could be compared in terms of delay, delay variability and multitasking performance. This work continues today.

Part of the benchmarking process had to address the issue of video quality, which in turn was complicated by the fact that different vendors used different error-concealment techniques. One of the reasons GSM voice calls are reasonably consistent in mobile to mobile calls is due to the fact that error concealment is carefully specified across the range of full rate, enhanced full rate and adaptive multirate codecs. M-PEG4, however, permits substantial vendor differentiation in the way that error concealment is realised. This introduced a number of additional quality management challenges.

Six years on, the story has moved on but the narrative stays the same. Headline clock cycle counts do not translate directly across into task speed and/or task consistency and stability. Take as an example Samsung's applications processor originally called Humming Bird but now called Exynos, a name apparently derived from the Greek words smart (expynos) and green (prasinos). This is an ARM-based device and competes directly with the Qualcomm Snapdragon device (also ARM based) and TI OMAP and Intel and AMD processors. Intel has a power optimised product codenamed Moorestown that at the time of writing is yet to be launched.

All these devices build on the reduced instruction set architecture used in the original Acorn Computer. It is reassuring to know that rather like acorns and oak trees big ideas can still originate from small beginnings. Both the Snapdragon device and the Samsung device (used in Samsung's Galaxy tablet) are optimised for high speed and low power. but achieve that in different ways. Snapdragon chips are optimised to achieve more instructions per cycle and integrate additional functionality including an FM and GPS receiver, HSPA Plus and EVDO modem and Bluetooth and WiFi baseband. The Samsung device has a logic design that can do binary calculations.

Both devices can be underclocked or overclocked and the 1-GHz version of the chip sets used in present products can process about 2000 instructions per second. However, the number of instructions needed to perform a task can be very different. This can be due to either how well the software has been compiled or hardware constraints dictated, for example, by memory and interface constraints (referenced earlier) or usually a combination of both.

This can mean that certain tasks such as floating-point calculations and 2D or 3D graphics may run faster on a particular device and in general it can be stated that software design rather than hardware capability is usually the dominant performance differentiator. A processor rated at 1 GHz will not necessarily be run at 1 GHz.

Code bandwidth has increased in parallel with processor and memory bandwidth.

In '3G Handset and Network Design' (page 122) we had a table showing how code bandwidth had increased between 1985 and 2002 from 10 000 lines of code to one million lines of code that implied that code bandwidth had increased by an order of magnitude every eight years.

Today (nine years later) an Android operating system in a smart phone consists of 12 million lines of code including 3 million lines of XML, 2.8 million lines of C, 2.1 million lines of Java and 1.75 million lines of C++, suggesting that the rule still more or less applies.

---

[5] http://www.eembc.org/home.php.

The impact of this code bandwidth expansion on income and costs is still unclear. Almost inevitably more software complexity implies more software faults, but additionally the Android applications store is open to any developer. This means that a handset vendor cannot test applications supplied by third parties in terms of their quality and efficiency either in terms of user experience or consumption of processor and memory bandwidth. This can result and has resulted in an increase in product returns on the basis of poor performance, a source of frustration for a handset vendor who has had no direct income from the supply of the application but is having to absorb an associated product support cost.

There may be a rule in which an increase in processor bandwidth, memory bandwidth, code bandwidth and application bandwidth can be related to an increase in offered traffic volume and value but the effect is composite. To say that an order of magnitude increase in code bandwidth results in an order of magnitude increase in per user bandwidth by the same amount over the same period of time is not valid or supported by observation.

The present rate of increase for example is significantly higher with some operators reporting a fourfold increase annually and present forecasts are generally proving to be on the low side (as discussed in earlier chapters). However, if the opportunity cost of delivering this data in a fully loaded network exceeds the realisable value then it could be assumed that price increases will choke this growth in the future, unless value can be realised in some other way. A compromised or inconsistent user experience including a short duty cycle may similarly limit future growth rates. Also, the value may be realised by the content owners or, as stated above, the application vendors rather than the entities delivering or storing the content.

Apple has managed this by establishing and then holding on to control of application added value. No other vendor has managed this including Nokia, who have made progress with their OVI application store though with a focus on developing market rather than developed market applications. The theory of course is that application value has a pull-through effect both on product sales and on monthly revenues and per subscriber margin.

This may well be true but the counterbalance is that application value incurs application cost at least part of which shows up in product-return costs or product-support costs. The product-return costs as stated above will generally fall on the handset vendor. Product-support costs for example customer help lines, may well fall on the operator. Although deferred, these costs can be substantial.

Present smart-phone sales practice has at least partially translated these costs into additional revenue. This has been achieved by implementing tariffs schemes that are so impenetrably complicated that most of us end up paying more per month than we need to. Many of us also have smart phones that have functions that we never discover or use, which supports the view that the most profitable user is the user who pays £25 a month and never uses the phone. If the user pays £8.00 a month to insure the phone and never leaves the house that is even better.

It is, however, reasonable to state that the functionality of device-centric software and the application value realisable from that software is intrinsically determined by the hardware form factor of the device that in turn determines uplink and downlink offered traffic value and delivery cost.

Consider a power-optimised multitasking multithreaded media processor in a mobile broad-band device. An example is shown in Figure 7.1.

The core has to be capable of handling multiple inputs and outputs that may combine text, voice, audio (including *high-quality* audio), image, video and data. For conversational variable bit rate traffic (a small but significant percentage of the future traffic mix), the processor has

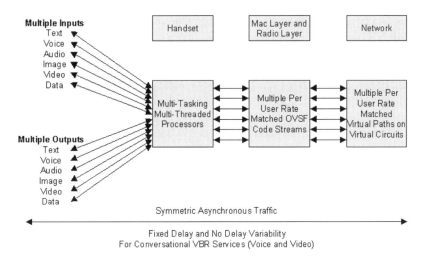

Figure 7.1   Multithreaded processor functional block diagram.

to maintain the time interdependency between the input and output data streams. In effect, the processor must act as a real-time operating system in the way that it handles multiple media streams into, across and out of the processor core.

The same level of determinism then needs to be applied to the MAC and physical radio layer, preserving the time interdependency of the multiple per user rate matched OVSF code streams and then on into the network with multiple per user rate matched virtual paths and circuits.

Note that together, the handset media processor, MAC layer and radio layer and network transport have to deliver tightly managed fixed delay and no delay variability, a requirement inconsistent with many existing processor cores. Processor loading is also generally being considered in terms of asymmetric loading – for instance, a higher processor load on the downlink (from network to mobile).

This assumption is really not appropriate for conversational variable bit rate services that are and always will be symmetric in terms of their processor load – a balanced, though highly asynchronous, uplink and downlink.

Note also that a balanced uplink/downlink implies a higher RF power budget for the handset. Most RF power budgets assume that the handset will be receiving more data than it is transmitting, which is not necessarily the case. A higher RF power budget in the handset places additional emphasis on processor and RF PA power efficiency. Given this, one of the most useful benchmarks for a video handset should be its milliamp/hour per megabyte performance under specified channel conditions. Given that the milliamp/hour per megabyte metric effectively determines network loading and margin, this could turn out to be one of the more important measures of phone (and processor) performance.

7.4   Audio Bandwidth Cost and Value

There should also at least in theory be some relationship between quality and value and there is certainly a relationship between quality and delivery and storage cost. Audio bandwidth for example is a product of the codec being used.[6] The adaptive multirate narrowband speech codec (AMR NB) for example has a sampling frequency of 8000 Hz and maps to a constant rate channel rate of 4.75, 5.15, 5.90, 6.70, 7.40, 7.95, 10.2 or 12.2 kbit/s. The adaptive MultiRate Wideband Speech Codec (AMR-WB) samples at 16 000 Hz and maps to a constant rate channel rate of 6.6, 8.85, 12.65, 4.25, 15.85, 18.25, 19.85, 23.05 or 23.85 kbit/s. The MPEG4 advanced audio codec samples at 48 000 Hz and can be either mono or stereo mapping to a channel rate of 24, 48, 56 or 96 kbit/s.

Any of the above can be made variable bit rate by changing the bit rate switch.

7.5   Video Bandwidth Cost and Value

For video coding the mandatory H263 codec supports a QCIF frame size of 176 by 144 pixels or sub-QCIF 128 by 96 pixels at up to 15 frames per second. MPEG4 is similar. Both map to a maximum 64-kb channel.

In 3G Handset and Network Design we suggested users might be prepared to pay extra for better quality images and video. Image and video quality is usually already degraded as a result of compression with the loss of quality being determined as a Q factor.

In J-PEG (image encoding), 8×8 pixel blocks are transformed in the frequency domain and expressed as digital coefficients. A Q factor of 100 means that pixel blocks need to be identical to be coded as 'the same'. A Q factor of 50 means that pixel blocks can be quite different from one another, but the differences are ignored by the encoder. The result is a reduction in image 'verity' – the image resembles the original but has incurred loss of information. In digital cameras, 90 equates to fine camera mode, a Q factor of 70 equates to standard camera mode.

These compression standards have been developed generally to maximise storage bandwidth. You can store more Q70 pictures in your camera than Q90 pictures. However, choice of Q also has a significant impact on transmission bandwidth. An image with a Q of 90 (a 172-kbyte file) would take just over 40 seconds to send on a 33 kbps (uncoded) channel. The same picture with a Q of 5 would be 12 kb and could be sent in under 3 seconds. The delivery cost has been reduced by an order of magnitude. The question is how much value has been lost from the picture. The question also arises as to who sizes the image and decides on its original quality, the user, the application, the user's device or the network. Similar issues arise with video encoding excepting that in addition to the resolution, colour depth and Q comprehended by J-PEG, we have to add in frame rate.

Perceptions of quality will, however, be quite subtle. A CCD imaging device can produce a 3-megapixel image at 24 bits per pixel. Consider sending these at a frame rate of 15 frames per second and you have a recipe for disaster. The image can be compressed and/or the frame rate can be slowed down. Interestingly, with fast-moving action (which looks better with faster frame rates) we become less sensitive to colour depth. Table 7.1 shows how we can exploit this. In this example, frame rate is increased but colour depth is reduced from 10 bits to 8 bits (and processor clock speed is doubled).

Note also how frame rates can potentially increase with decreasing numbers of pixels.

---

[6] http://library.forum.nokia.com/index.jsp?topic=/Design_and_User_Experience_Library/GUID-D56F2920-1302-4D62-8B39-06C677EC2CE1.html.

Table 7.1   Pixel resolution and frame rate

	Frame rates at 10 bits versus 8 bits per pixel output	
Example	10 bits (@ 16 MHz)	8 bits (@ 32 MHz)
1280 × 1024	9.3	18.6
1024 × 768	12.4	24.8
800 × 600	15.9	31.8
640 × 480	19.6	39.2
320 × 240	39.2	78.4

This gives us a wider range of opportunities for image scaling. The problem is to decide on proportionate user 'value' as quality increases or decreases.

Table 7.2 shows as a further example the dynamic range of colour depth that we can choose to support.

The M-PEG 4 core visual profile specified by 3GPP1 covers 4- to 12-bit colour depth. 24-bit is generally described as high colour depth and 32-bit colour depth is true colour. 32-bit is presently only used for high-resolution scanning applications.Similarly handset hardware determines (or should determine) downlink traffic. There is not much point in delivering CD quality audio to a standard handset with low-quality audio drivers. There is not much point in delivering a 24-bit colour depth image to a handset with a grey-scale display. There is not much point in delivering a 15 frame per second video stream to a display driver and display that can only handle 12 frames per second.

Intriguingly, quality perceptions can also be quite subtle on the downlink. Smaller displays can provide the illusion of better quality. Also, good-quality displays more readily expose source coding and channel impairments, or, put another way, we can get away with sending poor-quality pictures provided they are being displayed on a poor-quality display.

These hardware issues highlight the requirement for device discovery. We need to know the hardware and software form factor of the device to which we are sending content. It also highlights the number of factors that can influence quality in a multimedia exchange – both actual and perceived. This includes the imaging bandwidth of the device, how those images are processed and how those images are managed to create user experience value.

Table 7.2   Colour depth

Colour Depth	Number of Possible Colours
1	2 (ie Black and White)
2	4 (Grey-scales)
4	16 (Grey-scales)
8	256
16	65 536
24	16 777 216

7.6   Code Bandwidth and Application Bandwidth Value, Patent Value and Connectivity Value

The developing narrative is that bandwidth can be described in different ways. As code bandwidth has increased, application bandwidth has increased.

Application bandwidth value can be realised from end users and/or from patent income. As this book was being finished, Nortel raised $4.4 billion from the sale of 6000 patents to Apple, RIM and Microsoft. Some of the patents were in traditional areas such as semiconductors, wireless, data networking and optical communications but many were application based.

As application bandwidth has increased, the way in which devices interact with the network and the way that people use their phones and interact with one another has changed. The result has been a significant increase in network loading though quite how effectively this is being translated into operator owned network value is open to debate. The risk is that the operator ends up absorbing the direct cost of delivery and the indirect cost of customer help lines and product returns but misses out on the associated revenue.

However, it is too early to get depressed about this.

A substantial amount of application value is dependent on connectivity. Connectivity value does not seem to be increasing as fast as connectivity cost, but this may be because operators are filling up underutilised network bandwidth and can therefore regard additional loading as incremental revenue. Also, we are fundamentally lazy and often do not bother to change from inappropriate tariffs where we are paying more than we need for services that we never use, always good for operator profits.

In the longer term it is also interesting to consider the similarities between telecoms and the tobacco industry. Both industries are built on a business model of addiction and dependency. In telecoms and specifically in mobile communications, as application bandwidth increases, addiction and dependency increases. This should yield increasing returns over time. Let us hope there are no other similarities.

8

User-Centric Software

8.1 Imaging and Social Networking

The ability to capture images on a mobile phone has either coincided with and/or facilitated the social-networking phenomenon. Either way, this provides a justification for treating the topic of image processing in this chapter under the heading of user-centric software, exploring the relationship of mobile-phone software and social networking.

In theory at least imaging bandwidth should translate into network value. If we take a picture or video on a mobile phone and send it to a friend or share it with multiple friends on an image and video sharing web site then the cost of connectivity should be realised as operator revenue. This is, however, premised on the assumption that the image/images will be uploaded over the mobile broadband network on the basis of this being the easiest and most immediate option available. However, we usually have the option to wait until we get home and upload over a fixed connection. This will be an appealing option if the cost difference between an upload over the mobile network and fixed network is higher than the point of indifference, whatever that happens to be. The reduction of solid-state memory cost, the increase in solid-state memory density and a reduction in memory power drain have made it easier both technically and commercially to provide phones with enough memory to support the home-upload option.

Imaging capabilities in mobile phones can, however, provide significantly more functionality than just taking pictures. For a start phones can potentially see in the dark better than we can and can see things in higher resolution and greater colour depth and in parts of the ultraviolet or infrared spectrum. Imaging functionality can include an ability to perform tasks that would be otherwise difficult, reading a bar code for example. The combination of imaging with pattern recognition and/or pattern matching combined with user context and location and positioning context potentially allows a mobile device to move from being a communications tool to a device that allows us to interact with the physical around us. Similarly combining imaging with other sensing capabilities may add value. Software has several functions in the image processing chain. Software can correct for distortions introduced by low-cost optical components and can be used to improve image quality including compensating for user limitations, shaky hands being an example. Software can then be used to sort images or search images or match images.

Making Telecoms Work: From Technical Innovation to Commercial Success, First Edition. Geoff Varrall.
© 2012 John Wiley & Sons, Ltd. Published 2012 by John Wiley & Sons, Ltd.

The imaging processing chain and the software used in the image processing chain therefore merits attention both in terms of what it can do today but also what it may be able to do in the future.

8.2   The Image Processing Chain

When considering the imaging processing chain, it is logical to start with the image. Our eyes see the world in terms of brightness (luminance, also known as luma), colour (chrominance, also known as chroma) and the shapes and patterns of objects within our field of vision. Our eyes have a remarkable dynamic range and can tolerate and process visual information in anything between direct sunlight (+100 000 lux) to a moonlit night (fractions of a lux).

There are various tricks that help us increase this dynamic range, for instance we just see things in black and white at very low light levels. We also have autowhite colour balancing as part of our natural visual tool set. White has a green tinge in daylight, a yellow tinge under fluorescent light and an orange tinge under incandescent light, but the tinge disappears courtesy of our natural image processing chain. These dynamic range and colour-capture and colour-correction capabilities have to be recreated in artificial image processing systems.

In addition, our natural image processing system is incredibly efficient at sorting out the entropy in the viewed image (the useful information) and redundant information that can be effectively ignored and discarded. This process is also adaptive. If you are hurtling down a hill on a mountain bike, your brain will just be processing the visual information relevant to the task in hand (staying on the bike and not hitting any trees in the process).

These compression capabilities and adaptive pattern recognition capabilities have to be recreated in artificial image processing systems. We are also naturally adept at pattern recognition, recognising a face in a crowd for example. These pattern-recognition capabilities have to be recreated in artificial image processing systems.

Each component in an artificial image processing chain has a required function and a 'wanted effect'. Unfortunately, most components also produce 'unwanted effects' that introduce impairments. We can, however, go some way towards cancelling out or at least concealing these effects, as described in the following sections.

8.2.1   The Lens

The wanted effect in a lens is optical quality, the ability to focus an image on the sensor array with minimum distortion. Optional functions include a good depth of field, focal length, angle of view and optical zoom.

Unfortunately, because of cost, size and weight constraints, most lenses used in camera phones are far from perfect. The unwanted effects that result include vignetting (not enough light getting to the edge of the sensor array causing edge/corner shading) and lateral chromatic aberration caused by rays of light being sent obliquely across the colour filter array. The effects can be cancelled out or reduced by using antivignetting/shading correction algorithms, antiblur algorithms and colour correction.

8.2.2   The Sensor Array

A sensor array is an array of photosensitive cells, the discrete parts of which are described as pixels. The photosensitive cells are more or less identical to solar cells in that their job is

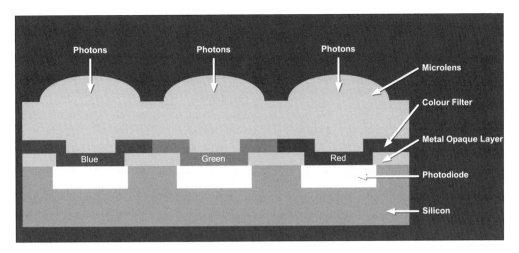

Figure 8.1   Microlens, colour filter and photodiode. Reproduced with permission of Aptina Inc.

to turn incoming photons from the lens into countable electrons. The efficiency with which they accomplish this task is described as quantum efficiency (QE) measured in bits per lux per second. Bigger pixels collect more photons, but take up more space and cost more. More pixels on a sensor array increase resolution but take up more space and cost more.

In the implementation pictured in Figure 8.1 a microlens is placed over each pixel. The photons pass through a colour filter and an opaque metal layer and are captured by a photodiode.

A sensor array is either a charge coupled device (CCD) or CMOS device. In a charge coupled device, the charge is transported across the chip and read at the corner of the array. In a CMOS device, each pixel has either 3 or 4 transistors, which amplify and move the charge across a wired backplane. Because each pixel on a CMOS sensor has several transistors located next to it, some of the incoming photons hit the transistors rather than the photodiode, so sensitivity is lower than a CCD device. The transistors also produce noise.

This noise (and the effects of transistor mismatch) can, however, be reduced by a process of double sampling and a combination of noise-cancellation and noise-reduction algorithms.

CCD sensor arrays produce better images particularly at higher (multimegapixel) resolutions. CMOS devices cost less, use less power and their ability to address individual pixels, individual rows of pixels and individual columns of pixels allows for more flexibility both in terms of image processing, pattern matching and image recognition. CMOS devices in 2005 typically had a resolution of the order of 300 000 pixels. Today (2011) some high end SLR still-image cameras have CMOS arrays that can resolve 24 megapixels or more. High-end CCD sensors can resolve over 40 megapixels. Usually, a small microlens is used over each pixel through which the light passes.

Noise compromises accuracy but as noise floors reduce over time, partly a consequence of volume-related yield optimisation, these devices enable algorithmic innovation that properly applied can yield algorithmic value.

This is what might be termed valuable value as it does not cannibalise other parts of the component value chain. In theory, people buy more phones at a higher price and do more things with them that indirectly or directly create additional income.

This is not the case with some other forms of innovation. In an earlier chapter we mentioned RF power-amplifier innovation that allowed four or five amplifiers to be replaced by a single component. Given that the expectation will inevitably be that the single device price point will stay the same this is not a compelling proposition for vendors with sunk manufacturing investment. In fact, generally it can be stated that an innovation that delivers to all parts of an industry value chain will face the least adoption resistance.

Camera phones more or less fall into this category. True, they may have cannibalised the low end of the digital camera market but as we shall see later in this chapter the higher-end digital camera market is alive and well and may well have benefited from people using camera phones wanting to upgrade to something that took better pictures and was easier to use.

This is a microscale effect with a macroscale impact.

But back to technology.

8.2.3   The Colour Filter Array

Colour images are a mix of red (400 nm), green (510 nm) and blue (700 nm). A pixel will have a red, blue or green filter and the readings from 3 pixels (a red, blue and green pixel) are combined together to produce the chroma information. This is why colour sensors are three times less sensitive than black and white sensors. This part of the component chain is known as the colour filter array.

Human eyes are less sensitive to green than they are to red or blue. The colour filters are therefore arranged in a Bayer pattern, originally invented and patented by Kodak. The Bayer 'checkerboard' has red and green in one row and blue and green in the next. This is shown in glorious black and white in Figure 8.2.

The Bayer pattern means that interpolation algorithms have to be used. The unwanted effects of interpolation include optical crosstalk (rays of light taking an oblique route through a red or blue pixel before landing on a green pixel) and blooming caused by charge overflows into neighbouring pixels. Antialiasing filters reduce these effects but the filters blur the image and obscure detail. Using edge-sharpening algorithms (known as high-frequency component gain algorithms) can put this detail back but these algorithms amplify noise in the flat regions of the image. This can be avoided by using edge-recognition algorithms.

The sensor array and colour filter array provide the information needed for the DSP and microcontroller to perform a number of system functions including autoexposure, stray-light compensation, autofocus, autowhite balance and gamma balancing (correcting for the nonlinear relationship between pixel value and displayed intensity on devices such as TV monitors). These are 'wanted effects' or required functions.

The unwanted effects introduced by the DSP are due to the fact that is has to digitise a complex combination of waveforms that have large variations in signal amplitude (wide dynamic range). When the input bandwidth exceeds the sampling rate, the DSP will produce aliasing effects (difference frequencies/false frequencies). False low frequencies caused by ADC undersampling create Moire effects (the distortion that you see when someone wears a check jacket on television), coarse quantisation causes fringing, ringing and shadowing.

Some of these effects ripple up and down the processing chain. For example, a series of flash guns go off at a press conference and put the DSP into compression, prediction errors fill the output buffer and trigger heavy requantisation, which produces tiling effects. If the system drops chroma coefficients in a desperate attempt to recover, then colour disappears (not that

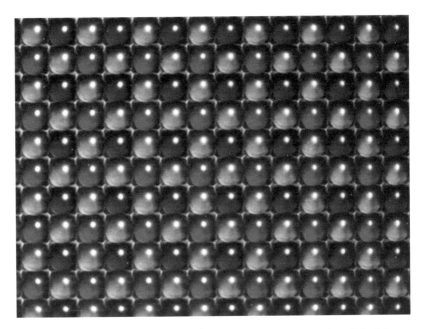

Figure 8.2   Kodak Bayer pattern.[1] Reproduced with permission of Kodak.

you will really notice by this stage). These are sometimes described as excess compression effects.

This brings us to the final stage of the processing chain, the JPEG and MPEG encoder.

8.2.4   The JPEG Encoder/Decoder

Entropy and redundancy are separated in still images by dividing the image into eight-by-eight pixel macroblocks. Each macroblock is coded in terms of its luma and chroma content divided into what are termed as Y, U and V planes in which Y is the luminance and U and V are colour difference channels. Colour difference is a more efficient way of describing the RGB content of each discrete part of the image. The JPEG compression algorithms are based on a discrete cosine transform (DCT) that separates out the luminance and chroma information in the spatial and frequency domain. These are expressed as DCT coefficients.

Compression can then be achieved by comparing one macroblock with other adjacent macroblocks and in effect just coding the difference from block to block. This is something of an oversimplification. There is actually a stage described as quantisation that removes perceptually insignificant data (and reduces the number of bits per DCT coefficient) and AC/DC prediction that uses filters to predict coefficient values from one or more adjacent blocks, but otherwise that's JPEG in a nutshell. There is also a version of JPEG known as motion JPEG – a succession of JPEG images used for video streaming.

---

[1] http://pluggedin.kodak.com/pluggedin/post/?;id=624876.

8.2.5   The MPEG Encoder

Generally though, moving images, i.e. video, are coded using MPEG (usually MPEG 4). JPEG is based on the principle of macroblock to macroblock comparison. MPEG adds in image to image/frame to frame comparison and motion estimation.

Image comparison is based on I frames, P frames and B frames. I frames are encoded as still images and are not dependent on any reference frames. P frames depend on the previously displayed reference frame and B frames depend on previous and future reference frames. I frame coding is sometimes described as intracoding (literally, coding within), P and B frame coding is sometimes described as intercoding (literally, coding between).

In MPEG 1 and MPEG2, the macroblock sizes are fixed. In MPEG4, the macroblocks can be subdivided or sliced and become part of a more complex encode/decode process involving shape coding, texture coding and motion-estimation coding.

Motion-estimation coding predicts the contents of each macroblock based on the motion relative to a reference frame. The motion vectors and the difference between the predicted and actual macroblock pixels are encoded. The way this search and comparison algorithm is implemented by different vendors can make a substantially impact on the performance of the whole encode/decode process.

Search techniques include cross-searching, step searching, diamond searching. Matching techniques include sum of absolute difference block matching, luma/chroma motion vector matching and methods based on deriving chrominance motion from luminance motion.

Shape coding is also known as object coding. The MPEG 4 standard describes how video objects can be described in terms of their shape. Video objects in an image stream make up a video object sequence (VOS) that is made up of video object layers (VOL). The video object layers grade the importance of the coded information and help to support 'graceful degradation' algorithms, sometimes also described as enhancement layer encoding.

Video object layers can be described within video object planes (VOP) that can be built into video object groupings (VOG). Shapes can be opaque or transparent and 2D or 3D. They can also be described in terms of their 'texture' (texture coding), in which picture gradients (consisting of AC/DC coefficients) can be combined with horizontal and vertical prediction algorithms to reduce the overall code rate.

Most of the remaining algorithmic effort then gets focused on eliminating the 'compression artifacts' – the unwanted effects of compression. These include 'blocking' in which the borders of each macroblock become visible in the reconstructed frame and 'ringing', which creates distortion near the edge of image features.

Blocking is reduced by using a low-pass filter. Deringing depends on being able to detect the edges of image features. A 2D filter is then applied to smooth out areas near the edges but with little or no filtering on the edge pixels (which would cause blurring).

Finally, there are various error-resilience techniques used to hide the fact that 'the channel' (which for a camera phone includes the radio channel) is highly variable with high and unevenly distributed errors and frame erasures.

Error resilience depends on the use of resynchronisation and motion markers, the use of extender header codes, data partitioning and reversible length coding also known as forward decoding/backward decoding, also known as bidirectional coding.

8.3 Image Processing Software – Processor and Memory Requirements

Each of the algorithms and filter functions discussed occupy processor clock cycles. Most of them also occupy memory space.

For JPEG, the DCT transform, quantisation, variable length coding and AC/DC prediction all have an impact on processor loading. For MPEG, motion estimation, shape coding and texture coding add to the encoder/decoder overhead. Quite often there are just not enough clock cycles to go round.

8.3.1 Image Processing Hardware

This is why most image processing chains in mobile phones use hardware coprocessors/ hardware accelerators to meet power-budget and latency constraints. The power-budget/ latency-budget issue has also prompted device vendors to suggest and sometimes implement a range of new memory and DSP and microcontroller architectures optimised for media processing/image processing applications. This creates a number of complex decision issues when designers are trying to decide on which algorithms to use on which hardware platform.

8.3.2 Image Processing, Voice Processing and Audio Processing

An additional complication is that it is not just the image processing chain that we need to deal with but also voice and audio. The inclusion of wideband high-quality solid-state microphones in high-end cameras today is a sign that enhanced audio-capture capabilities will be an inherent part of future camera phone products – the addition of hear what i hear (HWIH) to see what i see (SWIS) as part of the user experience. From a mobile broadband revenue perspective the challenge is to make *'this is what I am looking at now'* have more value than *'this is what I was looking at then'*. This means that camera-phone connectivity to a mobile broadband network has to be easy and low cost and image quality has to be as good as you get with an SLR for fixed image capture or a high-end video camera for video capture. In practice there is a growing overlap between camera phones, SLR cameras and video cameras. Can a camera phone be as good as the other two options?

8.3.3 35 mm Cameras as a Quality and Functional Performance Reference

As users, most of us, consciously or subconsciously use 35-mm film cameras as a quality and functional performance reference.

The Leica A, the first commercially successful 35-mm camera, was introduced in 1925. It was small, light weight, had a high top speed for fast action shots (a 500th of a second), a 50-mm standard focal length lens, a focusing lever and infinity lock, a 50-mm viewfinder and built in frame counter. It could take 36 pictures without reloading. 57 000 units were built between 1925 and 1936 – a mass-market product by the standards of the day. Many camera phones today would struggle to match the image quality and functionality provided by this device and similar devices.[2]

---

[2] http://www.cameraquest.com/mguide.htm.

Table 8.1   Digital camera performance benchmark 2005

	Entry level		Midtier		High End	
Retail price	£100	£200	£300	£400	£600	£6000
Resolution	2.1 Mpixels	3.24 Mpixels	4.0 Mpixels	7.0 Mpixels	8.0 Mpixels	16 Mpixels
Digital zoom	3×	3.4×	3.6×	6×	10×	10×
Optical zoom	None	3.6×	10×	10×	10×	10×
ISO rating	200	400	400	400	800	1250

8.3.4   Digital Cameras as a Quality and Functional Performance Reference

It is only relatively recently, around about ten years ago, that the image quality achievable even from high-end digital cameras matched that achievable from traditional film-based cameras. This is a function of the performance available from the sensor array and the image correction and image processing algorithms used.

Digital cameras have replaced film on the basis that they are more convenient. They save on film cost and the preview facilities mean we can throw away all those vastly embarrassing pictures we took on holiday before they get developed. We have a choice today of low-cost entry-level fixed-focus digital cameras, midtier compact or digital SLR cameras and high-end digital SLR cameras.

Table 8.1 compares a cross section of cameras available in 2005 in terms of cost, function-ality and quality using resolution, zoom capability and ISO sensitivity as the chosen points of reference. Note that resolution is stated in effective pixels. This is the total number of pixels excluding the pixels used for colour balancing and the pixels outside the range of the lens.

Zoom capability is simply a function of the lens design (optical zoom) or the resolution available (digital zoom depends on reasonably high resolution so that you can throw most of the picture away and still have acceptable quality). ISO sensitivity is the sensitivity available from the sensor array and is equivalent to the sensitivity rating for 35-mm film.

Opinions might differ as to what constitutes a high-end camera so this has been widely defined above as anything that costs between £600 and £6000!

8.3.5   Digital Camera Performance Benchmark 2011

Six years on £300 buys a midtier camera such as an EOS Canon 1000D with a CMOS sensor providing a resolution of 10.1 Mp and an ISO rating of 1600, a seven-point autofocusing system with spot monitoring, the ability to shoot images at 3 frames per second and a 2 GB memory card. £1000 buys a higher-end version such as the EOS 50D with 15 Mp resolution, an ISO rating of 12 800, a nine-point AF (auto focus) system and the ability to shoot images at 6.3 frames per second. The higher ISO ratings allow for fast shutter speeds in low light conditions. An equivalent value Nikon adds in an ability to capture full HD (1920 × 1080) movie clips with sound. Nikon started adding WiFi to some models in 2005 but the market impact of this functionality has been limited. This reinforces the point made earlier that wireless connectivity is apparently not as compelling a proposition as anticipated by some vendors. If you have £10 000 burning a hole in your pocket then you could buy a Pentax 645D with a CCD sensor providing a medium format image with 40 MP resolution and a 77-point sensor.

There are a number of advantages to using an SLR camera. First, what you see is what you get (a hinged mirror behind the lens reflects the image up into the viewfinder), secondly the user can choose from a vast range of standard and specialist lenses. The disadvantage (for a digital SLR) is that the mirror only swings out of the way when the shutter is pressed, which means that the image does not reach the sensor until the moment of exposure. Only then does the camera have a chance to do autocorrection on the image.

Compact cameras, also sometimes described as lens shutter cameras, don't offer the optical choice or quality of an SLR camera but the fact that the sensor array itself is looking at the image before the shutter is depressed helps to improve the metering process. In other words, the image that comes out of a compact camera can actually look better than the image coming out of a digital SLR. The digital SLR demands more knowledge and skill from the user and may require more postediting to get the required colour and light balance effects.

In addition to the cost of the optics, a high-end SLR or compact digital camera also requires a substantial amount of storage bandwidth. To preserve image quality, images are either saved as raw data or as a TIFF (tagged image file format). A TIFF file uses lossless compression. The original image can be recreated after compression without losing any of the original source information. Even a TIFF of an 8-megapixel image still produces a 20-Mb file. This takes 30 seconds to save on to a compact flash card. High-end cameras get round this problem by having buffer memory – typically enough to support ten or twelve seconds of 'fast shooting' at say 2 frames per second. Midtier compact cameras still have good-quality optics – not perhaps as good as a high-end SLR but still good enough for most print applications, at least up to A4 size.

If pictures are just being taken to put on the web, then a GIF image (graphic interchange format) is used. Midtier and (most) entry-level digital cameras are therefore more than adequate for 'display online' applications.

8.3.6   The Effect on Camera Phone User Expectations

The rapid increase in digital camera quality over the past ten years and the rapid decrease in price has, however, had an effect on what users expect in terms of the quality available from a digital camera phone hardware platform. For reasons of cost, size and weight, the lenses used in camera phones are of relatively poor quality. The use of a CMOS sensor (for cost and power reasons) instead of the CCD sensor used in higher-end digital cameras also tends to result in higher image noise and lower sensitivity.

Most users of course just expect their camera phones to be as good as an equivalently priced digital camera in terms of optical quality. While this is possible, it does imply additional cost and complexity and the only way in which a digital camera phone can be as good as an equivalently priced digital camera is either by subsidising the product, or to make the product in much higher volume. By default, Nokia achieved this because so many of their phones include camera-phone functionality. Nokia became and still remains the largest vendor of digital cameras albeit with a phone included. Did Nokia sell more phones as a result of adding imaging functionality? Well yes almost certainly, but the offset cost includes the additional imaging component cost included in phones owned by users who never use the functionality and the additional customer support costs of sorting out users who cannot work out how to take a picture let alone send it to someone.

8.4   Digital Camera Software

Users also expect their camera-phone software to be just as good as an equivalently priced digital camera in terms of functionality. All digital cameras and camera phones use digital processing to improve or rebalance the captured image. High-end cameras, however, do the job better and faster than low-end cameras and better and faster than most camera phones.

Digital camera manufacturers set out to differentiate their camera products on the basis of what might be described as 'algorithmic value'. One example is the time it takes a camera to autofocus and the accuracy of the autofocus. To be achieved accurately, autofocusing depends on multipattern metering and averaging of the edge characteristics of the image. To be achieved quickly (within a few hundreds, or preferably a few tens of milliseconds), autofocusing depends on highly optimised and highly parallel processing algorithms. The same principle applies to almost all other functions including autoexposure (metering and averaging the luminance characteristics of the image) and colour balancing (metering and averaging the chrominance characteristics of the image. Today, midtier digital cameras can handle capture speeds of at least two frames per second with absolutely no compromise in image capture quality.

If camera phones aspire to replace digital cameras, or at least aspire to replace low-end or midtier digital cameras, then they have to provide equivalent functional performance.

8.4.1   Digital Compact and Digital SLR Cameras as a Quality and Functional Performance Reference

What does this mean in practice? Let's consider two digital cameras in 2005, one a digital compact, one a digital SLR, both are from the same manufacturer, both are priced at about £700 (Table 8.2).

Table 8.2   Digital compact cameras versus digital SLR

Specification	Digital Compact	Digital SLR
Resolution (Mpixels)	8	6.3
Sensor size (mm)	9 × 6	23.7 × 15.6
ISO range	50–400	200–1600
Lens (35 mm equiv)	35–350 mm	27–105 mm
Aperture range	F2.8–5.2	F3.5–4.5
Shutter speeds	8–1/3000 s	30–1/8000 s
Metering	Multisegment (256), centre weighted, spot	3D colour matrix, centre weighted, spot (AF point)
EV compensation	±/2EV in 0.3 EV steps	±/5 EV in 0.3 or 0.5 EV steps
Viewfinder	0.44-inch tft, 235 000 pixels	Optical fixed pentaprism
Continuous shooting speed	1.2 fps/2.3 fps (display off)	3 fps
Continuous shooting capacity	5 shots	4 shots (RAW), 9 shots JPEG fine
RAW file processing time	10 s	O s (up to 5 shots)
Start up time	3 s	0 s
AF speed	600 ms	300 ms
Battery life	240 shots	400 shots
Dimensions	116 × 85 × 121 mm	140 × 111 × 78 mm
Weight	600 g	595 g

Note the maximum sensitivity of the Compact is ISO 400, whereas the digital SLR has an ISO of 1600.

Six years on the market equivalent for the digital compact camera would be something like the Sony Cyber Shot HX 100V[3] but at a price point closer to £300 rather than £700. For this you get a 16.2-megapixel camera with thirty times optical zoom and an ISO sensitivity of between 100 and 3200. The camera has an inbuilt GPS receiver so that pictures can be geotagged. There is a 2D and 3D sweep panorama facility with autostitching. Still images can be captured at a rate of ten per second. Video can be captured at 30 frames per second at 1920 by 1080 (2 megapixel) resolution that equates to a data capture rate of 17 Mbps.

A high-end digital SLR such as the DSLR 850 from Sony, albeit at a slightly higher price (of the order of £1800) has an ISO rating of 100 to 6400 and a resolution of 24.6 megapixels. Bear in mind that at this kind of resolution a single JPEG image can easily be 20 MB or closer to 40 Mb as an uncompressed RAW file. If you want to absorb memory bandwidth this is a good way to do it.

Comparing compacts with SLRs at equivalent prices just on the basis of resolution may suggest that the compacts offer better value for money. However, compact cameras have smaller sensors, which means the pixels are smaller which means they collect fewer photons, which means they are less sensitive and have a lower signal-to-noise ratio.

The smaller the sensor, the more the magnification needed to produce a same size print. Smaller sensors also will tend to exhibit more severe chromatic aberration and sensor blooming (where a charge leaks from an overloaded pixel to its neighbours). The lens on a compact gives a longer focal length than the standard lens of the SLR, though the motor drivers can be quite slow.

Compacts and SLRs both typically have a wide aperture range. A wider aperture will deliver more photons to the sensor array but will reduce the depth of field (the range over which different parts of the captured image remain in focus). Other differences include maximum shutter speed, generally faster for SLRs1/3000 for the compact) and metering patterns, probably the one area (apart from size) that the compact has an advantage.

The time available to the compact for metering (given that the sensor array is looking at the image before it is captured) mean that the pictures coming out of the compact will probably look better than the pictures coming out of the digital SLR.

Digital SLRs generally have a wider dynamic range for EV compensation. EV (the exposure value) is the amount of shutter speed or aperture adjustment needed to double or halve the amount of light entering the camera. Generally, the digital SLR has an exposure system that will tend to preserve more highlight detail than the compact. Thus, although the image coming out of the SLR might look darker and muddier, the images will have a higher useable dynamic range and will respond better to photoediting. The viewfinder in the SLR gives a true representation of the image (you are looking through the lens). The compact relies on a miniature LCD that by default is only giving you its own version of the image.

The viewfinder and the LCD on the back of the compact camera both absorb power and processor clock cycles. The shows up both in terms of battery life (numbers of shots per recharge cycle) and when trying to shoot at faster frame rates. RAW file processing time is a function of the embedded memory included in the devices (more in the digital SLR).

---

[3] http://www.sony.co.uk/hub/cyber-shot-digital-cameras?campaignId=12003904&s_kwcid=sony%20digital%20camera%20review|7512510395.

Start up time for the digital SLR is generally immediate, compacts may take several seconds. Autofocus (AF speed) with an SLR will also usually be faster. This makes a difference in terms of how responsive the camera feels to the user. Although the compact is dimensionally smaller this actually makes it harder to hold steadily (to avoid camera shake). It also weighs the same as the digital SLR.

So although compacts generally have better resolution, image stabilisers and longer-range zooms, they do not have the speed, viewing systems or ultimate image quality of their digital SLR equivalents. The above examples also illustrate how high-end product features migrate towards low-end products over time just like the car industry and/or mobile phones where smart-phone functions migrate over time towards lower-end products.

Self-evidently, it is not practical to replicate the optical quality and functionality packaged in a 600-g digital camera in a 100-g cellular phone. However, users of digital cameras don't necessarily understand this and we cannot expect consumers to be particularly sympathetic to the cost and size and power budget constraints implicit in cellular phone design. All they will say is . . . . this phone doesn't take good pictures. Having said that, a modern smart phone generally produces some pretty impressive results.

8.5   Camera-Phone Network Hardware

Even so. it is sensible to concentrate on some of the other user benefits that come from packaging phones with cameras.

Optical quality and functionality are not just a function of the digital camera or camera phone but a composite of the components used to store, process and handle images once they leave the users device. Image stores, image servers and the way in which imaging storage is realised in hardware are all part of the 'quality and value' process.

For imaging, the amount of storage needed is partly driven by how often users take pictures but also by the imaging bandwidth. Given that high-end cameras are now capable of capturing 40-megapixel images, it seems churlish not to provide an equivalent increase in storage capability. Imaging bandwidth directly determines the amount of storage needed. Note we cannot automatically expect users to downgrade a RAW or TIFF image to a JPEG just for our convenience. What's the point of having a decent camera capability if you the throw 90% of the image in the bin. Other metrics include the time it takes to upload and download images and the stability of the storage medium (nonvolatility over time and temperature). This brings us to the next topic.

8.6   Camera-Phone Network Software

Image data such as the date and time a picture was taken and the exposure used can be stored in an EXIF (exchangeable image format file). Devices that have GPS can add location information to the file, which can be stored either in a camera's memory card or sent with the image to be stored remotely.

Camera phone network software addresses three requirements – image editing, image cataloguing and image management. These functions can, to an extent, be performed in the camera phone or digital camera but with either type of device, the expectation has to be that the images will at some stage be stored somewhere else.

A user can do image editing on a PC using PhotoShop or Paint Shop Pro – this is not software that you would want to load onto a portable device. These programmes typically take up at least 200 MB of RAM and 500 MB of hard-disk space. A cataloguing software programme (Photo Shop Album or Paint Shop Photo Album for example) will take up a similar amount of memory and you will also need to add some image management software. Microsoft Photo Premium sets up a browser so you can search images by date, keyword, folder or size but it takes another 300 MB of hard-disk space.

Some (possibly most) of this browser functionality can of course be packaged in the user's device. Browser presets are useful for directing users to the 'right' companies like Opera[4] are active in developing added value embedded browser capabilities.

An alternative is to use an online photo album like Kodak[5] or Snapfish[6] Kodak limit you to 700 JPEGs of any size in an album but you can have as many albums as you like. Snapfish provide unlimited (JPEG or zip file) storage space on condition that you buy a print or gift item (mugs etcetera) from the Snapfish service once a year. If pictures with a resolution of more than nine megapixels are uploaded, Snapfish have the right to downsize the image.

These online albums provide some fairly simple image-editing capabilities such as cropping and red-eye removal. Snapfish also provide some autocorrection including autocontrast. Both also provide view/review and share facilities. Companies such as Scalado[7] and Shutter Fly[8] provide similar capabilities though optimised for camera-phone applications supported via cellular network operators and integrated with existing MMS server topologies.

Just as camera phones have to at least aspire to have equivalent functionality to digital cameras, so these network-operator-specific service offerings have to be at least as good as existing and competitive online services. This is a challenge, particularly when competitive online services are free.

8.7   Summary

User expectations of camera-phone image quality and functionality are influenced by the performance now available from entry-level and midtier digital cameras. Optical quality is a function of lens quality, sensor quality and the effectiveness of the algorithms used to correct for optical and processing impairments. Functionality includes capabilities such as autofocusing and autoexposure. The accuracy and speed with which these functions are performed provide differentiation between low-end and high-end products.

Camera phones have at least to aspire to providing quality and functionality that is equivalent to similarly priced digital camera products. This can only be achieved through subsidy or by producing products in much higher volumes. There are also fundamental form-factor and power-budget issues to address.

The user experience is also determined by how images are stored and managed in the network and whether the quality of the original image is preserved. In terms of image editing, cataloguing and management functionality, user expectations are increasingly dictated by the

---

[4] http://www.opera.com/.
[5] http://www.kodakgallery.com/gallery/welcome.jsp.
[6] http://www2.snapfish.co.uk/snapfishuk/welcome.
[7] www.scalado.com.
[8] www.shutterfly.com.

capabilities of third-party online service providers such as Kodak and Snapfish who have no prior affiliation either to the cellular radio industry or to the traditional service provider community.

Image quality, functionality and interoperability (the absolute requirement to be able to interchange images between devices between networks) are obvious prerequisites for MMS-based image-capture and image-sharing platforms. This is both a challenge and opportunity but implies an increasing need to integrate camera-phone design policy with network design and specification, particularly the sizing and specification of image storage bandwidth and performance metrics.

Engineers with knowledge of digital camera design have to understand the practical cost and form-factor constraints of cellular-phone design. Cellular-phone designers have to appreciate the optical and functional performance expectations implied by present and future digital camera products. Network hardware and software engineers have to understand that image bandwidth and image quality expectations will determine access network, transport network and storage network performance requirements.

9

Content- and Entertainment-Centric Software

9.1 iClouds and MyClouds

In this chapter we review how content and entertainment software has changed over the past five to seven years and set this in the context of contemporary events (announcements that have occurred literally as this chapter is being written (June 2011)).

These include the announcement by Apple of iCloud,[1] a set of free cloud services intended to encourage people to store data and access applications on demand remotely (in the cloud) for access from any device but including the iPhone, iPad, iPod touch, Mac and PC. The assumption is that you might own all of these and if anything changes on any one device it changes on all of them.

The iCloud is serviced from three data centres the latest of which[2] is a 500 000 square foot facility in Maiden North Carolina five times the size of an existing 109 000 square foot facility in Newark California and represents an investment of somewhere between $500 million and one billion dollars.

At the same time Microsoft have announced upgrades to the X Box 360 including the ability to access live TV programs from Sky TV in the UK, Canal+ in France and Foxtel in Australia with presumably the intention of inking similar deals with providers in the USA and in parallel started marketing the Microsoft My Cloud.[3]

In May 2011 Microsoft spent $8 billion dollars buying the Skype telephone service and rumours in the Press started circulating about their intent to buy Nokia's handset business, now with a market value less than HTC, the Taiwan-based manufacturer of smart phones.

Five years earlier this would have seemed an unlikely scenario with Nokia consolidating its dominance in the handset market. Unfortunately for Nokia the company proved overreliant on Symbian, its home-grown operating system and failed to realise the scale of the impact that the introduction of the iPhone would have on the smart-phone market and Nokia's position in

---

[1] http://www.apple.com/uk/icloud/?cid=mc-uk-g-clb-icloud&sissr=1.
[2] http://www.datacenterknowledge.com/archives/2010/02/22/first-look-apples-massive-idatacenter/.
[3] http://www.microsoft.com/en-us/cloud/default.aspx?fbid=_7o8c0rKG39.

Making Telecoms Work: From Technical Innovation to Commercial Success, First Edition. Geoff Varrall.
© 2012 John Wiley & Sons, Ltd. Published 2012 by John Wiley & Sons, Ltd.

that market. Competition at the lower end of the market in developing countries further eroded market share and margin.

In 2010 Stephen Elop formerly President of Microsoft's business division and prior to that COO of Juniper Networks, Adobe Systems and Macromedia joined Nokia as President and CEO.

Microsoft were themselves trying to work out a response to the shift that had occurred in the last quarter of 2010 when smart phones started outselling personal computers. (Approximately 100 million smart phones against 94 million personal computers.)

This had and has two implications. The added value for Microsoft per PC shipped is close to $40, the added value for Microsoft in a mobile phone is $15 dollars and the market has to be shared with Apple and Android. Both are problematic. Apple has a first-mover advantage that at the time of writing appears unshakeable. Android is effectively open sourced and therefore will be likely to force down operating system realised value.

Buying Nokia might be a solution but that all depends on what happens next both in terms of handset hardware and software form factor and functionality. It also depends on the market mix. If everyone bought $500 smart phones then additional software value would be realised. However, what we are trying to establish is the mix between hardware and software value going forward and who will own that value. The answer is hidden somewhere in the hardware and software changes that have happened in the last five years, the hardware and software changes that are happening today, the hardware and software changes that are likely to happen over the next five years and the impact these changes will have on the smart phone market, TV market, games market and content and entertainment sector and application added value. We just need to know where to look. The journey takes us from camera phones via gaming platforms to a future that may be uncertain but in theory at least should be predictable. The analysis needs to include successes and failures. The accuracy of the prediction is dependent on the amount of detail that we can capture and the depth of the analysis that we apply to that detail so here we go.

9.2   Lessons from the Past

At CEBIT in March 2005 Samsung launched their V770 camera phone with the stated aim of proving that a ' midtier' digital camera phone capability could be packaged into a camera phone physical form factor. The 770 was a 7-megapixel phone with interchangeable lenses. In common with many other camera phones, it also supported basic camcorder capability. The product was enthusiastically hailed as a breakthrough by the technical press but relatively few were sold. The problem was that the product attempted to be a digital camera with a mobile phone added in as an extra.

Given that there was no real enthusiasm by the operator community to provide additional subsidy for the product it fell between two stools, being overtaken by digital cameras that worked better and mobile phones that worked better and cost less, albeit with less exotic imaging functionality. It also looked odd, felt heavy bulky and unbalanced and for all these combined reasons failed to set a trend.

Six years later the Samsung Galaxy[4] offers a similar resolution (eight megapixel) and LED flash, autofocus, touch focus, face detection, smile shot and image stabilisation with a video

---

[4] http://www.samsung.com/global/microsite/galaxys2/html/.

capability that can record full 1080p HD video and a preinstalled image and video editor. The success of this product suggests that early disappointments (in relative terms) can serve to calibrate later market introductions in a positive way.

Despite the fact that most high-end cameras will now also shoot HD video, digital camcorders still exist as a discreet product sector but as with still cameras functionality can be shown to be changing over time. A digital camcorder six years ago (2005) from Canon priced at around £600 to £700 typically had 800 000 or 1.2-megapixel CCD sensors. The default resolution (DV video) was usually 720 by 480 pixels (345 000 pixels) with the additional pixels being used for image stabilisation, digital zoom and still photography. Camcorders with 3- or 4-megapixel CCD sensors were also sold on their ability to produce quality still images in addition to video. Video images were typically recorded on to tape or disk and still images recorded on plug-in memory.

9.2.1   Resolution Expectations

The top of the range Samsung 770 seven-megapixel camera phone could 'only' handle QVGA 320 by 240 pixel video (76 000 pixels). Users of camcorders expected to have 720 by 480 pixel DV resolution (345 600 pixels) more or less equivalent to 640 by 480 pixel VGA (307 200 pixels). This gave adequate quality when displayed on a computer monitor.

The other format found in video phones in 2005 was 176 by 144 pixel QCIF (25 344 pixels) derived from 352 by 288 pixel CIF (101 376 pixels), which is one quarter the size of a standard 704 by 576 pixel PAL TV image known as 4CIF (405 504 pixels).

Note that CIF (common intermediate format) is a standard originally from the TV industry, whereas VGA (video graphics array) was introduced in 1987 as a standard for describing computer monitor resolution. The VGA standard, however, included 1920 by 1080 pixel high-definition TV (2.073 megapixel) that at that time was an aspiration but is now a common reality.

Today, HD video recorders are available for around £300 and will work in conditions of almost total darkness (around one and a half lux) and when required can realise a still image resolution of seven or eight megapixels. Just as digital cameras can now perform as video camcorders, video camcorders can also perform as a still image camera. A contemporary product such as Panasonic HDC SD90[5] is a 1080-pixel HD device that shoots at up to 50 frames a second with an optional additional lens for capturing 3D images.

9.2.2   Resolution, Displays and the Human Visual System

Just as in audio systems where you would not generally marry a £5000 amplifier to a pair of £5.00 speakers, there is not much point in capturing high-quality video and showing it on a low-resolution display.

From an end-user-experience perspective the useful most relevant measure of display quality is perceived picture quality. This can either be done by establishing a 'subjective quality factor' that is more or less equivalent to the mean opinion scoring used in voice. A more objective

[5] http://www.panasonic.co.uk/html/en_GB/Products/Camcorders/1MOS+HD+Camcorders/HDC-SD90/Overview/6828169/index.html?gclid=CMHi-rvMnKoCFcxzfAodzDpSyg.

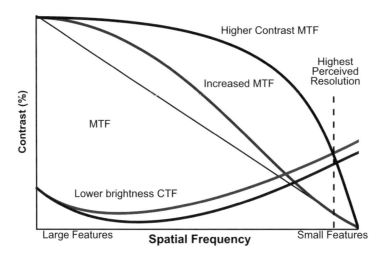

Figure 9.1 Contrast threshold function and MTFA. Reproduced by permission of Planar Systems Inc. (Clarity Visual Specialist Display Technologies).[6]

method is to use a method known as the modulation transfer function area. This is illustrated in Figure 9.1.

The quality of a video stream is partly a function of the frame rate and colour depth, but also a function of the contrast ratio, resolution and brightness of the display. Contrast ratio defines the dynamic range of the display (the ratio of the brightest white the system can generate divided by the darkest black), resolution defines the display's ability to resolve fine detail expressed as the number of horizontal and vertical pixels. Brightness is, well, brightness, measured in foot-Lamberts or candelas per square metre.

Brightness captures attention, contrast conveys information.

To be meaningful, brightness and contrast need to be characterised across the range of spatial frequencies being displayed. Spatial frequency is the ratio of large features to small features – the smaller the features in an image, the higher the spatial frequency. The overall number of pixels in the display determines the limiting resolution. The modulation transfer function is a way of comparing contrast to spatial frequency. As features get smaller, the contrast ratio will reduce. However, this measure does not take into account the limitations of human vision. We need a minimum contrast for an image to become distinguishable and this is measured using a contrast-threshold function.

Adding together the modulation transfer function of the imaging system and the contrast threshold function of human vision yields a crossover point that determines the highest perceptible resolution.

Increasing the brightness does not make much difference to the maximum perceived resolution, whereas increasing the contrast ratio significantly increases the amount of image content conveyed to the viewer.

---

[6] http://www.clarityvisual.com/.

The example in Figure 9.1 is applied to benchmarking quality in larger display systems but the same rule set applies to smaller displays. These performance metrics are important because they define the real-world experience of the user – a better quality image makes for a more immersive experience, the more immersive the experience, the longer a session will last and the more valuable it should be.

9.2.3   Frame-Rate Expectations

Six years ago the Samsung V770 produced 15 or 30 frames per second. 30 frames per second would normally be considered adequate for video (cinema film runs at 24 fps, PAL at 50 and NTSCC at 30 fps) but a number of streaming applications in 3G networks were being trailed in 2005 at 35 fps (the faster frame rate hides some of the quality issues). Some of the higher-tier camcorders could even then shoot at 240 frames per second. Camera phones have therefore not destroyed the compact camera, digital SLR or camcorder – these product sectors all remain alive and well partly due to continuous innovation and the pull-through effect of big-screen HDTV. All of these products have had an impact on memory footprint and memory technologies.

9.2.4   Memory Expectations Over Time

Six years ago the Samsung V770 had 32 MB of internal memory and an MMC Micro card slot. The Canon Camcorder had 11 GB of memory on its DV cassette plus 8 Mb on the SD card. The Hitachi camcorder had 9.4 GB of memory available on the DVD plus the SD card. Both the tape and DVD formats were capable of storing half an hour of DV quality video and audio.

9.3   Memory Options

9.3.1   Tape

Tape storage is rapidly becoming a historical curiosity, but back in 2005 it was a relatively mainstream option for camcorders. The Mini DV tape usually used had a number of advantages – relatively reliable and robust and low cost, less than 4 dollars for 11 GB of storage. There was, however, a mechanical process involved. The tape moved at 18.9 mm a second over a read/write head rotating at 9000 rpm. The head laid down tracks diagonally across the tape, 12 tracks per frame for PAL and 10 tracks per frame for NTSCC. This means that the tape was quite bulky (the cassette was 12.2 mm by 48 mm by 66 mm) and the recording sequential – that is, random access requires the tape to be mechanically wound forward or back. It is thus not suitable (if used on its own) for mixed media (video, audio and data) applications.

9.3.2   Optical Storage

DVDs in comparison can be configured for random access (DVD-RAM), or sequential access (DVD-RW and/or DVD+RW).DVDs come in three sizes, four and a half inch, three-inch mini DVDs and Sony's two and a half inch universal media disk. Data is stored in a spiral of pits (up to 48 kilometres long for a double-sided DVD) embedded within the disk. The reflectivity of the pits can be changed by the application of heat from a red (640 nanometre) laser that is also used to read the data. A DVD RAM can be rewritten about 10 000 times, a DVD-RW or DVD+RW can be rewritten about 1000 times (the whole disk gets erased on the write cycle). The disks rotate at between 200 and 500 rpm. Although DVD-RW and DVD+RW are optimised for sequential storage, there are zone storage and search techniques that mean they can be used for storing and reading a mix of video, audio and data. The spin speed changes to keep the read speed constant as the laser moves out from the centre of the disk. The transfer rate is typically 10 Mbps. The seek time on a DVD RAM disk is typically about 75 ms.

A standard four and a half inch DVD holds 4.7 GB of data, (9.4 GB on a double-sided disk), a mini DVD holds 1.4 GB and a universal mini disk holds 1.8 GB. And as stated in an earlier chapter moving to a blue laser at 405 nanometres rather than a red laser (a Blu Ray DVD) increases available storage to close to 30 gigabytes, enough for a two-hour HD movie.

The Sony Universal Media Disk was used (as a read-only device) in the original Sony PlayStation Portable released in 2004. The PlayStation Portable was significant in that it could deliver PS2 quality graphics on a high-brightness screen. PS2, the second generation of PlayStation produced by Sony had been released in March 2000 in Japan and competed with Microsoft's X Box, the Nintendo Game Cube and Sega's Dreamcast. The PS2 outsold all other consoles and reached 150 million units shipped by early 2011 with over 2000 titles available and over 150 billion titles sold since launch.

In 2007 Sony introduced a new portable device PSP 2000, also known as the PSP Slim as it was 33% lighter and 19% slimmer. In 2008 the PSP 3000 was introduced with an improved LCD screen with five times the contrast ratio and a built-in microphone. A redesigned and smaller model, the PSP Go, was launched in 2009. This removed the UMD drive and relied on WiFi connectivity for game downloading and Bluetooth connectivity for multiplayer headsets and handset connectivity.

The PSP2 also called the Sony NGP also called the Vita was introduced in June 2011 with an OLED[7] multitouch front display and multitouch pad on the back. The product included

Figure 9.2   Sony PlayStation Vita. Reproduced with permission of Sony Computer Entertainment Inc.

---

[7] Organic light-emitting diode display.

WiFi and 3G connectivity with a 'near' application to allow users to discover other Vita users nearby and 'Party' that allows users to voice chat or text chat while they are gaming.

The device has a front and rear camera with a frame rate of 60 or 120 frames per second, a six-axis motion sensing system (three-axis gyroscope, three-axis accelerometer and three-axis electronic compass, built-in GPS and Bluetooth). It is shown in Figure 9.2.

9.4   Gaming in the Cloud and Gaming and TV Integration

Taking the Sony portable offerings as a reference, over a period of five years this product sector has moved from a business model based on shipping products on an optical disk to a business model based on paying for downloaded games over the internet stored on local memory.

The assumption must be that this model also now evolves into just downloading games on demand as and when needed, effectively cloud-based gaming. Similarly, the Microsoft X box is evolving into a dual-purpose entertainment platform that can either store record and play TV programmes and/or download games from the internet. The X Box therefore needs a substantial amount of hard-disk space of the order of 250 GB.

A desk-top hard disk is typically now a three and a half inch diameter device spinning at between 10 000 and 15 000 rpm. The capacity in 2005 was typically 100 GB. Six years later capacity is typically somewhere between 750 and 1000 GB (one terabyte). The power budget of these devices (8 to 10 W) precludes their use in portable products.

Lap tops typically use two and a half inch diameter hard disks offering 250 GB of capacity. A spin speed of 10 000 rpm produces a seek time of under 5 milliseconds but power consumption is still over 5 W. Reducing the spin speed to 5000 rpm increased the seek time to about 15 milliseconds but reduces the power to about 3 W.

In 2005 iPods used 1.8-inch hard disks with between 20 and 60 GB of storage and a variable 3990 to 4200 rpm spin speed. Power consumption was about 1.3 watts but as with other portable products solid-state memory has taken over. In 2005 vendors were promoting either 1-inch drives (Hitachi) or 0.8-inch 'microdrives' (Toshiba and Samsung) for cellular phones.

The one-inch drives fitted into a Type 2 PC card form factor (5.2 mm) and weighed 16 g. Microdrives had a depth of 3.3 mm and weigh 10 g. One-inch disks typically held about 8 to 10 GB and microdrives were either 2 or 4 GB. Both had a spin speed of 3600 rpm giving a seek time of 12 to 15 milliseconds and a transfer rate of 4 or 5 M/bytes per second. A one inch drive had a power consumption of just under one watt, a 0.85-inch drive, about 700 milliwatts.

Some cellular handsets became available with hard disks (the Samsung V5400 for example and the 3-GB SGH-i300 launched in the UK) and Toshiba brought forward the availability of a 4-GB 0.85-inch hard disk but in practice over the past six years solid-state memory has eclipsed all other options for mobile applications. A hard drive remains the most economic and space effective way to put a terabyte of storage on your desk top but is not optimum in a mobile product form factor.

9.4.1   DV Tape versus DVD versus Hard-Disk Cost and
Performance Comparisons

Tapes may well continue to disappear from the market but they still offer good value for money. A mini DV tape costing 4 dollars implies a storage cost of less than 0.04 of a cent per megabyte but the memory medium cannot be used for random access applications.

An optical disk could potentially offer similar cost/performance ratios with the added benefit of sequential storage and random access capability.

A 4-GB microdrive can provide sequential storage and fast random access (15 milliseconds rather than the 75 milliseconds of a DVD) but presently costs about 10 cents per megabyte and as we have stated is nonoptimum when used in portable products.

9.4.2   Hard-Disk Shock Protection and Memory Backup

A number of manufacturers (Toshiba for example) have MEMS-based accelerometers built in to hard-disk drives to detect when a product is being dropped. The head can then be kept clear of the disk to prevent damage. Analogue Devices (low-g iMEMS) and Microelectronics have digital accelerometers targeted at hard-disk protection.

This is standard now in most lap tops though if you have dropped a lap top your problems may be more fundamental than just a damaged memory. Also, lap tops increasingly are going to solid-state memory linked to cloud connectivity, what used to be called the thin client model.

9.5   Solid-State Storage

This thin client model has mixed implications for the solid-state memory market. On the one hand, there is an increasing need for short-term storage but thin clients linked to cloud connectivity imply that neither the application nor the content are stored in the device but are downloaded as required on demand.

The future of both embedded memory and plug-in memory is thus open to debate. The debate benefits from an understanding of the various solid-state memory options and their relative advantages and disadvantages.

9.5.1   Flash (Nor and Nand), SRAM and SDRAM

Flash-based solid-state memory can be configured to look and behave like a hard disk (given its ability to retain data in power down) and has the big advantage of not having any moving parts. It is however, more expensive and comes in a bewildering range of form factors.

Embedded nonvolatile (FLASH) memory in a cellular handset is typically a mix of NOR for code storage and NOR and NAND for data storage. NOR sets the input memory state at 0, NAND sets it at 1. NAND costs less than NOR because it takes a number of memory capacitors and matches them to a smaller number of switching transistors. NAND is more error prone but the cost benefits make it a preferred option for most embedded data storage applications. Wear-levelling techniques have also increased the number of write cycles to typically 100 000 or more even on extreme geometry devices.

Flash memory, originally developed by Intel, works on the basis of having two transistors separated from each other by a thin oxide layer – one transistor is known as the 'control gate', the other is the 'floating gate'. The floating gate can have its state changed by a process known as tunnelling. An electrical charge is applied to the floating gate transistor that then acts like an electron gun, pushing electrons through the oxide layer. A cell sensor measures the level of charge passing through the floating gate. If it is more than 50% of the charge, the state value is 1, if less than 50%, the value is 0. The gate will remain in this state until another electrical charge is applied. In other words it is nonvolatile. It will retain its state even when power is not applied to the device.

The embedded (nonvolatile) flash memory sits alongside (volatile but lower cost) SRAM (static random access) and DRAM memory (dynamic random access). Random access means just that, the memory is optimised for fast random access rather than steady sequential access.

SRAM offers very fast read/write cycles (a few nanoseconds) but the data is held in a volatile 'floating gate' (a volatile flip flop). It does not need refreshing as often as DRAM (hence the description static) but is still relatively power hungry. SRAM is used for temporary cache memory.

DRAM memory holds data in a storage cell rather than a flip flop (which is why it is called dynamic). DRAM offers the lowest cost per bit of the available solid-state memories. A typical DRAM device will store up to 64 MB (512 Mb). SDRAM is synchronous DRAM – synchronous means the device synchronises itself with the CPU (central processing unit).

DRAM products are inherently power hungry in that the storage cells leak and require a periodic refresh to maintain the information stored in the device. There are techniques such as temperature-compensated refresh, partial array self-refresh and deep power down that help reduce power drain. These products are sometimes described as mobile DRAM or mobile SDRAM. The volatility and related power-drain issues of these devices and their optimisation for random rather than sequential access makes them unsuitable for storing large amounts of video requiring sequential access.

It has therefore now become increasingly common to supplement embedded memory with plug-in memory. Most plug-in memory consists of a NAND flash device (or multiple devices) on a die with a dedicated microcontroller.

In the context of a cellular handset, plug-in memory can be summarised in terms of when the technology was introduced.

9.5.2   FLASH Based Plug-In Memory – The SIM card as an Example

9.5.2.1   The SIM Card

The SIM card (subscriber identity module) became an integral part of the GSM specification process in the mid to late 1980s. The original idea was that it should be used embedded on a full-size ISO card (the standard form factor for credit cards) but as phone form factors became smaller through the 1990s it became normal to use the device on its own. The device is 0.8 mm high, which apart from the Smart Media Card (see below) makes it the thinnest of any of the plug-in memories and one of the smallest (26 mm × 15 mm). Early SIMs consisted of an 8-bit microcontroller, 8 k of ROM and 250 or so bytes of RAM and were used primarily to store the user's phone number (IMSI/TMSI) and to provide the basis for the A3/A5/A8 challenge and response authentication and encryption algorithms. Early SIMS also had a limited amount of data bus bandwidth, typically 100 kbps.

A SIM in 2005 had the same mechanical form factor but used a 16-bit or 32-bit micro-controller. Most of the vendor product roadmaps suggested then that one gigabyte of data storage would be technically and economically feasible within 2 to 3 years coupled with a faster bus (up to 5 Mbits/s) and faster clock speed (up to 5 MHz). These devices were termed HC SIMS (high-capacity SIMS), or MegaSIMS or SuperSIMS depending on the vendor and were targeted at the image storage market. A SIM used with a UMTS phone was known as a USIM (universal subscriber identity module) with additional functionality in terms of device and service identification. An ISIM was a SIM optimised to work with the IP multimedia subsystem intended to provide the basis for IPQOS management and control.

Although the SIM remains a mandated component within UMTS and a mandated part of the 3GPP1 and 3GPP2 standardisation process, it's longer-term role in managing authentication, access and service control is being challenged by microprocessor and memory manufacturers.

ARM11 microprocessors for example were promoted in terms of their ability to manage user keys and as a mechanism for providing network virus and M commerce protection. ARM/TI dual-core microprocessor /DSPs were promoted in terms of their ability to provide end-to-end security, multimedia and digital rights management.

Intel were promoting their 2005 offering 'Bulverde' on the basis of its built-in hardware-based security capabilities (DES, triple DES and AES encryption) and conformance with IT industry (EAL4+/EAL5+evaluation assurance levels) rather than telco industry security standards.

The SD card and MMC cards available at the time also promoted enhanced security and authentication features as part of their future network operator focused value-added proposition. Smart card/SIM/storage card combinations such as MOPASS (a Hitachi-led consortium working on SIM card/memory card functional integration) strengthened this story. These products were sometimes described as 'bridge media' products.

Thus, it was possible that some of the functionality proposed for the USIM/ISIM would migrate onto competitive device platforms particularly in markets with no prior SIM/USIM experience. The 1-gigabyte capacity of the SIM also could have been assumed to be insufficient for many of the evolving imaging applications, particularly video storage and movie image management and/or any associated gaming applications.

And that's how it has turned out so far. SIM cards have not made significant inroads into new markets and still predominantly perform the functions that they have always performed.

9.5.2.2   Compact Flash[8]

In 1994, SanDisk introduced Compact Flash, a 4-MB flash card with a depth of 3.3 mm (type 1 PC card) by 36.4 by 42.8 mm and a weight of 11 g. Compact Flash was compatible with the Integrated Device Electronics standard, which meant that the flash looks like a (small) hard disk to the operating system. Capacities ranged from 16 MB to 12 GB and the products were widely used in digital cameras. The devices could withstand a shock of 2000 Gs equivalent to a 10-foot drop and were claimed to be able to retain data for up to 100 years. Specification revisions included an increase in data transfer rate from 16 MB/s to 66 MB/s.

9.5.2.3   Smart Media[9]

In 1995, Toshiba launched Smart Media, an 8-MB device. This was the slimmest of the plug-in memories with a depth of only 0.76 mm but had the disadvantage of being dependent on the

---

[8] http://compactflash.org/.
[9] http://eu.computers.toshiba-europe.com/.

host controller to manage memory read/writes (it had no built-in controller but is just a NAND chip on a die). This and a relatively high pin count, has limited its application in smaller form factor devices, although it found its way into a number of digital cameras and camcorders (for still-image storage).

9.5.2.4   Multimedia Cards[10]

Introduced in 1997 by Siemens and Sandisk, multimedia cards were targeted from the start at small form factor cell phones and (not surprisingly, given their name), multimedia applications. Standard MMC cards had a depth of 1.4 mm and are 24 mm long and 32 mm wide. An MMC microcard was 18 mm long.

There was an MMC Plus and HS (high speed) road map with a variable bit width bus (1 bit, 4 bit, 8 bit), a 52-MHz clock rate (a multiple of the GSM 13 MHz clock), a 52-M/byte per second transfer rate and a search time of 12 milliseconds (equivalent to a hard disk at the time). Storage footprints were from 32 MB to 2 GB. A standards group working with the Consumer Electronics Advanced Technology Attachment standards body was established to address how an MMC card would interoperate with disk-drive memory and looked at future optimisation of MMC devices for audio and video stream management.

9.5.2.5   Memory Stick[11]

The Memory Stick was introduced by Sony in 1998. It had a depth of 2.8 mm, a width of 21.5 mm and is 50 mm long (the same length as an AA battery). The 'top of the range' model in 2005 was the Memory Stick Pro providing 2 GB of storage with a maximum transfer rate of 20 MB/s and a minimum write speed of 15 Mbps. A thinner version, the Memory Stick Pro Duo, had a depth of 1.6 mm and a capacity of up to 512 MB. These devices were optimised for video (and other types of media including images, audio, voice, maps and games). The high transfer speeds needed for video did, however, have a cost in terms of power drain – a memory stick used at full throttle used upwards of 360 milliwatts. Sony had a product road map showing 4 GB devices available by the end of 2005 and 8 GB devices by the end of 2006. Five years later in 2011 a 32-GB memory stick costs £40, a 2-GB memory stick costs £4.00. Most of us use memory sticks given to us as free promotional items at conferences, which was probably not the business model that Sony originally anticipated.

9.5.2.6   SD (Secure Digital) Cards[12]

Introduced by a group of manufacturers including Sandisk, Toshiba and Panasonic in 2000/2001, the SD card was developed to be backwards compatible with (most) MMC cards but with a reduced pin count (2 rather than 9 pins) and (as the name implies) extended digital rights management capabilities.

---

[10] http://www.mmca.org/.
[11] http://www.sony.co.uk/product/rec-memory-stick.
[12] http://www.sdcard.org/home/.

The standard SD card was the same width and length as the standard MMC card (24 by 32 mm) but is thicker (2.1 mm) and therefore able to support slightly more storage with a road map of 2- to 8-GB devices. The device weighed 2 g and had a smaller cousin known as a Mini SD card that was 1.4 mm by 21.5 mm by 20 mm and weighed 1 g and a Micro SD card which was 1 mm by 11 mm by 15 mm and weighs virtually nothing. As you may have noticed, Mini SD cards and Micro SD cards were not compatible with MMC microcards. Seek times were similar to MMC. Rather like the memory stick, the SD card had separate file directories for voice (SD voice), audio (SD audio) and video (SD video).

9.5.2.7   xD Picture Cards[13]

Introduced by Toshiba, Fuji and Olympus in 2002, Picture Cards had a depth of 1.7 mm, a width of 20 mm and a length of 25 mm, weighed 2 g and were presently available in capacities up to1 GB with a 2- to 8-GB product road map similar to the memory stick. A 1-GB card would store about 18 minutes worth of Quick Time movie at VGA resolution (640 by 480) at a frame rate of 15 fps including sound, or 9 minutes at 30 fps.

9.5.2.8   USB Flash Drives/Portable Hard Drive[14]

Slightly different in that they were specifically designed for moving data files on to or from a hard disk, USB flash drives plugged straight into the USB port. This did not mean that you had to have a 12 mm by 4 mm USB port in the host device but delivered flexibility in terms of plug in I/O functionality (hot swapping, USB hubs and all those other nice things that USB ports deliver). The objective was to have devices that were compatible with the USB2.0 standards[15] for high-speed USB that included transfer rates of up to 480 Mbps. This became standardised as USB 2.0 and remains a ubiquitous method of providing connectivity with computing devices and of delivering power for devices that need recharging).

The lessons from this are, however, salutary. The adoption of plug in solid-state memory has not been helped by the bewildering choice of formats and form factors that resulted in a loss of scale economy and an offer to consumers that was completely incomprehensible to the majority of the buying public. Incompatibility between formats and an inability to export and import data between formats compounded the problem.

This has almost certainly helped build the case for cloud computing on the basis of simplicity and ease of use and cost, irrespective of whether the application is content or entertainment centric.

The iCloud may be the next evolution in this cloud connectivity story presumably with an iGame console as yet (at time of writing) to be announced.

---

[13] http://www.fujifilm.co.uk/.
[14] http://usbflashdrive.org/.
[15] http://www.usb.org/home.

10

Information-Centric Software

10.1   Standard Phones, Smart Phones and Super Phones

In December 2009 Microsoft and NAVTEQ, a digital map and location company, signed a technology agreement on the development of 3D map data and navigation software including the street-level visuals that we now take for granted when we look at Google maps. 18 months earlier Nokia had acquired Navteq for $8.1 billion dollars and started introducing phones such as the 6210 optimised for navigation applications.

Over the next two years smart phones with GPS positioning and associated navigation capabilities became increasingly common but the unique differentiation that Nokia presumably aimed to achieve through the acquisition failed to materialise and it is hard to detect any particular market advantage achieved despite the substantial investment. This is arguably a question of timing with a general rule that is not a great idea to invest in a specialist technology provider servicing a market that is being aggressively commoditised.

In 2006 we undertook an analysis of the potential of the mobile metadata market. This gives us a benchmark against which market expectation and market reality five years on can be judged.

We premised the analysis on the assertion that three types of mobile phone hardware and software form factors would evolve over time, standard phones, smart phones and super phones. Standard phones are voice dominant and/or voice/text dominant. Standard phones change the way we relate to one another. These phones have relatively basic imaging and audio capture capabilities. Smart phones change the way we organise our work and social lives. They have more advanced imaging and audio capture capabilities than standard phones. Super phones, also known as 'future phones' in that they did not exist in 2005, would change the way in which we relate to the physical world around us.

Super phones would have advanced audio and image capture capabilities equal to or better than human audio and video processing systems, have equivalent memory resources but significantly better memory retrieval than human memory systems, a better sense of direction than most humans and an enhanced ability to sense temperature, gravitational forces and electromagnetic fields.

Making Telecoms Work: From Technical Innovation to Commercial Success, First Edition. Geoff Varrall.
© 2012 John Wiley & Sons, Ltd. Published 2012 by John Wiley & Sons, Ltd.

10.1.1 The Semisuper Phone

In a sense the ingredients for a semisuper phone were already available. We could capture 20 kHz or more of audio bandwidth using solid-state microphones, capture low-g gravity gradients using 3-axis accelerometers and have low-cost phones that included a temperature sensor, digital compass and positioning capabilities and build software added value on all of these hardware attributes. However, a super phone has to be as good as or better than humans at more or less everything. This means having imaging bandwidth that is at least perceptually equivalent or preferably better than our own human visual system, able to see and hear the world in three dimensions.

This is why humans have two ears and two eyes in order to gauge distance and depth and in order to respond and react to danger or opportunity. Similarly, humans have a well-evolved optical memory. We remember events visually albeit sometimes rather selectively.

There are, however, still significant hardware constraints. Imaging resolution even with high-end CCD' sensors, the 40-megapixel CCD referred to in the last chapter for example, is still short of present human resolution (about 200 megapixels). Imaging processing systems are also still significantly dumber than humans when it comes to taking spatial decisions based on complex multichannel inputs.

For example, a racing driver driving a Formula One car uses audio, visual, vibrational and wind speed information and a sense of smell in subtle and significant ways that are presently hard to replicate either in terms of capture bandwidth or information processing in artificial intelligence systems.

However, leaving aside how audio and visual information is actually used, there are some areas where devices are better at 'seeing' the real world than we are. CCD and CMOS sensors for example are capable of capturing and processing photons well into the nonvisible infrared spectrum (up to 900 nanometres for CMOS and over 1000 nanometres for CCD) – the basis for present night-vision systems.

There are similarities between the image processing chain and the RF processing chain. In RF signal processing, the three major system performance requirements are sensitivity (the ability to detect wanted signal energy in the presence of a noise floor), selectivity (the ability to detect wanted signal energy in the presence of interfering signals) and stability (the ability to achieve sensitivity and selectivity over time, temperature and a wide range of operating conditions).

Noise is a composite of Gaussian white noise introduced by the devices used to process (filter and amplify) the signal and distortion with the added complication of nonlinear effects that create harmonic distortion. Image processing is very similar. The lens is the equivalent of the radio antenna but collects photons (radiant energy in the visible optical bands), rather than RF energy.

10.2 Radio Waves, Light Waves and the Mechanics of Information Transfer

Radio waves are useable in terms of our ability to collect wanted energy from them from a few kHz (let's say 3 kHz) to 300 GHz, above which the atmosphere becomes to all intents and purposes completely opaque.

In wavelength terms assuming radio waves travel at 300 000 000 metres per second, this means our wavelengths of interest scale from 100 kilometres (very long wave) to 0.0001 metre (300 GHz).

Optical wavelengths of (human) visible interest scale from 300 to 700 nanometres but let's say we are also interested in ultraviolet (below 300 nanometres) and infrared (above 700 nanometres. Let's be imaginative and say we are interested in the 100 to 1000 nanometre band. Assuming light waves also travel at 300 000 000 metres per second (of which more later), this means a frequency of interest from 0.000 001 metres to 0.000 000 1 metres or 3000 to 30 000 THz (a nanometre is one billionth of a metre).

10.2.1 Image Noise and Linear and Nonlinear Distortions

Having collected our radiant energy in a lens we subject it to a number of distortion processes partly through the lens itself (optical aberrations) and then via the sensor and colour-filter array.

The sensor array converts the photons into electrons (an obliging property of silicon) but in the process introduces noise. The combination of lens imperfections and sensor noise introduce ambiguity into the image pipe that needs to be dealt with by image-processing algorithms.

10.2.2 The Human Visual System

Interestingly this is not dissimilar to the human visual system where we sample the real optical analogue world with two parallel sampling grids consisting of cones and rods. Cones are centred around the middle of the retina and work in medium to bright light and are colour sensitive. Rods are concentrated in an area outside the cones. There are fewer of them. They are optimised for low-light conditions but are not colour sensitive. This is why we see in black and white at night. If you look at a distant dim object (a star at night) it will be easier to see if you shift your gaze off centre.

A sensor array emulates the functionality of the cones in the human visual system. It has to deliver similar or better dynamic range, similar or better colour accuity and perceptually similar resolution. Note that although the number of cones equates to 200 megapixels, no one seems quite sure why we need that much resolution.

So image processing is just like RF signal processing, except that it has to take into account the idiosyncrasies of the human visual system. Note also that although humans are superficially alike (two arms, two legs, two eyes) we are also intrinsically different in terms of optical performance. Some of us are colour blind, some of us have better resolution. Scientists have solved this by creating an 'ideal observer', a sort of 'model man' against which we can judge artificial optical and image processing systems.

The human visual system was first described in detail (in the western world at least) by Johannes Kepler (1571–1630) who codified the fundamental laws of physiological optics.

Rene Descartes (1596–1650) verified much of Kepler's work and added some things (like the coordinate system) that have proved indispensable in present-day imaging systems.

Isaac Newton (1642–1726) worked on the laws of colour mixture ('the rays to speak properly are not coloured') and did dangerous experiments on himself pushing bodkins (long pins) into his eye. He anticipated the modern understanding of visual colour perception based on the activity of peripheral receptors sensitive to a physical dimension (wavelength). All that stuff with the apple was a bit of a sideshow.

In 1801 Thomas Young proposed that the eye contained three classes of receptor each of which responded over a broad spectral range. These ideas were developed by Helmholtz and became the basis of Young–Helmholtz trichromatic theory.

This optical wavelength business was all a bit curious and still is. Anders Jonas Angstrom (1814–1874) decided quite sensibly that light should be measured in wavelengths and that given that the radiant energy was coming from the sun, then it would be a good idea to relate the measurements to a hydrogen atom.

Thus, an Angstrom was defined as being equivalent to the diameter of a hydrogen atom and equivalent to one tenth of a nanometre. This means that a nanometre is equivalent to 10 hydrogen atoms sitting shoulder to shoulder (assuming this is what hydrogen atoms like doing). This is remarkable given that in 1850, the metre was an ill-defined measure and dangerously French.

In the 1790s when Fourier was thinking about his transforms during the French revolution, his compatriots were deciding that the metre should be made equivalent to one ten millionth of the length of the meridian running through Paris from the pole to the equator on the basis that Paris was and still is the centre of the world if not the universe in general.

Unfortunately, this distance was miscalculated (a failure to account for the flattening of the world by rotation) and so the metre was set to an arbitrary length referenced to a bar of rather dull platinum–iridium alloy.

This was redefined in 1889 using an equally arbitrary and equally dull length of platinum–iridium, then redefined again in 1960 with a definition based on a wavelength of krypton-86 radiation (the stuff that Superman uses so at least slightly less dull).

In 1983 the metre was redefined as the length of the path travelled by light in a vacuum during a time interval of /1/299 792 458 of a second, which effectively determined the speed of light as 299 792 458 metres per second.

This is not the same basis upon which radio waves are measured and not the same way that Angstroms are measured. Anders Jonas Angstrom's work (1814–1874) was carried on by his son Knut (1857–1910) and formed a body of work that remains directly relevant today to the science of optical measurement.

10.3   The Optical Pipe and Pixels

10.3.1   Pixels and People

We have said that the sensor array emulates the functionality of the human visual system in that it takes point measurements of light intensity at three wavelengths (red, green and blue). A 4-megapixel sensor will have four million pixels most of which will be capturing photons and turning them into an electron count that can be used to describe luminance (brightness) and colour (chrominance).

Pixels behave very similarly to people. There are black pixels and white pixels and 254 shades of grey pixel in between. Pixels of similar colour tend to congregate together. A white pixel in a predominantly black pixel area stands out as being a bit unusual (an outlier). There are good pixels and bad pixels. Some bad pixels start life as good pixels but become bad later in life. Noisy pixels can be a big problem not only to themselves but also to their neighbours. Pixels see life through red, green, blue or clear spectacles but lack the natural colour-balancing properties of the human visual system. Their behaviour is, however, fairly predictable both spatially in a still image and temporally in a moving image.

10.3.2   The Big Decision

In RF signal processing, we have to take a final decision as to whether a 0 is a 0 or a 1 is a 1 or whether a bit represents a +1 or −1 (See Chapter 3 section on matrix maths).

In the same way, a decision has to be taken in the image pipe – is this a black pixel or a white pixel? (or one of 254 shades of pixel in between) or a green, red or blue pixel of a certain intensity and should that pixel be there or actually moved along a bit. Given that pixels are arranged in 4 by 4 or 8 by 8 or 16 by 16 macroblocks this is like a game of suduko, or if preferred a slightly complicated form of noughts and crosses (diagonals are important in the pixel world).

Pixels that do not fit a given set of rules can be changed to meet the rules. These decisions can be based on comparing a point pixel value with that of its neighbours. Simplistically, it is this principle that is applied to image enhancement in the image pipe (noise reduction, blur reduction, edge sharpening) and to image manipulation either in or after the image pipe (colour balancing, etc.).

10.3.3   Clerical Mathematicians who Influenced the Science of Pixel Probability

This gives us an excuse to introduce three mathematicians, the Reverend Thomas Bayes (the clerics of middle England and their role in cellular phone design), George Boole and Augustus De Morgan.

The Reverend Thomas Bayes (1702–1761)[1] is arguably the father of 'conditional probability theory' in which 'a hypothesis is confirmed by any data that its truth renders probable'. He is most well known for his 1764 paper 'An Essay Toward Solving a Problem in the Doctrine of Chances' and his principles are widely used in gambling (of which Thomas Bayes would have disapproved) and image-processing and image-enhancement algorithms.

For instance, conditional probability theory is the basis for context modelling. For example, an area of blue sky with occasional white clouds will obey a certain rule set prescribed by the context (a sunny day), a dull day sets a different context in which greys will dominate and contrast will be less pronounced. Adaptive algorithms can be developed that are spatially and/or context driven, conditioned by the image statistics and/or any other available context information.

George Boole (1815–1864) developed Bayes work by introducing more precise conditional language, the eponymous Boolean operators of **OR** (any one of the terms are present, more than one term may be present), **AND** (all terms are present), **NOT** (the first term but not the second term is present) and **XOR** (exclusive **OR**, one or other term is present but not both).

Every time we use a search engine we are using a combination of Boolean operations and Bayesian conditional probabilities but Boole is also present in almost every decision taken in the image-processing pipe.

---

[1] http://www.york.ac.uk/depts/maths/histstat/bayesbiog.pdf.

Augustus De Morgan (1806–1871) was working in parallel to Boole on limits and boundaries and the convergence of mathematical series (or at least the description of convergence in precise mathematical terms). He is probably most famous for De Morgan's Law that applied to people and pixels goes something like this:

In a particular group of people

Most people have shirts.

Most people have shoes.

Therefore, some people have both shirts and shoes.

In a particular group of pixels

Most pixels have chrominance value x.

Most pixels have chrominance value y.

Therefore, some pixels have both values x and y.

More specifically, the work of Augustus De Morgan is directly applied to the truncation of iterative algorithms so that an optimum trade off can be achieved between accuracy and time – the longer the calculation the more accurate it becomes but the more clock cycles it consumes. The modern image processing pipe is full of these rate/distortion, complexity/distortion, delay/distortion 'trade off' calculations.

So the Reverend Bayes, George Boole and Augustus De Morgan have had and are still having a fundamental impact on the way in which we manage and manipulate image signals that have particular statistical properties and exhibit specific behavioural characteristics. They allow us to pose the thesis that 'this cannot be a black pixel because . . .

10.3.4   The Role of Middle England Clerics in Super-Phone Design

Their role in super-phone design is, however, far more fundamental.

Early super phones will be capable but dumb. They may be able to see and hear as well as we can but they cannot think like us. This requires the power of reason and logical analysis to be applied.

Smart phones declared role in life is to make us more efficient. Some of us, many of us, do not want to be made more efficient. We may own a smart phone because buying one seemed like a good idea at the time but many of us will only use a small subset of the phone's potential capabilities.

Most of us, however, would like some practical help in the way that we relate to the real world around us – knowing where we are, where we need to go, seeing in the dark, hearing sounds that are below our acoustic hearing threshold and having a device that can help us make intuitive and expanded sense of the surrounding physical world of light, sound, gravity, vibration, temperature, time and space.

This is the promise of the smart super phone – the future phone that will deliver future user and telecom value. The value, however, is enhanced if placed in a context of position and location, which brings us to the topic of mobile metadata management.

10.4   Metadata Defined

Metadata is information about information, data that is used to describe other data. It comes from the Greek, meaning 'among', 'with',' beside' or 'after'. Aristotle's 'Metaphysics' was a follow onto his work on physics – it was consequential to the original information or, in Aristotle's case, body of existing work.

Metadata can be spatial (where), temporal (when) and social (who). It can be manually created or automatically created. Information about information can be more valuable than the information itself.

Semantic metadata is metadata that not only describes the information but explains its significance or meaning. This implies an ability to interpret and infer (Aristotle).

We are going to stretch a point by also talking about 'extended metadata'. 'Extended metadata' is information about an object a place or a situation or a body of work (Aristotle again). The object, the place, the situation, the body of work is 'the data'. An object can be part of a larger object – part of an image for example.

The combination of metadata and extended metadata, particularly in a mobile wireless context, potentially delivers significant value but is in turn dependent on the successful realisation of 'inference algorithms', 'similarity processing algorithms' and 'sharing algorithms' that in themselves realise value (algorithmic value).

This might seem to be ambitiously academic. In practice, metadata and, specifically, mobile metadata is a simple mechanism for increasing 'user engagement' that by implication realises user value.

10.4.1   Examples of Metadata

Higher-level examples of metadata include electronic programming guides used in audio and video streaming. These can be temporally based (now showing on channel number) or genre based (sport, classical, jazz). The genre-based indexing of iPod music files is an example.

Lower-level examples include audio, image and video metadata descriptors that allow users to search for particular characteristics, an extension of present word and image search algorithms now ubiquitous on the World Wide Web. These descriptors also allow automated matching of images or video or voice or audio clips.

Manual metadata depends on our ability to name or describe content. We may of course have forgotten what something is called – a tune, perhaps. If we can hum the tune, it is possible to match the temporal and harmonic patterns of the tune to a data base. This is an automated or at least memory-jogging naming process that results in the automated creation of metadata, putting a name to the tune. This may not realise direct application value (the application will probably be free) but provides an opportunity to realise indirect value (sheet music or audio download value).

The example in Figure 10.1 will be visually recognisable to some because of the pattern combination of the time signature (3/4 waltz time) and the opening interval (a perfect fifth). If you just used the perfect fifth it could be a quite large number of tunes, Twinkle Twinkle Little

Moon River

Mancini   Mancini

Figure 10.1 Tune recognition.

Star, Theme from 2001, Whisper Not, and many more. However, only one of these examples is a waltz (clue to the tune, think, River and Moon). Congratulations. You have just performed an inference algorithm based on temporal and harmonic pattern recognition.

10.4.2   Examples of Mobile Metadata

Any of the metadata examples above are potentially accessible in a wide-area wireless environment. However, they are not unique to wireless. Mobile metadata combines metadata value with mobility value. Mobile metadata is based on our ability to acquire and use spatial knowledge of where a user is at any given time combined with other inputs available to us.

Spatial knowledge may be based on cell ID or observed time difference or GPS (macropositioning) possibly combined with heading information (compass heading), possibly combined with micropositioning (low-G accelerometers), probably combined with time, possibly combined with temperature or other environmental data (light level or wind speed).

Even simple cell ID can add substantial metadata value, for example images can be tagged and/or archived and/or searched and/or shared by cell ID location.

This is an area of substantial present research, for example by the Helsinki University of Technology[2] under the generic description of *'automatic metadata creation'*. The research highlights the problems of manual metadata. For example, when people take a picture it is boring and usually difficult to manually annotate the image. There will generally be a lack of consistency in manual notation that will make standardised archiving and retrieval more problematic.

In comparison, automated spatial and temporal metadata can be standardised and unambiguous. We know where the user is and what the time is. Areas of potential additional value are identified. These are based, respectively, on 'inference algorithms' also known as 'disambiguation algorithms', 'similarity processing algorithms' also known as 'guessing algorithms' and 'sharing algorithms' also known simply as 'group distribution lists'.

10.4.3   Mobile Metadata and Geospatial Value

For example, if it is known that a user has just emerged from Westminster Tube Station (macroposition), has turned right and then left into Parliament Square and has then turned to face south (compass heading) and up (micropositioning) and then takes a picture, you know

---

[2] http://www.cs.hut.fi/~rsarvas/publications/Sarvas_MetadataCreationSystem.pdf.

that the user is taking a picture of Big Ben. Theoretically this could be double checked against the typical image statistics of Big Ben.

This is the basis of geotagging and smart tagging familiar to us in present mapping systems. Smart tagging was standardised in Windows XP and has become progressively more ubiquitous over time.

'Big Ben' can then be offered as a metadata annotation for the user to accept or reject.

However, having automatically identified that it is Big Ben that the user is looking at, there is an opportunity to add additional extended metadata information to the image, for example, see below.

10.4.4   Optional Extended Image Metadata for Big Ben

Big Ben is one of London's best-known landmarks and looks most spectacular at night when the clock faces are illuminated. You even know when parliament is in session; because a light shines above the clock face.

The four dials of the clock are 23 feet square, the minute hand is 14 feet long and the figures are 2 feet high. Minutely regulated with a stack of coins placed on the huge pendulum, Big Ben is an excellent timekeeper, which has rarely stopped.

The name Big Ben actually refers not to the clock-tower itself, but to the thirteen ton bell hung within. The bell was named after the first commissioner of works, Sir Benjamin Hall. This bell came originally from the old Palace of Westminster; it was given to the Dean of St. Paul's by William III.

The BBC first broadcast the chimes on the 31st December 1923 – there is a microphone in the turret connected to Broadcasting House.

10.4.5   Mobile Metadata's Contribution to Localised Offer Opportunities, Progressive Mapping and Point and Listen Applications

There are of course various related localised 'offer opportunities' associated with this 'engagement moment', a Macdonald's 24-hour 'Big Ben Burger' or the latest geospecific downloadable Big Ben ring tone. Of course, if the super phone has decent solid-state microphones, then the user can record his own ring tones and use them and/or upload them with a suitable spatial, temporal and personal metadata tag.

Geospatial value is of course closely linked, or could be closely linked, to progressive mapping applications (where only relevant parts of the map are downloaded as you need them) and wide-area 'point and listen' applications. Point straight ahead in the Big Ben example and you could be listening to an audio download of Parliamentary question time.

If you wanted a 3D overview of the area for local navigation or local information reasons or a 'virtual 3D tour' of the inside of Big Ben then this could be downloaded from a server.

10.5   Mobile Metadata and Super-Phone Capabilities

So there are a number of extended opportunities that are specific to the extended audio and image capture and play back/play out capabilities now being included in new-generation phones and becoming increasingly familiar to us today.

This is important because many of our mobile metadata applications do not intuitively require a connection to a network. A Nikon camera with GPS and a compass capability will know where it is and what it is looking at and could have embedded reference data available. A top of the range Nikon SLR camera, for example, has 30 000 reference images stored that are used for autoexposure, autowhite balance and autofocus so it's a small step to add another few gigabytes of embedded reference data. The purpose of scene recognition in this example is limited to improving picture quality but the same principles can be applied to face recognition and other applications such as number-plate recognition.

It is therefore important to identify functionality that can be made unique to a phone, particularly a camera phone, connected to a network. The ability to connect to the web to access smart tag or geotag information is an important differentiator but not the only one.

We suggested that phones could be divided into three functional categories, standard phones that help us communicate with one another, smart phones that help us to organise our social and business lives and super phones that help us to relate to the physical world around us. Smart super phones combine all three functional domains. The mobile metadata that you get from network-connected smart super phones builds on these functional domains.

We have said that smart super phones by definition will have advanced micro- and macro-positioning and direction sensing capabilities combined with extended audio (voice and music) and image-capture capabilities combined with extended audio and display and text output capabilities plus access to server-side storage (the web) plus access to network resident rendering, filtering and processing power.

Returning to our Big Ben example, Big Ben is an object with a unique audio signature. Other objects and places have similar but different signatures, the sea side for example, a concert hall, an airport. We mentioned earlier that audio signatures make potentially good ring tones – Big Ben or the Westminster Peal.

If you fancy church bells as ring tones you will be interested in the relevant object metadata tag.

Consider the effect of bells rung together in a peal rather than singly, in English changes – with each bell striking a fixed interval, usually 200 to 250 milliseconds after the previous one. Peals of bells can sound very different in character and the effect is influenced by many factors; the frequency and amplitude of the bells' partials, considered individually and as a group; how they are clappered; and rather critically, the building in which they are being rung' (from a guide to campanology – the harmonic science and mechanics of bell ringing).

A paper on combinatorics, a branch of discrete mathematics with bell ringing as an example explores this topic further.[3]

Objects and places and spaces similarly have unique visual signatures. These can be object specific, The Statue of Liberty for example, or general, the seaside. A seascape will generally consist of some sky (usually grey in the UK), some sea (usually grey in the UK) and sand (usually grey in the UK). The image statistics will have certain recognisable ratios and certain statistical regional boundaries and there will also be a unique audio signature.

Object and place and visual signatures come 'for free' from the autoexposure, autocolour balancing and autofocusing functions in higher-end camera phones. Autoexposure gives you

---

[3] http://www.studyblue.com/#note/preview/632416.

a clue though not definite proof that it is day or night, autocolour balancing and autoexposure together give you a clue though not definite proof that the image is being taken in daylight outdoors on a sunny or cloudy day or indoors under natural, tungsten or fluorescent lighting. Autofocus gives you a clue though not definite proof of distance and spatial relationships within the image. Two cameras mounted a couple of inches apart and two microphones six inches apart would of course greatly increase the spatial relationship capture capability, a function beginning to be included now in 3G smart phones.

People have social signatures, these are specific (voice and visual) and physical (my lap top has fingerprint recognition which I am impressed by but have never used) and general, for example behaviour patterns.

10.6   The Role of Audio, Visual and Social Signatures in Developing 'Inference Value'

Now these are interesting capabilities and necessary for higher-end camera phone functionality but do they have a wider application value when combined with network connectivity?

Well yes in that they can add value to 'inference algorithms'.

'Inference algorithms' are algorithms that combine spatial, temporal and social information. Their job is to *'disambiguate'* the *context* in which, for example, a picture has been taken or an observed event is taking place.

So the fact that it is 4.00 a.m. (temporal) on the 21 June (temporal) and that you (personal signature) are somewhere on Salisbury Plain (cell ID spatial) looking at the sun (autoexposure and autowhite balance) rise over some strangely shaped rocks (image statistics and object recognition) at a certain distance away (autofocus) listening to some chanting (audio signature) tells the network with a very high level of confidence that you are a Druid Priest attending the mid summer solstice at Stonehenge. Does this particular example illustrate realisable value? Probably not, but more general applications of inference algorithms will – this is mechanised automated, mobile metadata inference value.

10.7   Revenues from Image and Audio and Memory and Knowledge Sharing – The Role of Mobile Metadata and Similarity Processing Algorithms

Inference algorithms potentially create new application value opportunities and add value to existing applications. Downloadable Druid ring tones spring to mind from the above example – a rather specialist market.

'Similarity processing algorithms' are related to but different from inference algorithms. The job of similarity processing algorithms is to detect and exploit spatial, temporal and social commonalities and similarities or occasionally, useful and relevant dissimilarities (the attraction of opposites). The relationship with inference algorithms is clear from the example above.

If the network knows I am a Druid (I am not but would not mind if I was) then the presumption could be made that I have common and similar social interests to other Druids. I am part of the Druid community that in turn has affiliations with other specialist communities. We may live in similar places. Before long, an apparently insignificant addressable market has become more significant. Not only does the network know that I am a Druid, it also knows that I attended the 2011 summer solstice and may want to share memories of that moment with other members of my social community. This is the cue moment for the suggested recipients list for any image or video and/or audio recording I may have made of the event or for the sharing of particular cosmic thoughts that occurred to me at sunrise. This is sometimes described in the technical literature as 'context to community' sharing. The suggested recipients sharing list might include other potentially interested and useful parties. A parallel example is the solar eclipse that happened in April 2006 in Africa and Central Asia. This spawned a swathe of specialist web sites to service the people viewing and experiencing the event either directly or indirectly including a NASA site.[4]

This is an example of a contemporary web cast of the event

Tokyo, Mar 25 2006: The national astronomical observatory of Japan and other organisations held an event to webcast a total solar eclipse from Africa and Central Asia when it occurred on Wednesday evening Japan time.

The event titled "Eclipse Cafe 2006" was the first of its kind organized by the observatory.

Participants viewed live images of the solar eclipse webcast at eight astronomical observatories and science museums across Japan, including Rikubetsu astronomical observatory in Hokkaido and Kazuaki Iwasaki space art gallery in Shizuoka prefecture.

Of the eight locations, participants in Hiroshima and Wakayama prefectures had opportunities to ask questions to astronomical researchers in Turkey using a teleconferencing system.

The webcasting service was offered with webcast images of solar eclipses in the past. On Wednesday, the images were transmitted from Egypt, Turkey and Libya.

The solar eclipse was first observed in Libya at around 7:10 pm local time and lasted four minutes, which is relatively long for a solar eclipse.'

Similarity processing algorithms would automatically capture these spatial (where), temporal (when and for how long) and social (who) relationships and create future potential engagement and revenue opportunities from a discrete but significant special interest group.

In the case of Big Ben, it is self-evident that there are potentially millions of spatial, temporal and socially linked images, memories and 'advice tags' available to share that can be automatically identified.

10.8   Sharing Algorithms

So automated sharing algorithms can be built on automated inference algorithms and automated similarity algorithms. Sharing algorithms are based on an assumed common interest that may include a spatial common interest (same place), temporal common interest (previous, present or possible future common interest) and a social common interest. Most of us are offered the opportunity to 'benefit' from sharing algorithms at various times through the working day. Plaxo and Linkedin, Friends Reunited and Facebook are examples. This is algorithmic value in the corporate and personal data domain. All are examples of extended social metadata tagging where the tagged object is . . . us. This is generally only acceptable if we have elected to accept the process. It also suggests a need to establish that we are who we say we are. There is an implicit need for social disambiguation.

---

[4] http://eclipse.gsfc.nasa.gov/SEmono/TSE2006/TSE2006.html.

10.9   Disambiguating Social Mobile Metadata

10.9.1   *Voice Recognition and the Contribution of Wideband Codecs and Improved Background Noise Cancellation*

We have said that spatial and temporal metadata are implicitly unambiguous. Social metadata is potentially more ambiguous. How can we be sure the user of the phone/camera phone is the owner of the camera phone? Voice recognition helps in that it can uniquely identify the user. Wideband codecs and improved background noise cancellation will significantly improve present voice-recognition performance. Note that these capabilities inherently depend on a user's willingness to be identified that in turn is dependent on the user's perception of value from the function. Higher-quality codecs and better background noise control will also improve general audio capture quality, making voice tags and audio tag additions to images (manual metadata input) and user-captured ring tones (a form of mobile metadata) more functionally attractive.

10.9.2   *Text-Tagging Functionality*

The same principal applies to speech recognition in that accuracy should steadily improve over time making speech-driven text tagging more functionally effective (manual metadata input).

10.10   The Requirement for Standardised Metadata Descriptors

A combination of factors are therefore improving the functionality of automatic metadata capture and manual metadata input over time. Automatic spatial metadata capture improves as positioning accuracy and object recognition/image recognition/image resolution improves; automatic social metadata capture improves as voice recognition improves. Manual metadata input (text tagging) improves as speech recognition improves.

However, the benefits of these improvements can only be realised provided that standardised descriptors have been agreed to provide the basis for interoperability. Interoperability in this context means the ability to capture, process and share metadata from multiple sources.

10.10.1   *The Role of MPEG 7[5] in Mobile Metadata Standardisation*

Even with improved speech recognition, text-based metadata input is problematic in that it is nonstandardised. Each user will tend to use a different vocabulary and syntax. This makes it difficult to implement inference and similarity algorithms other than slightly haphazard word and phrase matching.

It is easier to standardise automatically generated metadata. For example, to mandate prescriptive methods for describing the harmonic and temporal structures of a voice file or audio file or the colour, texture, shape and gradient of an image file or the structure and semantics (who is doing what to whom) of a video file.

Images, video and audio files can usually use Fourier descriptors, given that the majority of image, video and audio content will have been compressed using Fourier-based JPEG

---

[5] http://mpeg.chiariglione.org/standards/mpeg-7/mpeg-7.htm#3.1_MPEG-7_Multimedia_Description_Schemes.

or MPEG compression algorithms. The purpose of MPEG7 is to standardise these audio and visual descriptors to allow the development of standardised search, match and retrieval techniques, including metadata-based inference and similarity algorithms.

Taking imaging as an example, descriptors can be region specific or macroblock specific. For example, it is possible to search for a man in a red woolly hat in a particular area of an image in multiple images by looking for the frequency descriptor for the colour red and the 'closest match' texture descriptor.

Generically, all images are converted into a common unified format in which image features are identified based on the wavelength of the colours making up the scene, expressed as a standardised 63-bit descriptor. Some of the more complex algorithms, for example automated face-recognition algorithms, require more accuracy and resolution and use a 253-bit descriptor.

Video descriptions are based on a differentiation of simple scenes and complex scenes together with motion estimation using vectors (direction and magnitude). In a perfect world, this all works with present and proposed MPEG-4-based object coding, though in practice it is still problematic to extract meaningful video objects out of a high frame rate high-resolution video. These are nontrivial processing tasks not to be attempted seriously in embedded software and are presently better performed as a server-side function (potentially good news for network operator and server added value).

Cameras and camera phones equipped with depth sensors and/ or twin camera stereoptic capture makes it easier to capture 3D video and segmented objects that in turn make object-based coding more useful. These at present represent a specialised but emerging application sector. Thus, MPEG 7 marks a useful start in terms of a standards process but there is much practical work still to be done.

10.11   Mobile Metadata and the Five Domains of User Value

The fact that we have not arrived at an end destination does not mean we cannot enjoy the journey and mobile metadata already has much to offer in terms of 'interconnected user domain value'.

To summarise:

> **The Radio system domain,** mobile metadata can extract value from cell ID, direction and speed of movement and all of the other (many) user-specific behaviour metrics that are potentially available from the radio air interface. This value can be realised by integration with parallel mobile metadata inputs including:

> **The Audio domain** mobile metadata can be associated with voice capture and voice recognition, speech capture and speech recognition and wideband audio capture and audio listening.

> **The Positioning domain** – positioning and location value can be associated with (low-G accelerometer) and macropositioning (GPS or equivalent) and location systems.

> **The Imaging domain** – mobile metadata as an integral part of the image and video sharing value proposition.

> **The Data domain** – the integration of mobile metadata into the personal and corporate information management proposition.

10.12 Mathematical (Algorithmic Value) as an Integral Part of the Mobile Metadata Proposition

Realisation of mobile metadata value in all five user value domains is, however, dependent on an effective implementation of standardised 'descriptive maths' (MPEG7 descriptors or other standardised equivalents) and comparative algorithms – inference and similarity algorithms.

As stated earlier, some though not all of the algorithmic decisional techniques used in this space can be traced back to the Reverend Thomas Bayes, George Boole and Augustus De Morgan. Similarly, some though not all of the descriptor techniques can be traced back to Joseph Fourier (1768–1830) and Carl Friedrich Gauss (1777–1855).

But life moves on and it is important to realise that mathematical techniques and algorithmic techniques continue to evolve. In the telecoms industry, we have tended historically to focus on engineering capability (infrastructure deployment) and technology capability (product R and D) to provide competitive differentiation. This may be less relevant over time.

Modern mobile phones and modern mobility networks including mobility networks based on mobile metadata constructs are strategically dependent on mathematical and algorithmic value. This includes the 'descriptive maths' used to describe signals and systems and the 'decisional maths' used to respond to changing needs and conditions. An example of descriptive maths used in the signal space is the Fourier transform and present work on wavelet transforms. Descriptive maths helps us to capture and process the analogue world around us. One example of decisional maths applied at the system level would be an admission control algorithm.

Some of the mathematical value both in pure and applied maths is 'heritage value', the legacy of several thousand years of mathematical study and inspiration. Some of the mathematical value is 'contemporary value', the contributions presently being made by practicing mathematicians. So it is relevant to consider the areas in contemporary mathematics that are most likely to prove useful in terms of differentiating the future 'user experience proposition'.

10.12.1 Modern Mathematicians and Their Relevance to the User Experience Proposition

One way to do this is to find out what work, or more accurately, whose work is winning awards. For years there was no mathematical equivalent to the Nobel Prize. Rumour has it that the Swedish mathematician Gosta Mittag Leffler had an affair with Alfred Nobel's wife. This effectively shut out all future mathematicians particularly Swedish mathematicians from the Nobel award and recognition process.

There is a Field award for mathematicians but this is only awarded every 4 years and is restricted to mathematicians under 40. In 2002, the Norwegian government decided to fund a yearly award known as the Abel Prize[6] to mark the double centenary of the birth of Niels Henrik Abel. Niels Henrik Abel died in 1829 at the age of 27 after contracting TB following an ill-advised sleigh ride. In his short life he had, however, developed the foundations for group theory, also worked on in parallel by Galois.

Group theory is essentially an integration of geometry, number theory and algebra. Abel worked specifically on the commutative properties of group operations, arithmetical processes

---

[6] http://www.abelprisen.no/no/.

like addition where it does not matter in which order sums are performed. These came to be known as abelian groups.

Strangely, but perhaps not surprisingly, group theory is increasingly relevant in many areas of telecommunications including mobile telecommunications and mobile metadata management.

Geometry is important in vector maths (mathematical calculations that have direction and magnitude), number theory is important in statistical processing (for example, the processing of image signal statistics) and algebra crops up all over the place.

A popular description of group theory is that it helps discover and describe what happens when one does something to something and then compares the results with the result of doing the same thing to something else, or something else to the same thing. Group theory has been and is used in a wide cross section of physical problem solving including the modelling of turbulence and weather forecasting. Our interest in a wireless telecoms context is to study the role of group theory in the management and manipulation of complex and interrelated data sets that change over time and space at varying rates.

That's why it is useful to keep track of who is winning the Abel Prize each year. They are all mature contemporary working mathematicians and as such provide an insight into evolving areas of mathematics that are potentially strategically intellectually economically and socially important. This of course assumes that the Norwegian Academy of Science and Letters knows more than we do about the work of contemporary mathematicians, which for the majority of us is probably a valid assumption.

In **2003**, the Abel Prize was awarded to Jean-Pierre Serre for his work on topology (place and space) and group theory.

In **2004**, the Abel Prize was awarded jointly to Sir Michael Francis Atiyah and Isadore M Singer for their work on the eponymous Atiyah–Singer index theorem, a construct for measuring and modelling how quantities and forces vary over time and space taking into account their rate of change.

In **2005,** the Abel Prize was awarded to Peter Lax for his work on nonlinear differential equations and singularities, the modelling of odd things that happen at odd moments, breaking the sound barrier for example. Dedicated readers of RTT Hot Topics might remember our venture onto this territory in our March 2003 Hot Topic 'Turbulent Networks'. One reader hurtfully assumed the Hot Topic was an early April fool. This was of course untrue. System stability is going to be a major focus in multimedia multiservice network provision including multimedia multiservice networks with integrated mobile metadata functionality. Peter Lax wittingly or unwittingly is building on work done by Benoit Mandelbrot in the 1980s that in turn was based on the work of Lewis Fry Richardson (uncle of the late Sir Ralph Richardson) on aeronautical turbulence prior to the First World War.

In **2006**, the Abel Prize was awarded to Lennart Carleson for his work on harmonic analysis and his theory of smooth dynamical systems.

In **2007** Srinivasa S.R. Varadhan was awarded for his work on the maths describing rare chance and probability events and a unified theory of large deviation.

In **2008** Professor John Griggs Thompson, of Cambridge and Florida Universities and Professor Jacques Tits, of the Collège de France were awarded a joint prize for their work on algebra and in modern group theory.

In **2009** Mikhail Gromov received the prize for his revolutionary contributions to geometry.

In **2010** John Tate received the prize for his work on number theory.

In **2011** John Stony received the prize for work on topology, geometry and algebra.

Surprisingly, there are no Chinese mathematicians yet on this list, but presumably this will change in the future and it is likely that the specialisms may be similar.

From the Abel list, Lennart Carleson conveniently brings together two story lines that we have explored.

His work on harmonic analysis developed Fourier's work, but specifically applied to *acoustic* waveform synthesis. The popular press describes Lennart as the 'mathematician who paved the way for the iPod'. This is perhaps a rather journalistic interpretation. In practice, his work in this area in the 1960s marked an important advance in musical set theory, the categorising of musical objects and their harmonic relationships. These have subsequently proved useful in audio system simulation and design and automated audio metadata tagging. Robert Moog was developing his Moog synthesiser at the same time.

Lennart's genius is that he has done this work together with his work on dynamic systems, the branch of modern mathematics now dedicated to the modelling of large systems like financial markets and the weather – systems that change over time.

Multimedia multiservice network provision including multimedia multiservice networks with integrated mobile metadata functionality increasingly exhibit large-system behaviour. This is not surprising given that they are large systems with multiple inputs, many of which are hard to predict.

As with the weather, it sometimes seems hard to predict human behaviour, particularly long-term behaviour. This is, however, a scaling issue. We do not have enough visibility to the spatial, temporal and social data sets or the vector behaviour (direction and magnitude) of the data sets over a sufficiently wide range of time scales.

The mechanics of mobile metadata potentially allow us to capture and manage this information and ultimately, provided user elective issues can be addressed, to realise value from the process. At this point, data on data (metadata) becomes more valuable than the data itself.

11

Transaction-Centric Software

11.1 Financial Transactions

Transactions are normally thought of as financial transactions. Many of them are, but a transaction can cover almost any exchange between two or more people or two or more devices and are normally enabled by some form of identification and trust-establishment protocol.

However, let us start first with financial transactions and the role of mobile phones in the financial transaction process. Economic theory suggests that economic activity expands in line with population growth, particularly in developing countries. This may be true up to a point but at some stage growth becomes limited by resource and/or environmental constraints, and politics often gets in the way as well.

The disputed election at the end of 1997 in Kenya resulted in 1500 deaths and the displacement of 600 000 people. No shops or banks were open for 5 days and no air time resellers were available to top up prepay accounts. Safaricom[1] the local operator continued to allow calls and texts but more significantly a few months earlier a service called M-PESA had been established.

The service was disarmingly simple. A sender wishes to send money to a friend or pay for a product or service. He/she enters the recipient's phone number, the amount and a PIN number. The SIM-based software (either downloaded over the air or factory loaded) encrypts the SMS message. A centralised accounting system then transfers the funds and a confirmation message is sent. If the recipient owned an M-PESA account then this is a simple SMS receipt. If not, then a voucher is sent that is redeemable through a network of nominated agents.

One year on there were more than two million registered M-PESA users in Kenya, totalling 20% of Safaricom's installed base. Another one million people had received money through the system and 10% of Kenyans had used the service with an estimate that as much as 20% of Kenyan GDP was passing through the system. A similar service was launched in Afghanistan and Tanzania with India and Egypt following later.

---

[1] http://www.safaricom.co.ke/index.php?id=250.

Making Telecoms Work: From Technical Innovation to Commercial Success, First Edition. Geoff Varrall.
© 2012 John Wiley & Sons, Ltd. Published 2012 by John Wiley & Sons, Ltd.

Globally, only 1 in 7 people have a bank account, whereas (as we know) there are more mobile phones in the world than people, we just need to share them more evenly. Put another way, five billion people own a phone, one billion people own a bank account.

Just on its own and/or counting partner and associate networks Vodafone reaches a billion people. The Vodabank (our name not theirs) is therefore potentially the largest bank in the world in terms of addressable customer numbers, if not by value. This looks good for Vodafone but let us suppress the cynicism for the moment and at least give credit (no pun intended) where credit is due.

A simple SMS-based application has made it possible for a new generation of relatively low income users to pay bills or buy goods or send money securely and safely to friends and family without the need to handle cash. Employers can also pay employees and contractors through the system. Less cash means less crime (less incentive for mugging and theft and extortion).

So SMS saves lives, allows governments to communicate and allows individuals to spend and send and receive financial transfers. Not bad for an almost accidental add in to the GSM specification and living proof that simple things just occasionally can deliver spectacular value.

Case studies of the impact on other developing economies can be found at the Safari Com web site,[2] the Vodafone web site[3] or the Iceni Mobile web site.[4] The principle is being extended into agribusiness and medicine and health care.

11.2 The Role of SMS in Transactions, Political Influence and Public Safety

The short message service has been the unsung hero of the mobile telecoms industry. The first SMS message was sent in December 2002 with the first SMS phones available in 1993. 15 years later (2008) over 80 billion texts were sent and received in the UK making the UK, the second highest texting nation in the world, though dwarfed by the US that sent and received over 800 billion.[5]

SMS is a story that could fill several books of which at least two have been published by John Wiley[6] so repeating the story here would be duplication but a summary would be that SMS has been eye-wateringly profitable with a per bit revenue three orders of magnitude (1000 times) higher than voice but has also made a substantial political, social and economic impact.

So, for example, it has been widely reported that the revolutionary movements that swept through Tunisia, Syria and Libya in 2011 were enabled to a significant degree by SMS texting combined with picture messaging and video feeds to U tube. If this is true then apparently mobile phones can dictate to dictators. It can also save lives.

Over the past ten years the telecoms industry has had to respond to a series of natural disasters, Katrina, The Asian Tsunami, the Chinese Earthquake and unnatural disasters including 9/11. These events disappear from the daily news but have a long-term impact on the

---

[2] http://www.safaricom.co.ke/index.php?id=745.

[3] http://www.vodafone.com/content/index/about/about_us/money_transfer.html.

[4] http://www.icenimobile.com/.

[5] http://consumers.ofcom.org.uk/2009/12/uk-consumers-embrace-digital-communications/.

[6] SMS statistics from Short Message Service (SMS) The Creation of Personal Global Text Messaging by Fred Hillebrand. http://eu.wiley.com/WileyCDA/WileyTitle/productCd-0470688653.html and also 'GSM and UMTS the creation of mobile communication', edited by Fred Hillebrand, also John Wiley.

telecommunications industry. Conversely the telecommunications industry has had a vital role to play in responding to the events both in the immediate aftermath and in the longer-term recovery and rebuilding process.

In the specific context of the role of SMS, the following announcement is from May 2008.

'Earthquake Collapses China's Unicom's Two Networks
    May 13, 2008
    China Unicom (CHU) says that due to the earthquake, its G net and C net in Wenchuan, Sichuan Province have both broken.
    About 200 base stations of the company's G net and C net in the Aba area of Sichuan have reportedly been paralyzed. And because of busy traffic, the company's two networks in Chengdu have been congested, thus resulting in a slowdown of SMS communications.
    As a result of the earthquake, the communications of four counties in south Gansu Province have also been interrupted and about 500 base stations in Shaanxi Province halted.
    China Unicom says it has initiated an emergency plan and sent technicians to the disaster areas to recover the communications. There is no firm timeline on when the systems will be fully operational again'.

The news item highlighted two issues, one that cellular networks are not immune to natural disasters, a fact already well proven by Katrina and other relatively cataclysmic events, the second that SMS, (the short-message service) one of the simplest parts of the cellular protocol stack, continues to expand its role as a fundamental enabler for emergency communications and a mechanism for getting on with life in troubled times.

In practice, the Chinese networks were restored quickly but SMS had provided the most bandwidth efficient and link budget efficient mechanism for getting in touch with friends and family affected by the disaster. Also, because an SMS text is more robustly channel coded and has a high energy per bit than voice the text will get through in conditions in which a voice call would fail.

The Chinese government has also shown a developed awareness of the political value of SMS. The following is an extract from a blog just after the earthquake

Earthquake: China uses text messaging to assure public
    Mon, 05/12/2008 – 11:48 am
    The full extent of the damage caused by the 7.8 magnitude earthquake that hit China's Sichuan Province on Monday afternoon is just starting to become clear. The quake was felt in Beijing and Shanghai and in places as far reaching as Taipei, Hanoi and Bangkok.
    In order to reassure people and to squelch false rumors, the Chinese government is using *SMS text messaging* to mobile phones as well as internet postings to inform people that the areas where they live are not in the seismic zone. *Over a million such messages were sent in nearby Guangxi Zhuang Autonomous Region and Guizhou Province.*
    The government plans to use text messaging not only for emergencies, but for various situations relating to the public interest. The plan is part of the government's new openness in information regulations which it says will promote "openness as principle, being closed off as the exception" *in an effort to provide timely and accurate information to the public'*

As Western observers, we are conditioned to be cynical about Chinese politics and Chinese political motivation but to all intents and purposes the response by President Wen Jiabao to the crisis seemed heartfelt and genuine and considerably more open than the response of the Burmese leadership to the Cyclone Nargis[7] disaster that happened at the same time.

SMS broadcasting was used to a lesser extent after the Asian Tsunami[8] in 2004 but now appears to be becoming an alternative to radio as a mechanism for disseminating government information in an emergency.

11.3   The Mobile Phone as a Dominant Communications Medium?

Can you get to as many people by phone in China as you can by radio?

Well no. According to the Chinese State Administration of Radio there are 1.2 billion people in China listening to long-wave, medium-wave and VHF radio. According to our colleagues at The Mobile World, as at March 2008 there were 557.8 million cellular subscribers in China. Radio therefore still had the advantage in terms of numeric reach but this discounted two factors, cellular subscribers in China were being added at a rate of nine million per month and three years later had passed the one billion user mark.

By the time you read this, the mobile phone will be the dominant communications medium by user volume. Cellular users in China, anecdotally at least, are also more likely to have their phones on rather than their radios at any particular time. Mind you, many of the phones have an FM radio, which is a much used function both in China and particularly India (see also Chapter 21).

Whether it will also be a dominant payment and transaction medium is open to debate and this may be determined by how easy it is to use, which in turn may be determined by how successfully near-field communication is implemented as an enabling technology for the contactless smart card. NFC-enabled contactless smart cards are typically close-coupled applications in which the two devices (the reader and the smart card) are either touching or within a few millimetres of each other hence the description 'near-field communication'. The technology and the software associated with the technology is covered in Chapter 19, but many of us already use NFC enables devices to pay for bus and train journeys or going through passport control.

11.4   Commercial Issues – The End of the Cheque Book?

Legislation can also have an impact on adoption. The Payments Council in the UK[9] for example decided in June 2009 that cheque guarantee cards would no longer be valid in the UK from 30 June 2011, the start of a process to abolishing cheques introduced in Britain 700 years ago. This was justified on the basis that there had been a 65% decline in the use of guaranteed cheques over the previous five years. Only 7% of the one million cheques written in 2010 were guaranteed. The average transaction value per cheque was £392 but 88% of all cards had a guarantee of £100 or less. The announcement was packaged with a plan[10] to develop a mobile

---

[7] http://abcnews.go.com/Technology/Weather/story?id=4806331&page=1.
[8] http://news.bbc.co.uk/1/hi/world/4136289.stm.
[9] http://www.paymentscouncil.org.uk/media_centre/press_releases/-/page/1560/.
[10] http://www.payyourway.org.uk/.

payments plan using mobile phones – a good example of how slowly the regulatory process moves in the financial services sector (sometimes for good reason).

The point to make is that the technology and engineering issues are relatively straightforward to address, the commercial issues are more complex. Similarly there was no technical reason why phones cannot replace passports but the regulatory and political barriers to this would be substantial.

Essentially, as the value of a transaction increases the identification and security associated with that transaction should increase as well. Conversely, as the value of a transaction decreases the cost of the transaction also has to decrease. The interesting thing about mobile phones as a payment method is that potentially they combine both low transactional cost and a potentially robust authentication and security process.

The obvious point to make here is that mobile phones can only do this if they are connected to a radio network that in turn provides connectivity to a banking network that can fulfil the transaction. The radio network can become a banking network and has some, if not all, of the billing and financial management procedures in place to realise this, but this brings us on to the narrative of the next ten chapters, network hardware and software.

Part Three

Network Hardware

12

Wireless Radio Access Network Hardware

12.1 Historical Context

The next five chapters review the technology and commercial dynamics of network hardware. This first chapter looks at wireless radio access network hardware, but in practice the topic has to be broader. A modern radio base station in a mobile broadband network has a similar amount of software code to a digital switch a few years ago and has an ability to make decisions on admission control that would have once upon a time been taken by the central switch. This is useful because many of these decisions have to be taken so quickly that by the time the signalling had travelled to the centre and back again there would be too much decision-time latency.

Given that signalling travels more or less at the speed of light or 300 million metres per second this might seem odd, but in practice a six-kilometre round trip adds 20 microseconds to a delay budget and several trips might be needed to make a decision. Add a few million clock cycles to the process and it can be seen that distance becomes a nuisance, particularly for decisions that have to be made every few milliseconds.

Additionally, the once-clear distinction between computing networks and telecommunications networks irrespective of whether those networks are fibre, cable, copper or wireless or a mix of all four is disappearing, or has already disappeared. The process started some time ago.

Immediately after the Second World War, a decision was taken to capitalise on the work that Alan Turing had done at Bletchley Park on the Colossus code-breaking machine and build a general-purpose electronic computer.

The work was done at the National Physical Laboratory and was based on Alan Turing's conceptual discovery of the Universal Turing Machine, a computer that is not structured to carry out particular tasks. The result was a computer, the Automatic Computing Engine[1] that at the time was probably the fastest in the world.

---

[1] http://www.makingthemodernworld.org.uk/icons_of_invention/technology/1939-1968/IC.059/.

Making Telecoms Work: From Technical Innovation to Commercial Success, First Edition. Geoff Varrall.
© 2012 John Wiley & Sons, Ltd. Published 2012 by John Wiley & Sons, Ltd.

Thirty years later Seymour Cray delivered a Cray 1A supercomputer[2] to the Aldermaston Atomic Weapons Authority where it still remained in 1990, the last operating Cray 1A in the world. The machine cost £8 million pounds and had a Freon cooling system and an early form of vector processing that allowed the computer to achieve then unrivalled operating speeds.

Both of these machines can be traced intellectually back to the 1832 Babbage Difference Engine – the ancient ancestor of today's modern computing industry and therefore by default the ancient ancestor of today's modern telecommunications industry.

In fact, it might be intellectually more rigorous to describe these not as computing or telecommunications networks but as information networks, networks that have the potential to transform the world in terms of social and economic progress, education and health and possibly even environmental sustainability.

There are, however, fundamental differences between computing and communication in terms of cost and investment return that need to be factored into these global ambitions. It may also be that present telecommunications investment in developed economies will not result in networks that are more cost efficient or energy efficient than the networks being replaced.

This is not to say that these networks will be socially and economically unprofitable and/or that investment is not justified, but rather that the returns might take longer than anticipated and be more broadly based in terms of social rather than purely economic return, networks that become the connection engines of the information age, delivering benefits that with political will can be evenly distributed across developed, emerging and survival economies, a net gain for all.

12.2   From Difference Engine to Connection Engine

Babbage's design in the 1830s embodied almost all the conceptual concepts of the modern electronic computer. The Babbage engine never quite worked in practice due to tolerancing problems and Babbage's tumultuous temper but it was intended to be a general-purpose machine that could be programmed with punched cards to perform almost any function. Figure 12.1 illustrates the complexity of this device.

Similarly, the Automatic Computing Engine and Cray Super Computer were described as general-purpose machines but were applied to a very specific purpose, both were used to model and analyse the behaviour of atomic explosions.

Modern cellular networks are similar in that they have had a general purpose, providing person to person connectivity but are now evolving into being broader multipurpose delivery platforms. However, at heart a cellular network is still a connection engine.

Moore's law has powered the cellular industry to the point where a cellular or fixed line switch that would fill up a large room in the 1980s can now be fitted in a box that can shipped by lorry or air lifted to deliver instant emergency communication into disaster zones.

Although voice communication has to date been the predominant application, these networks are now beginning to do far more both in terms of imaging and video and data.

We have come to assume that each generation of computer is faster and more cost and energy efficient than its predecessor. Similarly, it might be assumed that each new generation of cellular network provides faster connection speeds at lower cost with ever better energy efficiency and that this provides the basis for an improved financial return. Both of these assumptions are unproven.

---

[2] http://www.makingthemodernworld.org.uk/icons_of_invention/technology/1968-2000/IC.110/.

Figure 12.1 1832 Babbage difference engine. Reproduced with the permission of the Science Museum/SSPL 10303264.

The first practical automatic telephone exchange based on electromechanical switching was introduced by Almon Strowger in 1888 and was first introduced to the UK in 1912. Strowger exchanges similar to the one illustrated in Figure 12.2 remained in use until the mid-1990s, proving the point that telecoms equipment when fit for purpose can have a remarkably long life cycle, particularly when compared to modern-day computing life cycles.

The development of electronic switching in the 1950s proved to be a surprisingly painful and expensive experience with practical deployment only being realised through the 1960s in the US and rather later in Europe and the rest of the world. These exchanges, however, remain today at the heart of every telecommunications network including every cellular network in the world and are completely fit for purpose for the job that they were intended to do, setting up, maintaining and clearing down telephone calls. Figure 12.3 shows an AXE digital switch manufactured by Ericsson.

Figure 12.2   Strowger exchange. Reproduced with the permission of the Science Museum.

We sometimes forget that exchanges have also been efficient at generating and distributing power and for many people the idea that an old-fashioned fixed phone still works in a local power outage provides a curious comfort. However, there is now a general assumption that telecommunication networks including cellular networks are moving away from centralised switching and transitioning to distributed architectures with IP-based addressing and prioritisation.

12.3   IP Network Efficiency Constraints

There is, however, no compelling evidence that this transition will either be cost efficient, energy efficient or directly environmentally efficient. IP networks were and are considered to be more resilient than traditional networks based on centralised switch architectures. As mentioned in the Introduction, they bear a close resemblance to the postal service, where packets can and often do travel by many and various routes from origin to destination.

Figure 12.3 AXE digital switch[3] with additional knitting. Reproduced with permission of Ericsson.

As with postal systems, packets can be stored (buffered) during the delivery process and as with postal systems packets can be lost or have to be deliberately discarded if the storage or buffer space overflows. This results in a variable customer experience. Post offices traditionally manage this by introducing differential pricing, first- and second-class stamps and guaranteed or signed for delivery. IP networks manage this by introducing multiple service levels based on multilevel packet prioritisation. In theory, this increases the multiplexing efficiency of the network, trading variable delay against network utilisation efficiency and trading delivery bandwidth efficiency against memory bandwidth occupancy. However, this requires routers to read and interpret IP addresses together with the priority labels embedded with the address. This information then provides the basis for a complex buffering and routing decision. This introduces variable delay into the delivery budget. The overall delay and the variability of the delay can be reduced by using hardware accelerators and high-speed memory.

These are costly energy-absorbent devices.

Even more frustratingly, just as in the traditional postal system, many of the packets delivered over the network are unwanted and have negative value – they annoy the receiver. These unwanted packets get in the way of latency-sensitive end-to-end services including voice communication. In a cellular network these costs are multiplied in several ways.

IP address overhead has to be accommodated within the RF link budget and the unwanted and unneeded negative value packets cost money to deliver. Wireless marketing teams have decided that users want and need to be offered high peak data rates even if this compromises average data throughput. High peak data rates, can only be achieved by using higher-level

[3] http://www.ericssonhistory.com/templates/Ericsson/Article.aspx?id=2095&ArticleID=1378&CatID=362& epslanguage=EN.

modulation schemes that are inherently noise sensitive and perform poorly most of the time resulting in user frustration and packet send retries that absorb network bandwidth and battery and network power. The higher-order modulation schemes require more linear and therefore less power efficient radio-frequency transmitters. The adaptation schemes needed to manage and mitigate these effects in turn introduce variable delay and add to the overall energy budget.

At this point you might question why the fixed and mobile telecommunications industry remains so set on IP network implementation. The idea that LTE networks will be more cost efficient and power efficient than the networks they are replacing is naïve. The notion that they can be profitable is not.

12.4   Telecoms – The Tobacco Industry of the Twentyfirst Century?

Telecommunications networks including cellular networks now provide access to instant information on a scale that would have been hard to imagine thirty years ago. Information is addictive and develops dependency, two core ingredients of a successful business model. Even better, the capital and energy costs of storing and managing the information in thousands of server farms around the world are paid for by other third parties.

For 150 years (from 1850 to the year 2000), the telecommunications industry has made money out of people's desire and need to interrelate to one another – the telegraphy and telephony age. Telegraphy and telephony provided a basis for social and economic progress based on social, economic and business efficiency, profitability with a social purpose and associated emotional value.

Information networks extend that model by helping us to interrelate with the physical and intellectual world around us. Earlier chapters discussed the role of the super phone and the interrelationship with the physical world around us. The next ten chapters shift this focus to this interrelationship between information networks and the intellectual world.

Information can have emotional value. It can also be misused, but in general it has to be assumed that informed decisions and actions are better than uninformed decisions and actions. Broader and more efficient access to information should at least theoretically enable wiser more broadly collective decision making, the world should become more intellectually efficient.

However, the tobacco analogy would suggest there may be longer-terms costs attached that are not presently visible and might only become evident in the distant future. It has taken Europeans 500 years to wake up to the health risks of tobacco but if the world was to worry about unknowable long-term risk then the world would never move on. In the meantime, the telecommunications industry appears to have reinvented itself more by accident than intent. The economic implications are, however, neither simple nor straightforward.

12.5   Amortisation Time Scales

Electromechanical switches were expensive to develop and slow to be deployed. In the UK depressed wage rates after the First World War meant that it was more economic to use human operators to set up a call. The financial returns from the investment were substantial but were based on an extended deployment cycle. The development cost of the Strowger exchanges still in use in the 1990s had been amortised over more than 100 years and the capital costs would have typically been amortised over forty years or more.

It would be surprising if the electronic switches installed over the past thirty years did not have a similar long tail in terms of investment return. Old technology takes a surprisingly long time to die. The counterargument is that the computing revolution has changed the economic rule book and that R and D and capital amortisation now has to be achieved in months not years and certainly not over fifty years or more. The fact that a Cray computer was still being used fifteen years after it was delivered is considered curious.

But communications is not computing.

There are similarities, both are dependent on hardware and software but this is where the similarity ends. The mechanics of connection are more complex than the mechanics of calculation. Telecommunication networks are dependent on signalling systems and protocols that took decades to develop and that absorbed thousands of man-years of international standardisation effort. SS7 signalling is one example.

Cellular networks also have to develop and standardise the radio access part of the network. Adding IP protocol to fixed or wireless networks introduces additional cost and complexity. It is no surprise therefore that the companies that have enjoyed the most success in the telecommunications industry and that are presently the most resilient are those that have managed to support long-term investments that realise long-term returns bearing in mind that the definition of long term in this context is thirty years or more. In Europe, that includes companies that have ownership and shareholder structures that give them a measure of protection against institutional shareholders focused on shorter-term returns.

Telecommunications companies in command and control economies similarly can find it easier to justify thirty- to fifty-year investment returns. Whether US companies can still do this must be open to question.

12.6   Roads and Railways and the Power and Water Economy – The Justification of Long-Term Returns

If we are defining telecommunications as just another transport system along side roads and railways and power and water utilities then it is reasonable to assume that the model of the last one hundred and fifty years will carry on for the next 150 years. Change occurs over decades rather than days and the return on investment is on a similar time scale.

There seems to be a growing assumption that universal broadband connectivity irrespective of how it is delivered will drive future economic growth. In common with other nation states, the US and the UK have ambitious plans in place to roll out next-generation networks that will allow us to work rest and play at a new level of efficiency and intensity, assuming that this is want we want or need to do. The release of the Digital Britain[4] report in 2009 by the UK government articulated the ambition but was nonspecific on where the funding would be found.

12.6.1   Historical Precedents – Return on Infrastructure Investment Time Scales

There are many past examples where infrastructure investment has been considered as critical to achieving economic, social and political progress.

---

[4] http://webarchive.nationalarchives.gov.uk/+/http://www.culture.gov.uk/what_we_do/broadcasting/5631.aspx/.

When technology innovation has coincided with an ability to raise capital and an appetite for risk then the change can be dizzyingly fast. The canals and railways of Victorian Britain were an extreme example. Over the next one hundred and fifty years, telegraphy, telephony and telecommunications grew out of a similar mix of invention combined with private and public investment. The return on that investment can be very long term. Passenger numbers on the railways for example are presently increasing in the UK. Most train journeys cross a nineteenth-century bridge or embankment.

Investment benefit time scales of this order are difficult to reconcile with conventional investment horizons. Railway planning is best done over thirty- to fifty-year time scales and telecommunications should arguably be the same.

Conveniently thirty to fifty years is coincident with the average working and earning life span, a fact that makes or should make infrastructure investment interesting and relevant to pension-fund investors. Australian pension funds have embraced this opportunity with some enthusiasm but most other countries remain focused on shorter-term returns.

12.7   Telecommunications and Economic Theory

Economic theory when applied to telecommunications unearths other puzzling contradictions and apparently underexploited opportunities. For example, it could be argued that speed equates to value. This of course depends on the fiscal cost of achieving the speed. This may include environmental cost – Concorde comes to mind. It also depends on the additional value achievable from additional speed. Canal journeys may take longer but may be more efficient and in some instances the net gain over faster alternatives, for example trains and lorries, may be greater. If the commodity being transported is increasing in value over time, the longer it takes to get there the more it will be worth.

But surely the canal analogy cannot be applied to the economic theory of telecommunications? The benefits of broader-band telecommunications must surely outweigh investment cost? Well, that depends on the economic scale of the benefit and the time scale required for a return on the investment. If the cost of delivering broadband to rural areas could be amortised over 50 years, effectively what happens with telegraph poles, then almost certainly a net gain will be achieved. If the cost has to be amortised over five years then almost certainly it will not.

This explains the present lack of appetite for private sector broadband investment. It also implies a need for a longer-term plan.

12.7.1   Telecom Politics in Malaysia – An Example of Long-Term Planning and Mixed Results

In 1991 the Prime Minister of Malaysia Dr Mahathir bin Mohamad produced a social and economic plan that would make Malaysia a self-sufficient industrialised nation by 2020 based on an annual growth of 7% between 1990 and 2020.

The building of a multimedia supercorridor stretching from Kuala Lumpur to Penang was promoted as a key enabler in the industrialisation process, alongside ambitious plans to upgrade the roads, railways and utility sector (power and water), transforming Malaysia into an Asian economic tiger.

The multibillion dollar telecoms plan was formalised in 1998 as a connectivity corridor 15 kilometres wide and 50 kilometres long, a Malaysian Silicon Valley to include two smart cities that would have cutting-edge communication and facilitate 'electronic government', the later being a touch sinister given Dr Mahathir's propensity to imprison political opponents.

Dr Mahathir resigned in 2002 after 20 years in office blaming the failure of the plan on the failure of the Malay community to adapt to an industrialising society and later stating that the education system had failed to deliver as well. In practice, Malaysia had been achieving year-on-year growth of 8% from 1987 up until the currency crisis of 1997, so blaming the Malay population might have been disingenuous though they may well have had well-founded doubts about the overall aims of the plan. Additionally, the policy was credible only as a result of positive Asian investment sentiment, which proved in the case of Malaysia to be unsustainable at least in the shorter term.

The policy did, however, achieve a noticeable if not radical cultural shift that included greater emphasis on electronics engineering education and a gender-neutral approach to engineering recruitment, an example that the UK and other countries could usefully follow.

12.7.2   The UK as an Example

Harold Wilson attempted something similar after the 1964 British election. The White Heat of Technology would rescue Britain from industrial decline. The British political process more or less guaranteed that this ambition would be frustrated. Autocracy may have been the missing ingredient here.

Forty five years on the question we seem to be asking ourselves is a reworking of the Wilson mantra – will the White Heat of Telecommunications Technology rescue Britain from industrial decline. After all, if Britain can be set on the road to recovery simply by putting in some broadband investment then just think what could happen elsewhere in the world. The answer of course is that investment in broadband connectivity is only part of the solution.

There is a parallel need to invest in the education of a new generation of telecommunications engineers and a need to invest in the R and D and manufacturing capabilities needed to bring new telecommunication technologies and techniques to market. These are needed in order to reduce costs and increase added value. This is needed to reduce return on investment time scales – a virtuous circle within a virtuous cycle. The same principles almost certainly apply to other national economies. The only difference is that it is challenging to create a nationally specific source of telecommunications engineering expertise from a virtually zero start point.

12.7.3   The Advantage of an Autocracy?

Taking Malaysia as a reference example, Dr Mahathir's political power base and initial lack of democratic accountability allowed for a number of ambitious projects to be undertaken of which telecommunications investment was one, but given that these were substantially financed by inward investment implied that they would be vulnerable to shifts in international investment sentiment that in turn would prompt political change and so it was. The pattern has been repeated in many countries since, both in the Middle East and in Europe (Portugal and Ireland).

Similarly, China is creating a potentially world-beating home-grown telecommunications R and D and manufacturing capability at a speed that is probably only possible in a closely coupled command economy. India could potentially do the same though political and industrial policy making in India is traditionally elliptic, which tends to slow things down.

At a recent conference in Cambridge one of the speakers[5] with experience of selling into China and India pointed out that regulatory policy does not always align with political policy. In China, as an example, the political system remains communist, but the regulatory system is capitalist.

12.7.4   UK, US and ROW (Rest of the World) Implications

What does this mean for the telecommunications industry in the UK, US and rest of the world? It seems unlikely that either the UK or US will adopt autocracy as an acceptable method for accelerating technology investment, though given the state of present UK politics you can never be too sure. Liberal democracies are not immune to the occasional miscarriage of justice but a whole scale abandonment of the democratic process might be a step too far. The end does not always justify the means to the end.

Collaborative initiatives that bring common interest countries together are one alternative, but recessions tend to increase national self-interest – the car industry being a present stark example.

So we are forced back to seeing what we already have and whether we can make better use of what we have. Theoretically, the UK should have a legacy advantage. In radio communications for example, the UK was one of the first countries to implement digital mobile GSM networks, the first country in the world to implement digital broadcasting with the launch of DAB in 1995 and DVB-T in 1997, and one of the first countries to introduce digital two-way radio. The UK presently hosts the world's largest TETRA-based public safety network. The UK will be one of the first countries to implement DVB -T2 broadcasting and has a leadership role in LTE standardisation. A UK company directly supports nearly 300 million cellular subscribers internationally and has a track record of technology innovation that has delivered global competitive advantage. We have several UK companies that are world leaders, or at least potential world leaders, in radio and telecommunications silicon. The UK undertakes internationally acclaimed fibre optic research. The UK is surprisingly good at designing and making and occasionally launching low-cost satellites. Depressingly, this is probably not enough.

It is remarkable how easily national technology resource, the engineering collateral of a nation, can be squandered as a result of political paralysis. Political inefficiency may be the price that we pay for living in a democracy and if so we should just accept the fact that free will is not free but comes with a cost attached. Even discounting the direct and indirect 'cost of democratic governance', developed nation states have high running costs that are a consequence of quality of life expectations compounded by other factors such as a perceived need to defend themselves or a usually misplaced desire to invade other nation states. In this context the concept that any developed nation state can survive as a predominantly service economy must be open to question.

By implication the UK needs a broadband technology economy rather than a broadband service economy. A broadband technology economy should be intrinsically efficient at realising a return on legacy R and D and manufacturing investment.

---

[5] Mr Kanwar Chadha speaking at the Future Wide Area Wireless Conference 27 January 2011.

So back to the assumption that universal broadband connectivity will drive future economic growth. This is almost certainly true but the problem is that we have no obvious way of getting to where we need to be. The return on investment is not sufficiently attractive to bring in private investment and public cash is not presently a viable alternative.

The UK has the advantage of a one hundred and fifty year legacy of telecommunications R and D and manufacturing investment. It therefore makes obvious sense to couple broadband connectivity investment policy with telecommunications industry investment policy. Countries such as Germany, Canada, France, Italy and the US are in a similar position. An economic model that somehow aligns and combines these interests is probably the best and possibly only way forward.

12.7.5   The Role of Telecommunications in Emerging Economies

In emerging countries, telecommunications has a different though no less profound role to play. The provision of microtrading and microfinance over cellular networks discussed in Chapter 11 is one example of the potentially transformative power of low-cost and not necessarily broadband connectivity.

12.7.6   The Telecommunications Economy – A Broader View of Broadband

It is therefore useful not to get overfocused on a broadband debate but to take a broader view of the role that telecommunications and the telecommunications industry can play in developed and developing economies.

Developed economies often have legacy telecommunications R and D and manufacturing investment that should be nurtured and nourished as a national resource. Telecommunications R and D and manufacturing in these countries is not the problem but the solution to the problem.

Emerging economies do not have this advantage but have the potential to make rapid economic progress with telecommunications as the essential facilitator – a telecommunications economy rather than a broadband economy. They can and do insist on in-country R and D and manufacturing as a precondition for overseas companies wishing to do business in the country. The Digital Britain Report referenced earlier justified broadband investment as a precondition for economic and social progress. The thinking in the report reflected a global debate on the contribution that broadband connectivity is expected to make to developed and emerging economies.

Irrespective of whether broadband is delivered over fibre, cable, copper or wireless or any mix of these, it comes with a high price attached. From an industrial policy perspective, the provision of broadband connectivity is an obvious business opportunity but only if the technology and engineering economics make sense. If the technology and engineering economics do not make sense it becomes an expensive obligation. The issue of course is whether wireless is more or less economically efficient as a delivery platform when compared with other options including fibre to the curb and fibre to the home.

This is in turn dependent on the definition of broadband, both in terms of average and peak data rates and contention ratios, the demographics and topology of the country into which broadband is being deployed and the efficiency with which wireless is integrated technically and commercially with other delivery options.

For example, fibre arrived in our street in 2009 as an upgrade to the existing cable network and we were duly leafleted with the offer of 50 Mbit broadband, all you can eat TV and telephone connectivity at £50 per month. Problematically for the fibre supplier, we have acceptable ADSL2 connectivity and minimal appetite for all you can eat TV already adequately supplied without a subscription fee via a £100 Freesat satellite receiver that we fully own and are still happy with two years later.

This highlights that even in urban areas in developed economies (the UK can just about still claim to be a developed economy) the time taken to achieve a return on broadband connectivity investment may not be consistent with short- or long-term shareholder expectation. Local zoning restrictions preclude cellular as a broadband access option even if we needed it, which we don't.

12.7.7   Australia as an Example

Simultaneously, the Australian government decided the Australian nation should have its own national broadband network delivering 100 Mbit fibre to 90% of the population over the next eight years. The cost is an estimated 43 billion dollars and is billed as Australia's largest ever infrastructure project

The proposed funding is made up of an initial government investment of 4.7 billion dollars with the balance as a mix of 51% government bonds and 49% private funding. It is claimed that the scheme will create 37 000 jobs.

The cost equated to a public funding investment per household of 1650 Euros. New Zealand has a similar programme that equates to 600 Euros per household. The equivalent UK national commitment is three Euros.

Whether the Australian and New Zealand initiatives prove to be economically efficient or politically popular is always going to be dependent on how much of the promised funding materialised within the political election cycle, how well it is spent and the terms on which private sector funding is secured. The latter is the detail that has been problematic. As one example, in Brisbane a private sector group called the i3 group had a plan to start work on a new fibre optic network in early 2011 using the underground sewer system. Unfortunately, the company became embroiled in an investigation by the Serious Fraud Office following a failure to deliver on similar projects in the UK. So far public funds have been spent, private capital has been raised and a fibre optic network has failed to appear, not the most promising of starts for a new funding model designed to transform broadband connectivity. The promised jobs have appeared elusive as well.

These projects are perhaps better attempted in command and control economies with a more deterministic political purpose. Singapore for example has a government plan to spend 500 Euros per household and it can be imagined that failure there will not be an option.

The justification for these investments is based on an expectation of improved industrial, social and political efficiency with additional benefits such as lower carbon emission based on an assumption that the need to travel will reduce. The underlying argument is that emerging economies need broadband in order to become developed economies and developed economies need broadband in order to remain internationally competitive.

The combined investment, however, is likely to be of the order of hundreds of billions of dollars and it is fair to question whether a better overall return could be realised by providing connectivity to bottom-end rather than middle- (emerging) or top-end (developed) economies.

12.7.8   Universal Narrowband Access in Survival Economies

A bottom-end economy is a survival economy in which the average annual wage is less than ten dollars per day ($3260 dollars per year). Emerging economies have incomes between $3260 and $20 000. Developed countries have median incomes greater than $20 000.[6]

Day to day life in survival economies is dominated by the need to find food, water and shelter (though TV is popular too). Many of the four billion people trapped in survival economies do not have bank accounts and have no way to exchange goods and services other than through basic bartering or cash, which can be dangerous to handle. When a natural disaster happens, survival economies are dependent on food aid brought in by oversees agencies by truck or air. The cost per calorie of this food aid once administration, transport and security costs are included is ridiculously high.

The Safari Com M-PESA scheme described in Chapter 11 has been an effective antidote to these problems. administered through a network of 10 000 local agents. Transaction volume is presently doubling every four months. The host operator, Safaricom, enjoys low churn rates and income from six million SMS messages a day. The publicity from the scheme helped with a stock exchange flotation which included a significant percentage of small shareholders to whom Safaricom has just sent a five-dollar dividend by SMS transfer – a transaction that would have been uneconomic using traditional banking.

From a user perspective menu prompts are in English or Swahili. An unexpected side benefit of the project has been to increase basic literacy, numeracy and memory skills (remembering a PIN number).

In the postelection riots subscribers were encouraged by Safaricom to share prepay credit in order to keep communication going. The riots meant that substantial food aid had to be brought in. The distribution of aid as cash instead of food parcels was piloted with SIMS issued to families who were then sent small amounts of money to allow them to buy food and supplies.

This reduced the cost per calorie and targeting the aid where most needed. The process also helped to establish local supply chains and sustain the local economy. A similar initiative was then started in Afghanistan, where recent events suggest there were other applications that could be usefully developed. For example, the cost of providing international observers in the Afghan election was estimated as being of the order of $300 million dollars. This does not include the loss of life that occurred and the political cost of evidential fraud. Spending this money on mobile phones and mobile infrastructure to enable voters to vote by phone is a viable alternative for elections in countries with limited democratic experience and would deliver the added benefit of a sustainable microeconomy on the Kenyan model.

Healthcare is an additional potential application based on the integration of M Health and M Wealth (poverty eradication using access technology).

12.7.9   Universal Narrowband Access in Survival Economies – The Impact on Developed and Emerging Economies

So if we diverted universal broadband infrastructure spending in developed and emerging economies and invested the equivalent amount, several hundred billion dollars, into universal *narrowband* access in survival economies what would happen?

---

[6] http://www.wri.org/.

We would suggest that globally there would be a net social, political and economic gain. Survival economies would move more rapidly to become emerging economies, in turn creating new markets for the developed world to serve. Attempts have been made to quantify this. The World Bank estimates that a 10% growth in broadband connectivity delivers 1.3% of GDP growth.[7]

In the wireless telecoms economy it is evident that profits in saturated developed economies are flat at best. Being forced into meeting universal broadband access obligations by wireless or other means will introduce additional cost and minimal profit. It is plausible that the taxpayer will absorb some of this investment pain but this hardly seems credible in today's international political climate. Given the growing interdependency of countries around the world it seems odd that the potential for survival economies to be the new engine of global social, political and economic progress has been largely ignored.

Partly, this may be due to fear of the unknown. For example, there is an argument that suggests that growth in emerging and survival economies will trigger large increases in carbon emissions. There are equally strong counterarguments to suggest this would not happen and that other positive metrics would outweigh any related carbon risk. Economic growth and a low carbon economy do not need to be mutually exclusive.

Note also that survival economies are not geographically specific and can be found within developed and emerging economies. Over 70 million Americans do not have a bank account.

12.8 The New Wireless Economy in a New Political Age?

The present political enthusiasm for broadband connectivity investment may be misplaced and is certainly too rooted in nationalistic competitive ambition. In a closely coupled global economy, future profits and business opportunity will be dependent on how well emerging countries perform and how fast survival economies can become emerging economies. Fortuitously, it would appear that mobile phones and mobile networks are important and perhaps essential to this transformation process.

We may be witnessing the birth of a new age of truly international growth and prosperity with the telecoms industry as a facilitator of global social, political and economic progress – the new wireless telecom economy. The fact that telecommunications networks have reinvented themselves as information networks does not alter this fundamental truth but should reassure us that the telecommunications industry continues to have a socially and economically profitable purpose in both emerging and fully developed economies.

The argument therefore is that narrowband access to survival economies should take precedence over broadband investment in developed economies. This would not be naïve altruism but a pragmatic way to maximise the social and economic dividend from global connectivity investment. Present broadband investment proposals in developed economies are justified on the basis of delivering nationalistic competitive advantage. This creates additional distance between the rich and poor on the planet, an unsustainable model for future global growth.

Access technologies, fibre, cable, copper and wireless, can, however, potentially deliver social and economic progress based on social and economic equality, the precondition for achieving at least a measure of political stability. Connectivity could be more broadly defined

---

[7] As reported by Bong Goon Kwah, Korea Telecom Cambridge Wireless Conference, 27 June 2011.

and delivered as a superutility in order to realise a step function reduction in connectivity cost and energy efficiency.

12.9 Connected Economies – A Definition

Connected economies can be interpreted on several levels.

There is a close interconnectivity between developed, emerging and survival economies and a growing recognition that South America, Africa and Greater Asia will be or will continue to be the engines of future global prosperity providing fast growth markets for the developed world to serve. The 2016 Olympics in Brazil will be a reflection of this global transition.

Additionally, these emerging economies are becoming increasingly able to undertake local technology and engineering development. Over six million young people graduated from Chinese universities this year of which about half a million are engineering graduates – an engine of innovation. India similarly is producing 400 000 new engineers per year.

The availability and cost and effectiveness of engineering resource is closely coupled with the world's ability to deliver connectivity. Additionally, telecoms connectivity is dependent on the availability of civil engineering skills and an ability to source and use energy as efficiently as possible.

The telecom companies that trace their heritage back to the Victorian era, Cable and Wireless, British Telecoms and Siemens being three examples, have built their business on the back of multidiscipline skills that include mechanical, civil and energy engineering.

Most telecoms companies today regard a capability to source and use energy efficiently as a core part of the connectivity proposition for both rural and urban deployment. Connectivity is, however, not a term that is exclusive to telecoms but includes transport, water and power.

To put this in an historical context, steam and coal, steel and oil, the motor car, the aeroplane, the valve and the transistor have created and shaped the world that we live in today – the discoveries and developments, inventions and innovations that have provided the foundation for today's modern connected economy.

However, technologies always come with a social and environmental price tag attached. As a general rule the costs and the benefits are not equally distributed geographically of demographically – technology does not have a good track record for delivering social or economic equality.

So, should technology and the engineering needed to exploit technology have an explicit social purpose and if so how should that purpose be measured and managed? *An* answer rather than *the* answer to the question can be found by analysing technology history in the developed world. The Science Museum[8] helps with the analysis process by dividing 'recent' (1750 to 2000) technology history into eight overlapping periods

12.9.1 Enlightenment and Measurement 1750 to 1820[9]

Henry the Eighth, a very modern politician, though with a marital record that would have not stood up well to modern media attention, laid the foundation for the age of Enlightenment by

---

[8] These extracts reproduced with permission of the Science Museum http://www.sciencemuseum.org.uk/.
[9] http://www.makingthemodernworld.org.uk/stories/enlightenment_and_measurement/05.ST.05/.

spending a disproportionate amount of money on the British navy. This allowed Britain to build its colonial empire generating the wealth that went partly into an expanded University system.

Two hundred years later, that same University system delivered Isaac Newton to the world and a generation of thinkers with the time and money to consider how the world could be understood and therefore managed through experiment, measurement and reason.

Most things that moved and many things that didn't were measured including rainfall, deaths and electric charge, an age of scientific curiosity and instrumentation. However, this was enlightenment for the few not the many – an important shift in intellectual thinking but with limited impact on the poor and disadvantaged.

12.9.2   Manufacture by Machine 1800 to 1860[10]

The era in which 'machines to make machines' produced the looms and spinning machines that powered the industrial revolution destroying the Indian cotton spinning industry in the process.

The industrial revolution resulted in a process of intense urbanisation and rural social deprivation. The net wealth of the nation increased but so did social and economic inequality. In parallel, the traditional education system failed to respond to the need for engineers. This was to sow the seeds of Britain's industrial decline in the twentieth century.

12.9.3   Age of the Engineer 1820 to 1880[11]

This underlying problem was hidden by a small group of Victorian engineering superstars, James Watt, Robert Stephenson and Brunel.

12.9.4   Industrial Town[12]

Between 1800 and 1900 Britain's population more than quadrupled and became predominantly urban. Production machinery made it possible for factories to employ hundreds of people. Towns with high-density working populations created new problems of housing, health and transport.

12.9.5   Second Industrial Revolution[13]

The first Industrial Revolution had been based on steam power, factories and railways. The second was based on new kinds of power – electricity, new chemicals, new plastics and new drugs – particularly from recently industrialised nations like Germany and the USA. The USA introduced mass-production techniques into the weapons industry – a change which was to have a profound influence on the world's-twentieth century political order.

---

[10] http://www.makingthemodernworld.org.uk/stories/manufacture_by_machine/02.ST.01/.
[11] http://www.makingthemodernworld.org.uk/stories/the_age_of_the_engineer/03.ST.03/.
[12] http://www.makingthemodernworld.org.uk/stories/the_industrial_town/06.ST.02/.
[13] http://www.makingthemodernworld.org.uk/stories/the_second_industrial_revolution/05.ST.01/.

12.9.6   The Age of Mass Production[14]

By 1914 Henry Ford's car lines in Detroit showed the world that production and machine repetition could transform the economics of personal transportation. Mass production generated mass employment, mass consumption and a World War in which technologically advanced mass-produced weapons enabled killing and wounding on an unprecedented scale, the age of mass destruction.

Wireless telecommunications started to have a substantial social, political and economic impact.

12.9.7   Defiant Modernism[15]

The atomic age had resulted in the age of mass destruction being taken to a new level of deadly efficiency, but as a small compensation, the Second World War provided the basis for inventions and innovations that could be potentially repurposed to realise social and economic progress.

The V2 rocket provided the basis for the space programmes of the 1950s and 1960s, the atom bomb was supposed to result in nuclear power stations that would produce electricity 'too cheap to meter' and emissions 'too small to measure', aspirations that significantly failed to materialise.

12.9.8   Design Diversity 1950 to 1965[16]

The availability of new materials and the dawn of the age of plastic provided opportunities to differentiate the design of everyday products in order to meet consumer expectations of choice, cost, quality and functionality.

12.9.9   Age of Ambivalence[17]

Organised environmental movements and environmental campaigns reflected a growing unease and failing confidence in the ability of technology to deliver social and economic progress at an acceptable environmental and social cost. This included concerns about genetic modification techniques.

In parallel, there was a growing recognition that the gap between rich and poor countries and the gap between rich and poor within those countries was getting wider over time both in terms of wealth and health – a recipe for economic, political and social instability.

---

[14] http://www.makingthemodernworld.org.uk/stories/the_age_of_the_mass/.
[15] http://www.makingthemodernworld.org.uk/stories/defiant_modernism/04.ST.02/.
[16] http://www.makingthemodernworld.org.uk/stories/defiant_modernism/04.ST.02/.
[17] http://www.makingthemodernworld.org.uk/stories/the_age_of_ambivalence/02.ST.06/.

12.10   Inferences and Implications

A number of *general* inferences can be drawn from this particular slice of technology history.

The social and economic benefits of technology and engineering innovation are generally not evenly distributed geographically or demographically nor are the direct and indirect costs. The direct costs are those associated with initial deployment and maintenance costs thereafter. The indirect costs are the social and economic costs that can be both short term, the loss of cotton spinning jobs in India being one example, or long term. Only now, 150 years later are we becoming aware of the full environmental cost of the industrial revolution.

By the end of the nineteenth century, Britain had become the world's most advanced connected economy based on a postal system, the telegraph, the railway, including the London Underground, a canal system, a steam-driven water system, the world's largest network of sewers, an underground hydraulic system, gas, electricity and a large merchant navy, a form of mobile connectivity.

This was infrastructure investment on a scale never seen before or since, built on the profits of the British Empire. Today, we still post letters in Victorian post boxes, travel on Victorian railways and the underground network, navigate and ship goods on Victorian canals, drink water delivered through Victorian pipes and flush effluent away through now leaky Victorian sewers. The water-based hydraulic system was retired in 1977 and the curtains of the London Palladium are now opened and closed using electric rather than water power.

A number of *specific* inferences can be drawn from this.

The infrastructure of a connected economy is expensive and has to be financed from somewhere else by someone else. The return on investment has a long tail extending to hundreds of years.

Telecommunications is only one of several bidirectional connectivity systems including the postal service with which it was successfully integrated for over fifty years. Water and sewage is of course bidirectional but modern electricity grids are also now being designed as two-way systems. There has been some debate as to the practicality of homeowners earning money from allowing passersby to access their home WiFi systems, an arrangement that could potentially result in an extra ten million access points becoming available. The snag here might be the occasional home owner who decided to increase power and uplink gain beyond conformance limits in order to attract more revenue.[18]

These caveats aside there may be merit in reconsidering the concept of the repurposed super utility. Superutilities were invented by the Victorians as a mechanism for delivering multiple services more efficiently. The Gas Light and Coke Company is one example; the General Post Office (Post and Telecommunications) is another.

Utilities are delivered in the developed world with bizarre inefficiency, the only possible justification being the employment provided to the trench-digging, hole-digging and meter-reading community.

It is inherently inefficient to have multiple organisations separately supplying electricity, gas, water, sewage and fibre, cable, copper and wireless connectivity.

If Britain was a dictatorship there would be a strong argument in favour of setting up two organisations, the British Broadband Corporation and the National Connectivity Corporation.

---

[18] Open forum discussion, William Webb, Future of Wireless Conference, 27 June 2011.

The British Broadband Corporation would take over responsibility for public service broadcasting from the BBC but would also provide broadband connectivity using fibre, copper, cable and wireless and would be responsible for developing new broadband and narrowband delivery technologies that could be applied into emerging and survival economies. The logo could stay the same as well.

It would have a social remit – 'entertain, inform and educate' would do nicely, a term apparently coined not by the BBCs Lord Reith but by the American broadcasting pioneer David Sarnoff in 1922.

The National Connectivity Corporation would provide all other forms of connectivity with a similar remit to develop connectivity technologies, particularly water and electricity, that could be applied cost efficiently and energy efficiently into emerging and survival economies.

Both corporations would be publicly owned but profit making with dividends only payable to pensioners. The two organisations would be mandated to develop mutually supporting collaborative delivery models.

This of course will not happen but does suggest that there may be merit in considering new more closely integrated ways of delivering water, electricity and telecommunications to remote communities in bottom-end economies.

12.11   The Newly Connected Economy

As in the Victorian age, connectivity efficiency is very much determined by how the energy needed is created and how efficiently that energy is used. In the nineteenth century water and sewage systems were powered by coal and steam. In the twentieth century telecommunications were powered and are still predominantly powered today from the electricity grid.

In the twentyfirst century, connectivity is much more likely to be powered from locally generated energy including wind and solar-powered base stations, and wind and solar-powered water pump and recycling systems all of which could potentially be community owned and managed. But let's consider the particular contribution that cellular phones can make to connectivity.

12.11.1   Connecting the Unconnected

By the time you read this the number of cellular phones in the world will equal the number of people (not counting the number of machine-to-machine devices that will be interconnected by then). This does not mean that everyone will have a cellular phone but is a reflection of the growth of multiple device ownership in developed economies. Cellular phones also need cellular networks or more specifically access to information networks in order to be useful.

However, achieving 100% penetration demonstrates that we have the *industrial ability* to provide a cellular phone to every human being on the planet and the *ICT ability* to deliver socially and economically useful personal and individual services to those phones.

This industrial and ICT capability is not concentrated in a single country or continent but spread across Europe, Greater Asia, the US and South America, a globally distributed deployment resource. China for example has very successfully leveraged its local market both to build home-grown vendors and attract inward intellectual investment in joint R and D programmes. Huawei in 2011, a company few people outside China had heard of ten years before, had a turnover of $28 billion, employed 110 000 people, half of them in R and D and had 5000 staff in Europe.

The cost and energy efficiency of deployment will be dependent not only on the capability and resources of these newer vendors and their older competitors but also on how well these services are integrated with copper, cable, fibre and other wireless delivery systems and the best social and economic returns may be from narrowband rather than broadband investment but without a doubt the modern world has sufficient technology and engineering and financial resource to connect the unconnected provided we have the political will to do it.

12.11.2   The Dawn of a New Age of Enlightenment?

Developed nations have visibility to a plausible model for providing affordable and sustainable informational, educational and economic connectivity to the survival economies of the emerging world, delivered physically side by side with water and power connectivity – a light at the end of a dark tunnel of technology-driven social and economic inequality, the potential dawn of a new age of engineering-based personalised mass enlightenment, enlightenment for the many not the few.

12.11.3   Enabling Wireless Technologies – Femtocells

And then I noticed that what I should really have been talking about in this chapter was mobile phone network topologies, but then this prompted me to look at what we had covered in Chapter 12 of 3 G Handset and Network Design only to discover that not only had this topic already been fully addressed but also very little had changed in the intervening eight-year period.

Base stations had become smaller, smart antennas had become a bit smarter, integration with WiFi networks had progressed a bit but not a lot.

For at least the last five of those eight years femtocells have been promoted as the new panacea. These devices never made much sense from a user-experience perspective. A femtocell in your home promised access speeds rather lower than the WiFi already installed that you now could not use as it was providing the backhaul for your new femtocell. The femtocell was available with only one operator providing the service so if other members of the family had service contacts with other operators, additional femtocells were needed that would then produce adjacent channel interference with other femtocells. Meanwhile, WiFi access points became cheaper and faster.

And this is probably quite a good though slightly negative point to end this chapter. Innovations only become widely adopted if they deliver added value to all players in the value chain. In the case of femtocells, they solve an operator problem (poor indoor coverage from outdoor cells) but as far as the user is concerned, why should that involve fixing a problem that is clearly the responsibility of the operator who already takes a large monthly fee but now appears to be expecting the consumer to provide free backhaul as well.

This brings us to our next chapter.

13

Wireless Core Network Hardware

13.1 The Need to Reduce End-to-End Delivery Cost

In Chapter 12 we suggested a brave new world in which telecom networks including mobile broadband networks would deliver social, political, intellectual, economic and possibly environmental value. However, this can only be achieved if the networks are financially sustainable.

This might seem back to front, but in Chapter 21 we review a series of forecasts including our own that suggest that one consequence of the transition from telecom networks to information networks is that there will be a 30-fold increase in data volume over the next five years over mobile broadband networks and a threefold increase in value.

This implies a substantial need to reduce end-to-end delivery cost. Backhaul costs presently represent 30% of this cost but are increasing over time. This cost of delivery is not just a function of volume but is also influenced by traffic mix and the degree of multiplexing gain that can be achieved.

13.1.1 Multiplexing Gain

Multiplexing gain is a function of single users having variable bit rate requirements that are then averaged over multiple users. Some of the bits are not latency sensitive and can be buffered. Additionally, there may be more than one point-to-point routing option that provides additional multiplexing gain. Mobile broadband offered traffic is highly asynchronous (bursty).

On the one hand, it could be argued that this makes multiplexing more rather than less effective. The counterargument is that bursty traffic remains bursty even after aggregation (persistent burstiness). What this means is that bursty traffic behaves in different ways and is affected by the characteristics of the paths over which it travels.

Generally, it can be stated that conversational, streamed and interactive traffic costs more to deliver than best effort and can be shown not to aggregate efficiently. The offered traffic can also be highly asymmetric. These characteristics together mean that backhaul has to be over provisioned in order to meet defined user QOS and QOE expectations. This increases capital and operational cost. However, we have also stated in previous chapters that although best-effort traffic can be used to increase delivery bandwidth utilisation this is only achieved by buffering that itself introduces cost.

Making Telecoms Work: From Technical Innovation to Commercial Success, First Edition. Geoff Varrall.
© 2012 John Wiley & Sons, Ltd. Published 2012 by John Wiley & Sons, Ltd.

Backhaul strictly defined is a delivery network not a buffered network. Buffering will be applied at the edge of the network. Delivery options include point-to-point microwave, point-to-multipoint microwave, multipoint to multipoint microwave also known as mesh networks, free-space optics, fibre, cable and copper.

Many markets are growing at a rate where traditional fibre or copper backhaul cannot be deployed fast enough. Point-to-point or point-to-multipoint microwave or free-space optic links are alternative options. These offer deployment time scale advantage and easy through-life reconfigurability but have different optimum fiscal- and energy-efficiency thresholds that are dependent on the technology and topology used and the characteristics of the offered traffic. In addition to the offered traffic mix, the trade off points for each option will be determined by network density, local geography, the mix between urban and rural coverage and local economic and business model considerations. The assumption is that future peak and average throughput requirements can be met either by more aggressive deployment of higher-order modulation schemes into existing bands and/or by implementing links in unlicensed or lightly licensed spectrum at 60, 70 and 80 GHz.

13.2   Microwave-Link Economics

Microwave-link economics and the choice of channel bandwidth and modulation schemes are also influenced directly and indirectly by spectral allocation and regulatory policy. A typical licensed spectrum product offer from a vendor such as Ericsson includes options at 6, 8, 11, 13, 18, 23, 26, 28, 32 and 38 GHz with a range of dish antennas from 0.2 to 3.7 M. The products use 4 to 256 QAM adaptive modulation and scale from 3.5 to 56 MHz channels. RF output power scales from +30 dBm (CQPSK at 6 GHz) to +16 dBm (256 QAM at 38 GHz), the receiver sensitivity thresholds for a one in 106 bit error rate scale from –59 dBm (256 QAM in a 56 MHz channel at 38 GHz) to –91 dBm for C QPSK at 6 GHz and the throughput scales from 4.1 Mbps to 345 Mbps. NSN having a matching portfolio though scaling to 450 Mb/s per carrier. These systems are usually frequency division duplexed.

Other option include 5-GHz systems using the unlicensed 5-GHz spectrum (802.11a) and/or the 60-GHz unlicensed band and/or the 'lightly licensed' 80-GHz band. These systems are usually time division duplexed.

As frequency increases bandwidth increases and antennas get smaller for a given gain. A one degree beam width antenna for example gives about 50 dBi of gain. Conversely at a given frequency, the bigger the dish the higher the gain. This translates into higher throughput but mast costs will increase, a function of weight and wind loading.

The problem (and opportunity) with microwave links is that as you move into the millimetre bands (30 to 300 GHz) the wavelength is the same length as a raindrop.[1] This causes non-resonant absorption that has to be accommodated in the link budget and/or by rerouting in exceptional but localised weather events (thunderstorms). There are also nonresonant absorption peaks for water at 22 and 183 GHz and oxygen at 60 and 119 GHz. The oxygen peak at 60 GHz is what allows this band to be made available as unlicensed spectrum as the absorption, of the order if 15 dB per kilometre provides protection against system-to-system interference.

---

[1] See 3G Handset and Network Design, Chapter 15, page 368.

Note that these are line-of-sight systems. With fibre, cable or copper you can turn a corner but with microwave you can only do that by installing another microwave dish and/or repeater.

Either way, line-of-sight point-to-point engineers argue that narrow beam width dish antennas are the most effective way to achieve link budget gain (and by implication, throughput) with the wide choice of available frequency bands, meaning that an optimum choice of system can usually be made.

13.2.1   Point-to-Multipoint

Another alternative option is point-to-multipoint microwave.

In point-to-multipoint a sectored antenna with for example a 90-degree beam width will service multiple sites, typically of the order of nine or so. The band options are ETSI standard 10.5 GHz, 26 GHz and 28 GHz. Bandwidths are presently 7, 14 and 28 MHz and throughput scales to 150 Mbps per sector with modulation options ranging from QPSK to 256 QAM. The systems are normally duplex with a duplex spacing of the order of 350 MHz at 10.5 GHz or 1008 MHz at 26 and 28 GHz. Output power is +20 dBm at 10.5 GHz and +18 dBm at 26 or 28 GHz. A 90-degree sectored antenna at the base station at 10.5 GHz provides 15 dBi of gain at 10.5 GHz and 17 dBi of gain at 26 or 28 GHz. A 0.3-m dish at the other end gives 28 dBi of gain at 10.5 GHz, 36 dBi at 26 GHz and 37 dBi at 28 GHz. Increasing the dish size to 0.6 m increases the gain to 33 dBi at 10.5 GHz, 40.5 dBi at 26 GHz and 42 dBi at 28 GHz.[2]

The spectral efficiency argument is based on the multiplexing and scheduling gain achievable from point-to-multipoint and its impact on backhaul efficiency and throughput. A microwave point-to-point connection achieves a similar scheduler gain to the one-to-one gain achieved in the RAN.

A microwave point-to-multipoint connection achieves the same one-to-one scheduler gain but has additional gain from multiuser diversity. Point-to-multipoint backhaul is therefore conceptually similar to a sectored node or e node B. Throughput in terms of joules per bit may also be lower, offering operational cost benefits, for example in solar-powered or diesel-powered applications. Point-to-multipoint could potentially be more broadly deployed but limited vendor support and limited spectral availability may constrain market adoption.

13.3   The Backhaul Mix

So how will the backhaul mix between fibre, cable, copper and wireless change over time? Let's try and answer the question initially in terms of mobile broadband access backhaul economics.

It seems likely that backhaul from femtocells will continue to be carried for at least part of the way over copper twisted pair with the caveat that fibre over utility poles may change this over time, a topic addressed in the next chapter.

Microwave vendors argue that microwave transport will meet the foreseeable capacity needs for mobile backhaul for many years to come using a combination of licensed, lightly licensed or unlicensed spectrum. The argument is based on the assumption that a 10-MHz LTE radio channel will require a backhaul of the order of 150 Mbit/s.

---

[2] Point-to-multipoint hardware examples, with thanks from Cambridge Broadband Networks.

The number of channels supported depends on the configuration of the base station on the site including the number of sectors and number of channels per sector, the channel spacing used (5, 10, 20 MHz or bonded channels up to 100 MHz) and cell geometry. Cell geometry is the term used to describe the distribution of users in a cell relative to the cell centre. The further away from the cell centre the weaker the link budget and therefore the lower the data throughput.

The IP packet throughput is therefore likely to be significantly less than the headline physical layer throughput figures would imply. A counterargument is that if operators start sharing the radio access network or if neutral host networks become more prevalent and if legacy GSM voice channels and HSPA and LTE data channels and ultimately presumably LTE IP voice get aggregated together then this implies several hundred MHz of instantaneously available bandwidth. A 13-band base station, if such a beast existed would have access to more than 1100 MHz of radio bandwidth, potentially available on each of three or more sectors with an average efficiency of between 2 and 3 bits per Hz. This is a gigabit rather than megabit discussion.

The offered traffic limit in this case would probably be the battery and heat-rise limitations of the user devices accessing the network and/or the number of users you would need to pack together to use the bandwidth in a small to medium-size cell. You may as well deliver television to each site as well adding a few more gigabits to the backhaul mix at least in one direction. Microwave vendors might argue that whatever the loading microwave can provide enough capacity just by adding more bandwidth. However, given that the dish antennas are band specific this implies a lot of antenna hardware on a site/mast that might be space- and weight-loading constrained. Line-of-sight problems would also shift the balance towards copper or fibre.

13.4   The HRAN and LRAN

The relative technology economics of wireless versus copper and fibre are therefore complex with many interdependencies. Ericsson makes things simpler[3] by dividing mobile broadband backhaul into two parts, the lower radio access network (LRAN) and the high radio access network (HRAN). The HRAN aggregates traffic from several LRAN networks using fibre or microwave. The LRAN provides the last mile of connectivity to anything between 10 and 100 radio base station sites normally, though not always, using microwave but with fibre and copper as alternative options.

They make the completely obvious point that microwave will generally have the lowest cost of ownership unless an alternative is already there or physically close. However, as you would expect from a vendor with very positive ambitions for LTE the argument is made that at some point in time some base stations may need gigabit backhaul links. The regulated licensed spectrum between 6 GHz and 38 GHz supports channel bandwidths up to 50 MHz, which with high-level modulation can just about deliver 500 Mbps under ideal conditions. Adding polarisation multiplexing and multiple input/output techniques increases this and adding channels multiplies the capacity.

---

[3] http://www.ericsson.com/ericsson/corpinfo/publications/review/2009_02/files/Backhaul.pdf.

As at 2011 Ericsson are/were offering 14 MHz or 28 MHz channels for LTE backhaul with the longer-term option of migrating to aggregated 28 MHz channels. The lightly licensed 71–76 GHz and 81 to 86 GHz bands known as E band support 10 GHz bandwidths but using simpler modulation on a single carrier. The light licensing has been adopted in the US and UK, takes a few minutes online and costs about $100 dollars a year. The bands became available in the US in 2003 and in 2006 in Europe. Using these bands, however, depends on local weather conditions. A tropical rainstorm every afternoon will seriously affect availability and throughput, but then we don't get a lot of tropical rainstorms in the UK, though actually we had one yesterday. There are some applications where throughput becomes more important when it rains. For instance, there are generally more traffic accidents and cars tend to break down so calls to the emergency and breakdown services will increase. Snow storms can also be a problem at some frequencies as can sand storms, though we don't get a lot of those in the UK either. For all of these weather-related events availability can be provided in other ways, for example by having multiple routing options or by having other microwave links at other frequencies.

The copper offer (reviewed way back in Chapter 2) includes high bit rate DSL (HDSL), enhanced high bit rate DSL (HDSL2) and in some cases single pair HDSL (SHDSL).

If there is a fibre to the curb nearby then this allows enhanced very high speed DSL (VDSL2) to be used providing 100 Mbps of upstream and downstream bandwidth. The 100 Mbps can be increased by implementing a crosstalk cancellation technique called vectoring and line bonding (multiple twisted pairs dedicated to the link). The vector cancellation characterises the crosstalk from the adjacent VDSL twisted pairs by knowing the transmitted symbols and calculating the crosstalk coupling gains then predistorting the transmitted symbol stream on the wanted channel to eliminate the crosstalk.

Microwave vendors are, however, keen to point out that whatever you might do to copper it can suffer the same fate as fibre and be taken out by accident when roads are dug up and can take several days to restore and make the case that a combination of backhaul provides the best guarantee of availability.

The microwave economic case is, however, dependent on the link budget including the fade margin (see weather events above). Link budgets are determined by the system gain that is a composite of the transmitted output power and TX/RX antenna gains minus the receiver threshold. This is set as the minimum received power required to support a bit error rate of less than one in 10^{12}. The system gain for a commercial backhaul radio will be of the order of between 160 and 200 dB. This implies that a range of up to 3 km is practical, but not more. This in turn implies these systems can only be used in urban or suburban applications with systems in the lower bands being better suited to rural applications.

Excepting that caveat Ericsson provide an example of a radio that supports 1.25 Gbps in both directions effectively an optical rate interface with a latency of less than a microsecond, transmitting in the higher band (81 to 86 GHz) and receiving in the lower band (71 to 76 GHz) on 5 GHz channels using differential binary phase shift keying (DPSK). The system gain is 162 dB. This would support 1.5-km hop lengths in heavy rain.

The point-to-point microwave market is served both by Tier 1 vendors including Ericsson, NSN and Huawei and more specialist vendors such as Bridgewave, Proxim, Aviat and Ceragon. The Tier 1 vendors have an advantage in that they have greater visibility to the future form factor of mobile broadband base stations, but essentially there should be room for both.

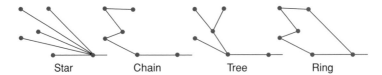

Figure 13.1   Network topologies. Reprinted with permission of John Wiley & Sons, Inc.

Ron Nadiv the VP of Technology and Systems Engineering at Ceragon has written a comprehensive book on Wireless Wireline and Photonic Networks published by John Wiley[4] that treats the topic of microwave backhaul in substantially greater detail than attempted here. A supporting White Paper[5] covers some of the same topics.

A summary of some of the points made about topology is, however, useful and relevant.

Most radio networks have a backhaul topology that is a combination of trees and rings, the tree being a combination of chain and star topologies, as shown in Figure 13.1.

Star topologies use a separate link from a hub to each site, which is simple but requires longer radio links and line-of-sight for each link that may or may not be possible. Interference is also more likely as all links originate at the same point so opportunities to achieve frequency reuse are limited.

Chain topologies don't have these problems but if one link goes they all go, just like those cheap Christmas lights you bought last year.

Putting the chain and star together produces a tree. In a tree topology fewer links will cause a major network failure and therefore fewer links require protection schemes (hardware duplication for example). Closing the chain produces the ring that provides the most protection.

Ceragon then develop a test case in which ten cell sites each requiring 50 Mbps aggregate to a total of 400 Mbps with every link that supports more than one site being protected. The options are shown in Figure 13.2.

The trade offs are as follows. The ring topology requires fewer microwave links but the links have to be higher capacity and use more spectrum, though since the ring has no more than two links to every node good frequency reuse can probably be achieved. The ring also requires some additional antennas but has inherently good availability due to the fact that a service failure would only occur when both paths fail. However, a ring will introduce additional latency and delay variation that might affect synchronisation signals and traffic latency. However, some of the tree links have lower bandwidth than those in the ring, which may introduce delay variation. Statistical multiplexing is more effective in a ring topology as the total capacity of the links in a ring is usually greater than the links in the branches of a tree. Rings are, however, more expensive to upgrade as each link needs an identical upgrade.

A practical example is then given of a Latin American operator transitioning from a TDM network to an IP network. Each cell site requires 45 Mbps of bandwidth for Ethernet traffic with additional E1 lines (2.048 Mbit/s supporting 32 by 64 kilobit voice channels). All wireless links were required to have hardware redundancy.

---

[4] http://onlinelibrary.wiley.com/doi/10.1002/9780470630976.fmatter/pdf.
http://eu.wiley.com/WileyCDA/WileyTitle/productCd-0470543566,descCd-description.html
[5] http://www.ceragon.com/files/Ceragon_Wireless_Backhaul_Topologies_%20Tree_vs_%20Ring_%20White%20Paper.pdf.

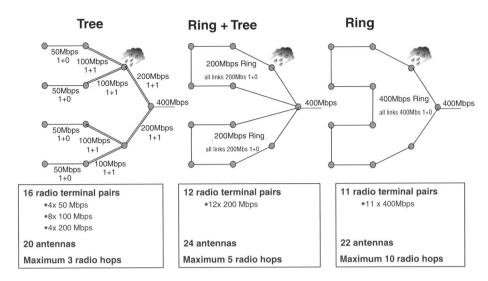

Figure 13.2 Aggregation topologies test case. Reprinted with permission of John Wiley & Sons, Inc.

An analysis suggested that the shift towards Ethernet improved the Capex and Opex economics of the ring topology compared to other options and offered better resiliency. The ring topology is combined with ring-based protection using the Rapid Spanning Tree Protocol. Apparently, this meets the Next Generation Mobile Network requirement of path restoration within a range of 50 to 250 milliseconds.

The other wireless option briefly referenced in Chapter 2 was free-space optics at 800 nm or 1550 nm with 1550 nm being the preferred option for safety reasons. These systems can support high bandwidth but are subject to scintillation and fog attenuation that precludes their use in many more mainstream applications.

13.5 Summary – Backhaul Options Economic Comparisons

In a voice-centric network a single T1 (1.544 Mbps) or E1 2 Mbit/s backhaul connection would have been sufficient to support a moderate size cell. Simplistically, an LTE channel with a peak data rate of 150 Mbit/ with three sectors needs up to 450 Mbit/s of capacity. Cell geometry reduces this. Adding channels per sector and/or implementing neutral host networks with shared backhaul increases this. Gigabit rather than megabit backhaul requirements are therefore at least plausible. Practical limits with copper are presently of the order of 100 Mbits but essentially this is a bandwidth-limited medium with propagation limitations and significant interference constraints (crosstalk).

It is usually very fully amortised but lease costs, for example, for an E1 or T1 line can be substantial both in terms of a set up charge and monthly rental and this cost increases linearly with capacity – you don't get a discount for multiple T1/E1 lines, which means that 10 Mbps of bandwidth cost ten times as much as 1 Mbps. So, for example, a T1/E1 circuit in 2008 would cost anywhere between $150 and $750 per month with a set up charge of over $500 dollars.

Fibre using higher-order modulation can deliver multigigabit throughput but is often not available as a legacy installation and therefore has to be aggressively amortised. A leased fibre capacity of 155 Mbps would be in the region of between $4000 and $7500 dollars per month and a set up charge of around $7000 dollars, assuming of course that the fibre is available, which it may not be.

The T1/E1 costs and bandwidth limitations and fibre cost and availability limitations together determine the market share of microwave systems in backhaul provision. After factoring in hardware costs, cable costs and licensing costs the payback for microwave compared to multiple T1/E1 lines is somewhere in the order of 4 to 12 months The availability of 10-Mbps Ethernet over copper over distances of up to 1.5 miles has changed this trade off point in some markets, but these services do not support TDM services.

The microwave versus fibre economics calculation is very much determined by distance and population density with the cost of digging the trench directly dependent on distance. The capex cost for fibre is definitely higher, up to about 2 Gbps, but the exact crossover point will be dependent on local factors. If the fibre can be shared amongst multiple operators the cost per bit will of course be substantially lower.

13.6   Other Topics

So now I have noticed that I had also said I would address core network hardware, router hardware, server hardware and cost and energy economics in this chapter. I am going to offer several excuses for not doing this at this point

Core network hardware has already been addressed in 3G Handset and Network Design and subsequently in more extensive and accurate detail by other more technically literate authors including Harri Holma and Antti Toskala's LTE for UMTS book (John Wiley ISBN 978 0 470 66000-3)

My own take on the changing shape of core network architecture eight years ago was that the overall complexity would increase rather than decrease over time but would become distributed and that cost would shift to routers that would be surprisingly hardware centric in order to realise throughput and latency metrics.

Lots of high-performance memory would also be needed in each router. The ability to buffer at router nodes meant that network bandwidth utilisation could be substantially increased but there were two associated costs, the memory and the bandwidth occupied by the signalling needed to avoid buffer overflow that would result in packet loss. The section on deep packet examination (Chapter 18 page 406, has proved particularly prescient).

Anyway, that's enough bragging for the moment.

The problem about server architectures is that I know very little about the subject but it is the sheer scale of the server farm phenomenon that is by far the most interesting thing about them. In an earlier chapter we mused on the fact that Google now have over one million servers humming away in order to ensure we can become instant experts on everything, something I have never had a problem with, even before the internet was invented.

The initial bright idea was to install server farms in places like Nevada until it was realised that the heat generated from the servers combined with the heat generated from the noon day sun was not an energy-economic proposition. So the latest trend is to choose places like Iceland that are both cold and have plenty of geothermal power to waste.

There are measures that can be implemented to make servers more efficient. This includes the use of the techniques used in mobile handset application processor hardware such as dynamic voltage frequency scaling, clock gating and improving utilisation. The concept is that power consumption should be proportional to the activity level. The devices are sometimes described as energy proportional machines.[6]

However, the argument that telecommunications networks are saving the planet needs to be approached with caution. Apparently we are travelling less on business due to effortless teleconferencing but somehow I think the ease with which we can now book a week end flight to Prague and back may have more than offset this trend.

Better just to accept that telecommunications makes us more socially, economically and intellectually efficient and leave it at that. But now it is time to do telegraph poles.

---

[6] The case for energy proportional computing, Barroso, IEEE Transactions, December 2007.

14

Cable Network and Fibre Network Technologies and Topologies

14.1 Telegraph Poles as a Proxy for Regulatory and Competition Policy

This chapter is nominally about telegraph poles also known as utility poles but telegraph poles are in effect a proxy for regulatory and competition policy, so the discussion will be rather broader.

Have a look outside your window at the nearest telegraph pole and have a look at what is on it, probably either power lines or twisted pair. Then consider how long the pole has been there, who owns the pole, maintains the pole and makes money out of the pole and then consider other uses for the pole, what else the pole could profitably be used for. It might also be useful to have a look at some of the other street furniture including telecommunication street cabinets, lamp posts and camera installations and to consider what goes on beneath our feet as well.

It does not take long to come to the conclusion that the way 'stuff' is delivered is not particularly efficient or effective. This is nothing new and before we grapple with those poles let's take a short journey underground and take a look not at telecommunications but at the sewage system.

14.2 Under the Streets of London

In the 1840s parliament ruled that all houses in London should drain into sewers that then ultimately ended up in the River Thames, which became too contaminated to drink. There were several outbreaks of cholera and in 1858 the smell of the Thames (The Great Stink) caused Parliament to go into recess.

As a result, Sir Joseph Bazalgette[1] was commissioned to turn 318 million bricks into the sewer system that still (just about) evacuates London today, a process started in 1859. Bazalgette constructed wide egg-shaped, brick-walled sewer tunnels rather than the narrow

---

[1] http://www.sciencemuseum.org.uk/broughttolife/people/josephbazalgette.aspx.

Making Telecoms Work: From Technical Innovation to Commercial Success, First Edition. Geoff Varrall.
© 2012 John Wiley & Sons, Ltd. Published 2012 by John Wiley & Sons, Ltd.

bore pipes previously used. This allowed the system to cope with subsequent increases in volume. By 1866 most of London was connected to the network.

The system was by no means perfect and still dumped tons of raw sewage into the Thames, which still happens today when it rains killing fish if not people by the thousand. When the Pleasure Boat Princess Alice sunk close to a sewage outflow in 1878, 640 passengers died, some from drowning but the rest were poisoned to death.

The supply side (water mains) also needed a major redesign and rebuild. This resulted in a sewerage infrastructure that is still more or less serviceable 150 years later, albeit in urgent need of a £4 billion upgrade. On the supply side more than half of the water mains are more than 100 years old. Around a third are 150 years old. About a third of our water in London is lost through leaking pipes before it gets to our home. This is inconvenient. London gets less annual rainfall than Istanbul or Rome and we are profligate in the way we use water. Many people in the world exist on 10 liters of water per day. We use this in one flush of the toilet.

Nevertheless, we owe a debt of gratitude to Mr Bazalgette. Apart from ensuring a reliable supply of clean drinking water he also managed to get the foul water from the legacy sewerage system into new low-level sewers built behind embankments on the river front. If you stand on the Albert or Victoria Embankments opened in 1870 or the Chelsea Embankment completed in 1874 you are standing above an intricate system of sewers and subways and railways that remain serviceable today. Mr Bazalgette by the way started life as a railway engineer.

Bazalgette went on to implement similar systems across the UK and in Budapest and Port Louis in Mauritius and trained a new generation of civil engineers who carried on his good work after his death on the 15 March 1891.

Some general points.

This was a massive undertaking but resulted in an infrastructure that has remained serviceable for 150 years. This was only possible because it was massively overdimensioned and paid for from government funds. A 50-year return might be considered acceptable to a pension fund, a 150-year return might be considered over long but the longer the infrastructure lasts the greater the social and economic return. There are still Roman aqueducts in use today.[2]

This is a two-way delivery system and therefore conceptually similar to telecommunications. The 360 000 miles of sewer in the UK have become host to fibre systems over the past ten years[3] but not all of the private sector initiatives have gone according to plan either in the UK or Australia, where as at 2011 legal disputes are in process.

This is almost inevitable considering the short-term investment horizon that the private sector has to work to or chooses to work to and the implicit long-term return of this sort of investment including a social return, in this case health, that cannot be reflected in a profit and loss account or corporate balance sheet.

This is a continuation of a conundrum that goes back at least to Plato and probably much further where public good and private ambition become hard to reconcile.

So what does this have to do with telegraph poles?

---

[2] http://www.romanaqueducts.info/q&a/11stillinuse.htm.
[3] http://news.bbc.co.uk/1/hi/technology/7104011.stm.

14.3   Above the Streets of London – The Telegraph

In 1837 the British physicists, William Cooke and Charles Wheatstone patented the Cooke and Wheatstone telegraph using the principle of electromagnetism to send and detect electrical pulses along a copper cable. The first public telegraph service in the world began in 1839, with two Cooke and Wheatstone machines either end of a 13-mile stretch of cable running 13 miles (21 km) alongside the Great Western Railway from London Paddington to West Drayton.

In parallel, Samuel Morse developed his system of dots and dashes used in the first telegraph from Baltimore to Washington in 1843. The initial idea was to bury the cable under ground but this proved unreliable and the telegraph pole was invented.

The extension of the telegraph system across the US and subsequently in other big places like the Australian outback was dependent on finding sources of local timber for telegraph poles preferably Douglas fir and pine, hard to find in desert areas.[4] The wood then needed to be pressure treated against rot and insect and woodpecker attack. Such apparently mundane material considerations had and still have a direct impact on ongoing operational costs (the replacement cycle of telegraph poles) and cost per bit or in this example cost per dot and dash.[5]

Fast forward 150 years. In March 2010, Virgin Media, a company started by Richard Branson announced[6] plans to trial a 50 Mbps fibre service over telegraph poles to a remote Berkshire village.

This followed a long battle with BT to get access to BT duct space and poles. The arguments for and against illustrate the ongoing dilemma posed by the question as to who really owns those poles. The answer at least morally if not legally is that we do.

14.4   Corporate Success and Failure – Case Studies – The Impact of Regulation and Competition Policy

BT point out on their web site[7] that they are the world's oldest telecommunications company with origins that can be traced back to the Electric Telegraph Company founded in 1846.

The companies and early private investors in the telegraph system mostly ran out of money in a process closely parallel to the railway boom and replicated more recently when the channel tunnel was built with private finance. The channel tunnel is now owned by the banks who have sold the debt to the pension funds that we depend on for our possible future retirement so by default we have paid for the channel tunnel even though we don't own it.

Anyway back to telegraph companies. The public sector had to take over the failed telegraph businesses and became by default a competitor to the surviving private sector companies. The takeover 'vehicle' as it would be described today was the General Post Office, after all this was just another two-way communication service. The GPO was run as a government department.

One of the longest private sector survivors was the National Telephone Company whose trunk telephone service was taken over by the GPO in 1896. In 1912 the GPO became the monopoly supplier of telephone services in the UK. A few local authority services survived

---

[4] More information on this is available from www.telegraphpoles.com.

[5] The Victorian Internet authored by Tom Standage, ISBN 13: 9780425171691 Published 2002, available from Amazon or Abe Books discusses the impact of the telegraph on Victorian society.

[6] http://news.bbc.co.uk/1/hi/technology/8562932.stm.

[7] http://www.btplc.com/Thegroup/BTsHistory/History.htm.

for a while, though the only long-term competitor proved to be the city telephone department in Kingston upon Hull, a town that has managed to successfully defy commercial logic both in telecommunications and bridge building[8] but not in fish.

In 1932 the idea was put forward that the GPO should become a nationalised industry but no progress was made on this until 1965 when a pipe-smoking Anthony Wedgewood-Benn raised the subject with Harold Wilson, then Labour Prime Minister. The result was one corporation split into two divisions: Post and Telecommunications, ratified in the Post Office Act of 1969 that then became two individual corporations with Post Office Telecommunications becoming British Telecom in 1980.

And then everything started changing again. The British Telecommunications Act of 1981 introduced the concept of competition, empowering the Secretary of State to license other operators to run telecommunications systems and accept user equipment supplied by other third parties provided the products complied with performance and quality benchmarks set by the British Standards Institute, a process described then and now as liberalisation. In 1982, Cable & Wireless were licensed to run a public telecommunications network through a subsidiary, Mercury Communications and the government announced its intention to privatise British Telecom, a decision ratified by the Telecommunications Act of 1984.

Two years earlier the UK had awarded two cellular licenses to consortia led by BT and Racal Millicom. The £25k license fee was considered to be expensive by the financial analyst community and Racal's prediction of 250 000 UK cell phone users by 1989, over three times more than the estimates of rival bidders was considered dangerously optimistic. By 1999 Racal (Vodafone) and BT (Cellnet) together had 1.1 million customers. Early business plans were predicated on high-priced mobile or transportable handsets, low penetration (under 2%) and low network cost to achieve profitability. BT and Securicor invested £4 million each in the original business plan.

In July 1999 British Telecom bought Securicor's 40% stake in Cellnet for £3.15bn making the mobile phone operator its wholly owned subsidiary. Cellnet had made £118 million profit the previous year. The business, by then rebranded as O_2, was sold to Telefonica in 2005 for £18 billion.

In that sense liberalisation was good for BT and to all intents and purposes for the government, who sold all remaining shares in the business in July 1993 raising £5 billion for the Treasury and introducing 750 000 new shareholders to the company.

In parallel, in 1991 the government had announced the end of the BT/Mercury duopoly of wire-line provision. Independent 'retail' companies were permitted to bulk-buy telecommunications capacity and sell it in packages to business and domestic users. Feeling squeezed in its local market, BT set off on a series of foreign adventures.

In June 1994, BT and MCI Communication Corporation, the second largest carrier of long-distance telecommunications services in the US, launched a $1 billion joint venture company providing advanced business services. BT subsequently acquired a 20% holding in MCI that was then sold to Worldcom for $7 billion, a profit of more than $2 billion on its original investment.

In July 1998 a 50 : 50 joint venture with AT and T was announced but wound down when the market turned down in 2002. AT and T has subsequently been modest in its international ambitions.

---

[8] http://www.humberbridge.co.uk/.

In December 2000 BT was forced to offer local loop unbundling (LLU) to other telecommunications operators, enabling them to use BTs copper local loops. By the end of August 2005, 105 055 lines had been unbundled but ten years on this remains an uneasy innovation with some hard to resolve technical problems (ADSL and VDSL cochannel interference – see Chapter 1).

In July 2003 The Communications Act introduced a new regulator, the Office of Communications (Ofcom) and a new regulatory framework in which companies could be given a general authorisation to provide telecommunications services subject to general and specific conditions including, for example, universal service obligations.

In the summer of 2004, BT launched Consult 21, an industry consultation for BT's twentyfirst-century network (21CN) programme. Billed as the world's most ambitious and radical next-generation network transformation it promised to transform the communication infrastructure of the UK by 2010 *'Using internet protocol technology, 21CN will replace the existing networks and enable converged multimedia communications – from any device such as mobile phone, PC, PDA or home phone, to any other device'.*

A business called Open Reach was set up to manage the UKs telecommunications infrastructure, treating the rest of BT on an equal basis as other operators. It was/is (at the time of writing) one of four businesses within the BT Group alongside BT Retail, BT Wholesale and BT Global Services. This is presented by BT as signifying its final transformation from 'a traditional telecoms company to a leading provider of converged networked services'.

So, did the UK tax payer get a good deal and has the process of liberalisation generated additional value and if so who owns that value. Well, the transformation of the UKs telecommunications infrastructure did not happen and it is still unclear how BT can finance this and still live with a Universal Service Obligation that includes the requirement to deliver broadband to remote rural areas.

On the 14 May 2009 Felixstowe TV (yes it really exists) had a gloomy announcement

> BT said today that it expects to cut a further 15,000 jobs over the next year, following 15,000 job losses already over the last year (including agency staff) and reported annual losses of £134 million.
>
> The company, which currently employs around 147,000 people, including 3,000 in research and development at Adastral Park in Martlesham Heath, says the losses are mainly due to BT Global Services, which provides IT network services to large companies. BT said that it aims to "work closely with unions" and that it has no plans for compulsory redundancies.

14.5   The Correlation of Success and Failure with R and D Spending

The Post Office had bought a disused airfield in Martlesham in 1968 as a new home for what had previously been the Dollis Hill Research Centre (in Dollis Hill in North West London). BT Martlesham built on the Dollis Hill legacy to create arguably one of the best telecommunication research centres in the world.

In 2006 BT was still able to claim a world leadership position in R and D[9] and was able to claim second position to NTT in terms of absolute spend though with this equating to 3.7% of

---

[9] http://www.btplc.com/Innovation/News/investment.html.

BTs turnover and 2.9% of NTTs turnover. The R and D spend had doubled from a 2001 level of £361 million.

To quote the 2006 DTI Press release;

BT director of research and venturing Mike Carr said: *"Our increased R&D spend is a clear indication of our success in transforming BT from a traditional telco to a networked IT services company through innovation."*

Mike said there were three reasons why BT has invested more in R&D: the increasing level of networked IT business it performs; the massive investment in the 21st century network; and the new products and services it develops – such as the digital TV service BT Vision, the wholesale mobile entertainment service BT Movio and the ground-breaking fixed-to-mobile service BT Fusion.

These statements are always a hostage to fortune.

This depressing Press Release on the 26 November 2010 tells the story for us;

The Department for Business Innovation and Skills R and D scoreboard, the twentieth and last version of the annual publication, was published on Thursday. BIS is scrapping the scoreboard as part of wider government cuts. In the words of science and innovation minister David Willetts, in the introduction to the 2010 report, the publication is no longer necessary because "today's companies better understand the importance of R&D to their long-term success".

In the fixed-line telecoms sector, investment was 48 percent less than its operating profit. Companies in the sector cut R&D spending by just over eight percent – this closely mirrored the decrease in BT's R&D expenditure – but its operating profits grew by 513 percent. In the same period, sales in this sector fell by 2.7 percent.

Both BT and Vodafone recorded an increase in operating profit in 2009, of 370 percent and 64 percent, respectively," the report added.

"Similarly to 2008, R&D expenditure in the UK software and computer services sector grew more quickly than sales in 2009,".

This suggests several things.

Privatised companies have to deliver short-term returns to shareholders and the management team are given substantial incentives to achieve these returns.

A short-term mechanism for achieving this is to reduce R and D expenditure. Additionally for BT, this R and D effort is now diluted across a far wider base that has to encompass a broad and unfocused ICT agenda rather than just the mechanics of telecoms delivery.

BT Movio and BT Fusion both turned out to be a poor use of R and D resources. BT Vision has, however, been a success[10] and combined with a return to financial health of the global services business explains BTs profit recovery. An additional 30 000 job losses over the two-year period to February 2011 have helped as well.

Thursday 3 February 2011 18.04 GMT

The group's global services division was forced to make huge write downs following cost overruns two years ago which pushed BT into the red during 2008/9.

However in the three months to end-December, the offshoot produced underlying profits of £141m and cut costs by 8% over the last 12 months.

---

[10] http://www.guardian.co.uk/business/2011/feb/03/bt-telecommunications-profits-rise.

BT added another 188,000 broadband customers in the third quarter bringing the tally to 5.5m, consolidating its position as Britain's biggest broadband operator, comfortably ahead of arch-rivals Virgin and Sky. Analysts say O2 and Orange may have lost market share.

In the third quarter to December 31, group pre-tax profit rose 30% to £531m, while free cash flow increased by 70%

Ian Livingston, chief executive said: "We have had a great few months with our highest share of broadband additions for eight years."

BT's broadband television service BT Vision added 40,000 customers in the quarter, compared to 24,000 additions in the previous three months. This brings the total to 545,000.

BT reduced its pension deficit from £7.9bn at 31 March 2010 to £3.7bn at December 31 2010, thanks in part to a soaring stock market.

Livingston said BT is making good progress in the introduction of its superfast broadband network; the company is spending more than £2bn to build a high-speed network to reach 15m homes by 2015.

In June 2011 BT Vision started providing access to BBC iPlayer as an IPTV service providing access to 400 hours of BBC TV and 1000 hours of radio programmes. This suits all parties reasonably well. The BBC gets a wider audience and another distribution channel that they don't need to pay for. A bigger audience helps make the case for a continuation of public service funding but also helps create income opportunities for BBC enterprise. BT Vision get to sell more higher-value subscriptions. So what does all this have to do with telegraph poles?

14.6   Broadband Delivery Economics and Delivery Innovation

The economics of broadband delivery are dependent on delivery innovation, which in turn is dependent on research and development. BTs research and development spending has at least in part been sustained by the profits made from a monopoly and then duopoly position in which the incumbent (BT) retained legacy advantage.

This advantage was then translated into a gain to the treasury through the privatisation process, though most of this money was then absorbed into higher priority spending including defence. A similar fate befell the income from the UK spectral auctions.

The grand plans to transform the fixed-line infrastructure by 2010 have slipped five years to 2015 and the government's exhortation that BT should deliver broadband to rural areas has minimal collateral given that the company's first duty is now to its shareholders who have no direct reason to want to undertake a loss-making venture. The only way the government could provide incentives to BT to bridge this 'digital divide' would be to trade a relaxation of competition policy against an extended broadband universal service obligation.

And that is where telegraph poles may come into play as key enablers. Installing fibre optic on telegraph poles requires specific changes to existing planning laws and will then require significant investment that BT is unlikely to want to make if forced to facility share with other third parties.

To reuse the well-worn motivated donkey aphorism, regulatory and competition policy is therefore the carrot and the stick in a new world of liberalised telecommunications provision.

Its effectiveness is yet to be established.

15

Terrestrial Broadcast/Cellular Network Integration

15.1 Broadcasting in Historical Context

In 1923, Captain PP Eckersley,[1] the BBCs first chief engineer launched a new era in broadcasting at Daventry in the Midlands. The first years of radio had individual stations transmitting programmes to individual cities across the country, from London to Glasgow.

The BBC transmissions in London for example were initially from Marconi House in the Strand, using a transmitter located in an attic room with aerials strung between towers on the roof adjacent to Bush House, home of the BBCs World Service. The transmission call sign was 2LO named after the transmitter illustrated in Figure 15.1.

Eckersley decided to have total coverage of the UK from one powerful transmitter located in the centre of the country. The site chosen had an ancient anglo saxon oak supposed to mark the exact centre of England, the Daven Tree, hence Daventry. The site was bought in 1924 and in July 1925 the world's first high-power low-frequency broadcasting transmitter was installed, known as Daventry 5XX and nicknamed the 'old gentleman' by its team of engineers. The 26-kW transmitter was the size of a tennis court and powered by two diesel generators each as big as a double-decker bus.

The transmitter achieved 85% population coverage, better than Eckersley expected. Within a few years smaller transmitters were built across the country but most radio receivers continued to have a mark on the dial saying 'Daventry'. Foreign short-wave transmissions began at the site in the 1930s and the engineers received Christmas presents from listeners, including tea from India and coffee from east Africa. If locals had an earth fault on their telephone system, they would hear the BBC Far Eastern Service in Malay or Thai when they picked up the

---

[1] Before being appointed as the BBC's first engineer, the multitalented Peter Pendleton Eckersley had been the producer, writer, and presenter of the first experimental half-hour programme that the Marconi company was licensed to broadcast each week from a studio in a former army hut in Writtle near Chelmsford. The programme was launched 'on air' on St Valentine's Day, 1922. The Captain had a gift for improvising on microphone. He became Britain's first radio star with an instantly recognisable call sign 'Two-Emma-Toc, Writtle testing' http://www.marconicalling.com/museum/html/people/people-i=17.html.

Making Telecoms Work: From Technical Innovation to Commercial Success, First Edition. Geoff Varrall.
© 2012 John Wiley & Sons, Ltd. Published 2012 by John Wiley & Sons, Ltd.

Figure 15.1   2LO transmitter.[2] Reproduced with permission of the Science Museum. From 2015 onwards the 2LO transmitter will be on display in the new Making of Modern Communications Gallery at the Science Museum in Kensington.

phone.[3] The Daventry site also saw the birth of radar, after transmitters there were used to test if aircraft could be spotted by bouncing waves from them.

The transmitters were retired in March 1992 after 67 years of service. Most of the land where the antennas had stood was sold off, reverting to farmland or being used to build houses. The few remaining buildings were initially used for storage, before being sold to Crown Castle in 1997 when the BBC transmitter sites were privatised.

These long-wave and short-wave systems were an example of high-power wireless, beautifully engineered with frequency synthesisers built into seven-foot tall 19-inch racks. To achieve an 85% coverage was remarkable given that Northants had (and still has) poor ground conductivity, the combination of sandstone and a low water table), Orfordness, set in salt water was in later years to prove more effective. However, the site is also home to a DAB broadcast

---

[2] The author is unsure of the provenance of the 2LO description. Possibly the transmitter had two local oscillators or perhaps the call sign was meant to be Hello Hello- answers on a postcard please.

[3] Thanks to Richard Lambley of Land Mobile Radio Magazine for sharing some of his early childhood memories of living close to the Daventry transmitter.

transmitter supporting local radio and localised infocasting and data casting, so may be playing a role in a newly developing radio network offer.

Similarly this bit of history might remind us that a rebirth of short-wave, medium-wave and long-wave long-distance transmission may be possible. Unlike the Voice of Russia, the BBC did not cut back on short-wave transmission and over the years, certainly up to a few years ago listeners steadily increased both in developing and developed countries. In 2010 the Foreign Office in the UK announced significant cut backs in World Service funding. A year later the decision was reversed and the Hindi, Arabic and Somali services were all reprieved. 'Political radio' and political broadcasting remains alive and well.

15.2   Digital Radio Mondiale

In the late 1990s, The Digital Radio Mondiale (DRM) Consortium[4] successfully lobbied for an international standard for digital AM for frequencies below 30 MHz. Digital AM is comparable to FM mono in terms of sound quality and can be broadcast with a footprint of 1000 miles from one transmitter. DRM can be used for a range of audio content, including multilingual speech and music, and has the capacity to integrate data and text. The DRM signal is designed to fit in with the existing AM broadcast band plan, based on signals of 9 kHz or 10 kHz bandwidth. It has modes requiring as little as 4.5 kHz or 5 kHz bandwidth but also includes modes that can take advantage of wider bandwidths, such as 18 or 20 kHz with applications that include fixed and portable radios, car receivers, software receivers and PDAs. Remember that 'wideband' is a relative term and 20 kHz bandwidths below 30 MHz are to all intents and purposes 'wideband' in terms of their application potential.

Whether DRM can survive competition from the internet in the longer term may be subject to debate but this is a supremely energy-efficient mechanism for delivering information globally and should not therefore be discounted. It is not just radio but can distribute text, pictures and html files. As we shall see in later chapters it may also have a role to play in public safety and disaster relief. Short-wave radio irrespective of whether it is analogue or digital will almost certainly continue to be important politically and educationally. The National Association of Short Wave Broadcasters[5] issues a monthly newsletter highlighting examples of the difference that short-wave radio continues to make particularly in countries where unrestricted access to the internet is not allowed.

15.3   COFDM in DRM

The DRM system uses COFDM with the number of subcarriers variable to suit channel allocations and required range (transmission resilience). There are three different types of audio coding. MPEG4 AAC audio coding is used as a general-purpose audio coder and provides the highest quality. MPEG4 CELP speech coding is used for high-quality speech coding where there is no musical content. HVXC speech coding can be used to provide a very low bit rate speech coder. DRM is therefore effectively a (very) wide-area version of DAB and DMB and was fully ratified as an ETSI specification (the ETSI ES 201 980 V1.2.2

---

[4] www.drm.org.
[5] http://www.shortwave.org/.

(2003/2004) DRM system specification and 101.968 V1.1.1 (2003–4) data casting standard. In parallel, the ITU ratified DRM for use in the medium-wave AM and long-wave frequency bands in Regions 1 and 3 (Europe, Africa, the Middle East, Asia and Australia/New Zealand).

15.4   Social and Political Impact of the Transistor Radio

Integrating broadcast receivers into small form factor hand-held devices is not particularly new as a design concept. There is some dispute as to who produced the first 'pocket-size' transistor radio. A product called the 'Regency' introduced in October 1954 came to market in parallel with early Texas Instruments pocket radio receivers. This triggered a 'form factor' race with Sony producing the 'world's smallest radio'[6] (March 1957). Two examples are illustrated in Figure 15.2.

Form factor and functionality is still directly dependent today on device geometry. In 2005 'scalability roadmaps' suggested that 95-nm devices would be replaced with 70-nm and 65-nm devices that would be replaced with 32-nm devices and this has duly happened. Device geometry determines DSP, microcontroller and memory functionality that in turn determines what we can do in a small form factor power limited handset. This in turn enables us to realise system value, which in turn realizes spectral value. Put another way, device bandwidth directly translates into application bandwidth, which directly translates into system value, which directly translates into spectral value.

Figure 15.2   Sony TR55 and TR63 pocketable transistor radios. Reproduced with permission of Sony Corporation.

---

[6] http://www.sony.net/SonyInfo/CorporateInfo/History/sonyhistory-b.html.
http://www.sony.net/SonyInfo/CorporateInfo/History/.

In developing countries it could be argued that mobile phones are replacing, or will replace, the transistor radio as the lowest common denominator way of accessing information and delivering information. Governments have started to use SMS texting to distribute political messages, a parallel delivery route to TV and national radio so it could be argued that this process of transition is already well underway.

On the other hand, transistor radios are, however, significantly simpler than cellular phones, for instance they don't have a transmitter. This means that they have a lower component cost and less battery drain. For someone earning a dollar a day or less even the most basic mobile phone remains an unaffordable luxury. Almost anyone can afford to own and use a transistor radio. Short-wave transmissions like satellite transmissions also travel effortlessly over geographic boundaries. Partially successful attempts to jam short-wave broadcasts in some countries suggest that some governments remain acutely aware of the potency of this often overlooked delivery option.

Earlier, we mentioned that the UK sold off the BBC transmitter sites in 1997 and in common with many other countries is presently preparing to auction TV spectrum released by the transition from analogue to digital terrestrial TV. In a sense this seems to be trading short-term treasury gain against a potential loss of control over national TV and radio broadcasting, with a risk that political influence shifts to third parties, owners of satellite broadcasting networks being one example. In the US the National Association of Broadcasters[7] is presently lobbying against the proposal by the FCC to take away an additional 120 MHz of TV broadcasting spectrum. 108 MHz, channels 52 to 69 had been taken away and auctioned in 2007. The additional spectrum would take out channels 31 to 51 that would mean (according to the NAB) that over 600 of the 1735 full-power radio stations would either need to close down or move to VHF, which would be rather impractical.

Looking at this globally, Table 15.1 provides an overview of the 500 MHz of broadcast (radio and TV) bandwidth available between long wave and 2 GHz. Table 15.2 lists the associated digital broadcast radio system options.

Table 15.1   Frequency allocations and digital broadcast system options

Radio Band	Frequency	System Options
Long Wave	3 kHz–300 kHz	DRM (Digital AM)
Medium Wave	300 kHz–3000 kHz (3 MHz)	DRM (Digital AM)
Short Wave	3 MHz–30 MHz	DRM (Digital AM)
VHF (FM radio)	88–108 MHz	DAB/DMB
VHF Band 3	174–233 MHz	DAB/DMB (218–230 MHz)
UHF Band 4	470–490 MHz	DVB-H, ISDBT, Media FLO
UHF Band 5	790–862 MHz	DVB-H, ISDBT, Media FLO
L Band	1452–1492 MHz	DAB/DMB
	1670–1675 MHz	DVB-H

---

[7] http://www.nab.org/.

Table 15.2 Broadcast receivers as at 2005 through to 2010

Broadcast receivers	Functionality
Analogue AM	Voice and low-bandwidth audio
Analogue FM	Voice, stereo radio, text and images (visual radio)
DRM	Voice, audio, text, data, images
DAB	Voice, audio, text, data, images, video
DVB-H	Voice, audio, text, data, images, video
ISDBT	Voice, audio, text, data, images, video
Media FLO (integrated with 1XEV)	Voice, audio, text, data, images, video
EDGE (Dual transfer mode).	Voice, audio, text, data, images, video
HSPA	Voice, audio, text, data, images, video

15.5   Political and Economic Value of Broadcasting

Broadcast spectrum has (always had) political value (Hitler, Mussolini, Churchill, Berlosconi), social value, evangelical value (Vatican radio) and economic value.

Economic value is a composite of license fee income (BBC in the UK), advertising revenue and (increasingly), participation bandwidth revenue. Participation revenues are a composite of text and voice value and (increasingly) image and video value, anything that works on a game show.

The realisation of fiscal value from spectrum measured in dollars per hertz or euros per hertz is dependent on delivering robust radio systems with adequate link budgets that ensure enough flux density to support consistent good quality reception of voice, audio and video content. This in turn is dependent on realising good receiver sensitivity, selectivity and dynamic range in TV and radio receivers. The earlier chapter on smart televisions (Chapter 8) also highlighted the value realisable from a return channel, in that example a broadband internet connection.

Note how participation revenues (which in some countries like Finland now exceed advertising revenues) are closely dependent on the robustness and consistency of this return channel. Participation revenues, in theory should increase as participation bandwidth increases. (This may not be true for participation margins where it will be hard to match the margin achievable on SMS voting and texting in terms of euros per hertz or euros per delivered megabyte.)

Note also the blurring of definitions between audio and video and text in these broadcast radio systems – visual radio with text is essentially competing directly with wider bandwidth digital TV. Any/all of the broadcast options are capable of triggering uplink bandwidth value.

So we are interested in the performance of the receiver in terms of sensitivity, application bandwidth and power efficiency and the performance of the receiver in the presence of locally generated transmit power (the return channel). We should also remember that we can also deliver radio and TV channels over existing cellular radio bandwidth.

This means that our choice of broadcast receivers could include existing analogue AM or FM radio, digital AM (DRM or equivalent), DAB/DMB, DVB-T or HSPA or EVDO mobile broadband products. Up until 2010 there was the additional option in the US of a system optimised for mobile reception known as Media FLO (forward link only) developed and promoted by Qualcom and an equivalent version of DVB T known as DVB H (handheld).

Both these systems were vigorously promoted from 2005 and both failed for reasons we need to examine.

A number of phones included and still include integrated FM tuners. These work as well as most FM tuners when used in mobile applications, that is quite well sometimes but not consistently well in weak-signal or high-mobility conditions.

There is nothing wrong in principle in adding digital subcarriers to existing analogue bandwidth and using these carriers for data and image transmission – an analogue and digital multiplex. This provided a perfectly adequate basis for first-generation 'visual radio' systems, for example in Finland.[8]

In practice, over the longer term, it should be possible to get better bandwidth and power efficiency and a more consistent user experience from a digital broadcast receiver. The same applies for analogue AM systems.

All of the digital broadcast systems should in theory and generally in practice provide a performance benefit in terms of quality, consistency and bandwidth and power efficiency. DRM would be (relatively) easy to integrate into present cellular handset form factors and dongle products were introduced[9] but failed to gain mass-market traction.

A DRM receiver works well under a wide range of operating conditions and is power efficient. It is, however, bandwidth limited and (at present levels of compression) incompatible with any existing TV content standards. There is a wide range of global programming available, Vatican Radio being one example. Up-to-date information on DRM receivers and DRM programming can be found on the Digital Radio Mondiale web site.[10]

15.6   DAB, DMB and DVB H

DAB/DMB is potentially useable across three radio bands – the VHF FM band, band 3 at 220 MHz and L band at 1.5 GHZ. It is MPEG2 TS and MPEG4 Part 10 compliant and supports useful features like multimedia object transfer, so it's definitely not now just an audio delivery system. At 2011 there are only 15 countries with DAB networks deployed, which is not enough to achieve global mass market adoption.[11] It does work well in the UK and despite the limited global uptake there are nice receiver products available.[12]

DVB-H has proved more problematic. The intention was that cellular phones would have mobile TV delivered via a DVB H physical layer either at Band 3 VHF, Band 4 or Band 5 UHF. Crown Castle purchased a 5-MHz bandwidth allocation in the US between 1670 and 1675 MHz and in 2005 started a DVB H trial in Pittsburgh.

Technically, DVB H was quite attractive. DVB-H increased the power efficiency of the receiver by time slicing. If multiplexed with a DVB T signal for example, a 2-Mbit burst would be taken out of the DVB T 15-Mbps stream and sent in a 146-millisecond burst. The receiver then powered down for just over 6 seconds then powered up for the next burst. The 2-Mbit burst was read into and out of a buffer at a constant 350 kbps.

---

[8] http://www.radioandtelly.co.uk/visualradio.html.

[9] For example, a product from Coding Technologies subsequently acquired by the Dolby Corporation www.dolby.com/professional/technology/broadcast/he-aac-dolby-metadata.html.

[10] http://www.drm.org/products.

[11] http://www.worlddab.org/country_information.

[12] http://www.pure.com/.

However, while this optimised the receiver power budget, it saved less power than might seem immediately apparent (the baseband processor still had to work pretty hard) and there had to be careful (and fast) synchronisation with the continuous and scattered pilots modulated on to the OFDM signal.

Extended receiver power down was, however, a well-understood technique and had been used in paging systems for at least 30 years, so it should have been possible to make this work satisfactorily. The time slicing also allowed for neighbour measurements and mobile initiated handover in multifrequency networks. This would have allowed a high-density DVB H network to be overlaid onto an existing cellular network.

There were, however, some practical issues when implementing a DVB H receiver in to a cellular handset. Producing a DVB H front end capable of accessing low band VHF, UHF (and L band) was challenging in terms of antenna design resulted typically in negative antenna gain of the order of between 5 and 10 dBi.

Equally problematic was the issue of GSM (or equivalent cellular) transmit power and wideband noise from the TX PA desensitising the DVB H receiver. This implied either a careful choice of DVB H channel allocation and/or some carefully designed (and potentially expensive and lossy) filtering.

There were various ways in which DVB H receiver sensitivity could have been improved, for example by using antenna diversity or additional time diversity (using the optional MPE FEC encoder) but this would not have overcome the problem of locally generated interference within the handset. Diversity gains within present handset form factors would also have been marginal. The other problem was that people didn't want to watch TV on their handsets, particularly if they had to pay for it and the battery went flat. Additionally there was scant incentive for operators to bear the cost of delivering content which they did not own and with no associated revenue stream.

ISDBT had similar problems to overcome. ISDBT, specified originally for the Japanese market, is arguably the most scalable of the present digital broadcast OFDM systems. It divides a 6-MHz channel into subchannel segments each of just over 400 kHz each segment has a variable OFDM multiplex, variable modulation (QPSK, 16 QAM, 64 QAM) and variable levels of convolutional coding. Table 15.3 summarises these capabilities.

Although the system promises substantial deployment flexibility, performance (as with DVB H) will be dependent on achieving successful (RF) integration into existing and future cellular transceivers and global scale. Updated adoption status by country can be found on the ISDBT web site.[13]

Media FLO was Qualcomm's proprietary offering for broadcast receivers. It shared some common techniques with DVB H in that it used a time division multiplex, with one or more traffic packets being transmitted within a reserved slot to all users in the service area. Users received the same packets from multiple cells. An OFDM multiplex was used to slow down the symbol rate and the symbols were soft combined to improve downlink performance.

Media FLO was deployed into Channel 55 in the UHF band. It had the merit of being intrinsically compatible with existing and future 1X EV DO networks but no inherent compatibility to the US TV ATSC system. The deployment of Media FLO prompted ATSC to develop a mobile optimised ATSC implementation. The market failure of Media FLO and subsequent

---

[13] http://www.dibeg.org/.

Table 15.3   ISDBT

Mode	Mode 1	Mode 2	Mode 3
Number of segments	One or three		
Bandwidth	432.5 kHz (one segment) 1.289 MHz (three segments)	430.5 kHz (one segment) 1.287 MHz (three segments)	429.5 kHz (one segment) 1.286 MHz (three segments)
Carrier spacing	3.968 kHz	1.984 kHz	0.992 kHz
Number of carriers	109 (one segment) 325 (three segments)	217 (one segment) 649(three segments)	433 (one segment) 1297 (three segments)
Modulation	QPSK, 16QAM, 64 QAM, DQPSK		
Number of symbols per frame	204		
Effective symbol duration	252 microseconds	504 microseconds	1.008 milliseconds
Guard interval	1/4 1/8 1/16 1/32		
Inner code	Convolutional 1/2 2/3 3/4 5/6 7/8		
Outer code	RS (204,188)		
Information bit rate	One segment 280.85 kbps to 1.7873 Mbps Three segment 0.842 to 5.361 Mbps		

sale of Channel 55 to AT and T in 2011 has not provided much motivation to move forward at all quickly on the mobile TV front.

Similarly, it would have been feasible in 2005 (though not necessarily economically attractive) to use EDGE as a broadcast radio bearer. A function known as dual transfer mode would have supported some 'ring-fenced' broadcast bandwidth within the existing slot structure. There are already broadcast packet channels for signalling bandwidth, so it would not have been a great leap to deploy broadcast packet channels for broadcast content and there were a number of receiver optimisation techniques such as joint detection (single antenna interference cancellation) that could be used to improve receiver performance.

15.7   HSPA as a Broadcast Receiver

HSPA was a more persuasive candidate. Release 7 included specific work items on receiver optimisation using advanced diversity and equalisation techniques. The pilot symbols on the channel can be used in a similar way to present OFDM-based broadcasting to provide active channel characterisation). The HSPA MAC (2-millisecond-based admission control) is also more IP friendly than the Rel 99 MAC and arguably more power efficient for broadcast reception). The less than enthusiastic market response to mobile TV by this time, however, meant that little was done in practice to commercialise this as a product offer.

There were also a number of deployment issues that arise from the inescapable fact that legacy cellular networks were not and are not optimised to support broadcasting applications (other than for signalling functions).

Traditionally (in the 1980s and 1990s), cellular networks were radio engineered to provide a balanced link budget. Given that handsets had less power available than base stations, it was usual to provide some additional link budget gain on the uplink. A typical cell site might provide for example 17 dB gain on the uplink using diversity and/or dual polarisation antennas and 9 dB of gain on the (sectored) downlink.

Similarly in the access network, the A bis and Gb interface was designed to support symmetric bidirectional (largely duplex voice-based) traffic. In contrast, digital broadcast networks used (and still use) ATM to deliver a one way traffic multiplex that is (completely) asymmetric and highly asynchronous (variable rate). The Rel 99 air interface addressed this issue by introducing an ATM-based radio layer (10-millisecond frame-based admission control) and an ATM-based radio access and core network.

The debate, however, then moved to focus on the future role of IP in broadcast networks. DVB-H for example was explicitly being positioned as an IP-based broadcast system solution. This meant that the IP subsystem at the transport layer had to behave at least as predictably as an ATM-based transport layer, or in other words, the IP subsystem had to replicate ATM end-to-end functionality (the ability to control a complex multiplex in the time domain). This was not something that IP networks were ever designed to do.

The integration of flexible network bandwidth provisioning and QOS based broadcasting was particularly challenging and needed to encompass transport and network design, server architectures and performance and content management systems.

In 2007 at the annual GSM World Congress trade show, mobile TV was everywhere. By 2011 it had all but disappeared. Is mobile TV dead or dormant and how will the success or failure of portable TV over the next few years impact the mobile TV offer?

Apart from users not wanting it, mobile TV market adoption has been frustrated by multiple standards illustrated by the muddle described above and a failure to harmonise spectrum on a regional and national basis. These together mean that scale economy could not and probably cannot be achieved.

15.8   Impact of Global Spectral Policy and Related Implications for Receiver Design and Signal Flux Levels

Global spectral policy is set every five years at the World Radio Congress, the next one, when this book was written, being WRC 2012. Five years earlier WRC07 produced policy on the following areas:

- The extension of Digital Radio Mondiale from long wave, medium wave and short wave to include the FM broadcast bands.
- The extension of the present T-DAB/DMB multiplex in Band 111 to include160 kbit enhanced audio transmissions and related European digital TV and DMB deployments.
- Agreement on additional HDD broadcast DTT DVB UHF transmissions and additional DVB H multiplex options.
- Finalised plans for allocating and auctioning 112 MHz of UHF between 470 and 862 MHz for cellular FDD transmission.
- Agreement on L band allocations between 1452 and 1492 MHz including potential DAB extensions and cellular FDD or TDD opportunities.

- The repurposing of S Band for DVB -SH and related hybrid satellite and terrestrial broadcast delivery options.
- International agreement on the 2.5-GHz cellular extension band from 2500 to 2690 MHz.

The emphasis was on producing harmonised band allocations that could deliver sufficient economies of scale to support new cost- and performance-competitive cellular transceivers with integrated broadcast receive capabilities. Some of this spectrum was made available at the expense of present users, including broadcasting and programme making and special events (radio microphones and professional wireless cameras).

The repurposing of this spectrum therefore implied significant engineering challenges and related business transition management opportunities. The repurposing of the upper end of the TV bands for mobile broadband has proved particularly problematic both technically and commercially. The difficulty revolves around the risk of hole punching. Hole punching is when a TV receiver is unable to receive a TV programme because it has been desensitised by a locally proximate mobile broadband device. One answer is to improve the dynamic range and selectivity of TV receivers but there are a lot of them already out there. Also, the problem is rather subtler than just dynamic range.

A DVB or ATSC UHF tuner will generally be designed to tune across the band. It has to do this because digital broadcast channels could be anywhere in the band depending on the country in which the device is being used or the part of the country in which it was being used.

The alternative is to produce country- or region-specific variants that can just work on the locally available channels. The disadvantage of producing country-specific or regionally specific variants is that it is difficult/ impossible to achieve economies of scale because addressable markets are too small.

The disadvantage of tuning across the band is that it is hard/ impossible to deliver good sensitivity from the device and the device is vulnerable to interference from other inband transmissions, for example from other mobile cellular handsets.

The difference is significant. For example, a digital broadcast tuner working across the whole UHF band would typically have a sensitivity of between -92 and -98 dBm. A UMTS handset working across a much narrower band, for example 25 MHz rather than 400 MHz could have a sensitivity of better than -120 dBm.

Admittedly this is not a like-for-like comparison as the UMTS handset is working in a 5-MHz channel and the digital broadcast channel is 6, 7 or 8 MHz, but the difference is in any case substantial. And the difference is important. One of the lessons learnt from the BT Movio experience in the UK and Modeo experience in New York, two examples of mobile TV trials that failed is that users with an integrated cellular/mobile TV and radio receiver expect to listen to digital radio and for that matter to listen and watch digital TV wherever and whenever they have cellular coverage. The present sensitivities and selectivity available from digital TV and radio tuners are not sufficient to meet that experience expectation.

One option is to increase broadcast transmission power levels. To an extent this is happening as digital switchover proceeds. In the UK for example digital TV signals were increased by 7 dB as analogue signals were turned off in a particular area.

The second option is to retransmit digital TV and digital radio signals from cellular transmitters. The objective with options 1 and 2 is to ensure that received signal strengths as seen by portable and mobile devices, in other words the flux density, are similar irrespective of whether the received signals are cellular or digital broadcast, a solution that has come to be termed as a cooperative network. An advantage of this is that portable televisions would also work and/or lap tops with DVB T or ATSC demodulators could provide users with cost- and energy-efficient linear TV programming. The link budget and bandwidth needed for HD, and in the longer term super HD, are particularly challenging for terrestrial broadcasting.

The broadcasting industry and cellular industry have been mast sharing and site sharing for over 30 years. The broadcasting and two-way radio industry have been mast sharing and site sharing for over 50 years. More recently, there has been a trend for cellular network operators to share mast space, Vodafone and Orange in the UK being a recent example. There is therefore an extensive body of knowledge and experience that can be applied to ensure that cellular, two-way radio and broadcast transmissions can coexist.

From a business perspective this is also good news for manufacturers of RF conditioning products. Mast sharing and site sharing has historically been based on the two-way radio and cellular industry being given, leased or sold space on broadcast transmitter masts. Increasingly we may now see the broadcast industry being given, leased or sold space on cellular macro- or microsites for local digital TV and local digital radio transmission.

The broadcast and cellular industries have a strong engineering common interest. Generally, when a strong engineering common interest exists there is also a strong commercial common interest, though sometimes this is not immediately apparent to the parties and vested interests involved. The broadcasting industry also has extensive experience of implementing OFDM-based physical layer transmission networks.

The consensus to emerge from WRC2007 was that mobile broadband would take three to five years to implement at the top end of the UHF band. This has proved overoptimistic, though commercial networks are now launched in parts of Europe including Germany and in the USA (AT and T and Verizon).

Engineering issues include managing coexistence between very different radio network topologies working at very different powers. In the days of omnipresent analogue TV, a country the size of the UK could be covered from 50 large high-powered transmitter sites and 1100 lower-power relay sites. This provided the statutory required coverage of 98% of the population including all communities with more than 200 people. Getting 90% coverage for a 3G network at 1900 MHz takes over 7000 base stations.

Major urban conurbations such as London were and still are covered substantially by one TV transmitter. For example, Crystal Palace in London broadcasts to a 'local' population of 8 million people. The analogue transmitter delivers one megawatt, cosharing mast space with DVB, DAB, FM and medium-wave transmissions. The station has a range of 60 miles for analogue TV and 30 miles for digital TV. HDTV broadcasts started from the site in May 2006.

Digital TV transmitters typically have effective radiated power outputs (the power that is delivered as a result of the transmitter power plus antenna gain) ranging from 2 to 50 kW. Crystal Palace is 20 kW. These output powers though relatively modest compared to analogue TV are still substantial when compared to cellular base stations. There is also a height-gain effect that increases effective coverage.

This provides economic benefits in terms of rural coverage and urban in building penetration and a benefit to users in terms of reduced DC power drain when viewing digital TV content. However, this in turn depends on network density.

15.8.1   Network Density and Cooperative Network Opportunities

There is a trend in digital TV broadcasting towards implementing an increasing number of lower-power sites, particularly in urban areas. Many of these potential sites are owned by cellular operators or administered, owned and/or managed by common third parties such as Crown Castle.

A faster than expected transition to high-definition TV including high-definition TV and an earlier than expected transition to super-high-definition TV strengthens the rationale for providing TV broadcast bandwidth from smaller sites previously optimised for cellular. The same principle could apply for local digital TV and digital radio broadcasts that could potentially be supported on interleaved spectrum.

Successful cosharing is, however, dependent on closer integration of broadcast and cellular technologies (DVB in Europe, ATSC in the US) and now LTE mobile broadband and legacy cellular networks and effective mitigation of shared spectrum interference.

15.9   White-Space Devices

This brings us to white-space devices.

Terrestrial TV networks are either implemented as single-frequency networks or more usually multifrequency networks. The reason for implementing multifrequency networks is that these are or at least have been high-power networks and the signals from a dominant mast can travel tens of miles and in some propagation conditions hundreds of miles. Compared to cellular networks (relatively low powered with transmit antennas installed generally at a much lower height than TV) TV networks are conservatively planned, which means that interleaved channels are often unoccupied in some places at some times. White-space devices are cognitive devices that are supposed to detect or be told when they can and cannot use this unused bandwidth.

On 4 November 2009 the US chose a new president and the FCC voted to allow white-space devices to be deployed into the US UHF TV band. This simple decision has potentially profound consequences for the broadcasting industry, the cellular industry and other communities with an interest in how regulated and unregulated and licensed and unlicensed spectrum is used in the future, both in the US, the UK, Europe and Asia.

The white-space ruling is the direct result of robust petitioning from the Wireless Innovation Alliance[14] supported by eight technology companies, Microsoft, Google, Dell, HP, Intel, Philips, Earth link and Samsung Electro Mechanics.

White-space devices based on spectrum-sensing principles already applied in the 5 GHz U-Nii band have been heralded as the birth of a new age in fixed, portable and mobile broadband access. In addition to the core companies in the WIA, other companies such as Motorola are actively promoting white space for municipal wide area WiFi. Alternative suggestions include the use of the spectrum for low-cost cellular wireless backhaul. The deployment of these devices is, however, controversial. Existing users in the band including TV and wireless microphones contend that interference issues remain unresolved.

---

[14] http://www.wirelessinnovationalliance.org/.

Cellular operators who have either already invested in the band (AT and T and Verizon in the US) or are preparing to bid in future 700- and 800-MHz UHF auctions question the impact that white-space devices will have on present and future mobile broadband business plans.

Additionally, white-space devices could be deployed in other bands in other countries. This could be perceived either as a threat or an opportunity for cellular operators and the service provider community.

The assumption is that as this is unlicensed spectrum, wireless access can be provided on a more cost economic basis, for example for free. In comparison, broadcasters using this spectrum have to cover their costs through licence fee or advertising income. Cellular operators using this spectrum have to recover their costs by charging an access fee.

The assumption that unlicensed spectrum is inherently more cost economic than licensed spectrum is only partly true and dependent on certain preconditions. The 2.4-GHz and 5-GHz ISM spectrum is 'free' in as far as there are no license fees that need to be paid, but there are still access costs that need to be recovered.

In domestic and SME applications, costs are limited to buying a wireless router and paying for an ADSL connection. Hardware costs are low because the ISM band is, more or less, a global allocation and this has allowed vendors to realise substantial economies of scale.

However, in other applications, for example public WiFi in hotels, airports, trains and hospitals, significant hardware installation and real-estate costs can be incurred that can in turn prompt aggressive access pricing. Similar constraints would apply to white-space devices.

For white-space devices to be successful they must have similar scale economies to existing WiFi and WiMax and cellular handsets.

The devices must support equivalent or faster data rates than existing cellular products and/or cellular handsets and when used as portable devices should use similar amounts of DC power. However, the receive and transmit functions have to work across operational bandwidths that are far wider (when expressed as a percentage of the centre frequency of the band) than all other existing mainstream radio systems. Additionally, the receiver dynamic range needed from these devices is substantially more than cellular or WiFi handsets.

This introduces a number of practical performance and cost issues.

15.9.1   Spectrum Sensing

White-space devices need to be able to detect when signals are present and not present. The signals to be detected are either high-power TV or low-power wireless microphones

One of the points when discussing the detection of TV signals is how reliably the pilot tone in an ATSC signal can be detected at a range of flux densities.

Companies such as Adaptrum,[15] founded by Robert Broderson the cofounder of Atheros, claim that time-domain matched filter techniques working across the whole of the 6 MHz ATSC channel provide robust signal sensing techniques. Motorola have added in geolocation to provide additional protection.

Broadcasters, however, remain sceptical. Even if sensing an ATSC signal can be achieved consistently down to low flux densities it is still feasible for devices to be in areas shadowed by buildings or hills – the hidden node effect. If the devices need to work in countries with

---

[15] http://www.adaptrum.com/.

DVB TV (with 7 or 8 MHz channel spacing) they need to detect a signal with a much more complex (and by implication hard to detect) pilot symbol structure.

Wireless microphones are different – these devices use FM modulation and are inherently low power.

The requirement to detect two different types of signals has to be realised in the presence of relatively high incident received energy from other adjacent channels. These signals potentially desensitise the receiver front end, thus potentially compromising the sensing function of the device.

15.10   Transmission Efficiency

To date, most technical work has focused on receiver performance parameters but white-space devices are supposed to deliver two-way wireless connectivity. The difficulty here is the bandwidth over which the devices will need to operate. In the US this would be somewhere or anywhere between channels 21 and 51 (512 to 698 MHz) or potentially from channel 2 upwards (54 MHz). The maximum power of these devices is relatively modest, of the order of 100 milliwatts, but difficult to deliver efficiently across these operational bandwidths.

Producing a device that would work in other markets, for example Europe, would mean devices would need to work somewhere or anywhere in UHF band 4 extending up to 790 MHz. Designers of cellular phones know how hard it is to deliver efficient RF power amplifiers at lower frequencies over extended operational bandwidths. Getting an RF power amplifier to work efficiently across 40 MHz at 800 MHz is challenging, let alone 100 MHz or more at 600 MHz.

As a comparison, the widest cellular bandwidth as a percentage of centre frequency is presently GSM 1800. 75 MHz at a centre frequency of 1747 MHz is equivalent to 4.3%. The 80-MHz ISM band at 2.4 GHz is 3.3%. White-space devices could be of the order of thirty to 40% – an order of magnitude greater.

15.11   Scale Economy Efficiency

One solution would be for white-space devices to be market specific that would mean white-space devices for the US market (Region 2) and rebanded devices for Region 1 (Europe) and Region 3 (Asia).

However, this would make it impossible to achieve sufficient scale economy. This would be particularly true if only deployed in the US. By 2014 the US will constitute less than 10% of the global market for cellular devices. RF development today is typically amortised over hundreds of millions of devices. Such volumes would be inconceivable from the US white-space market alone.

15.12   Signalling Efficiency

White-space devices will have to measure and continuously remeasure bandwidth occupancy and share these measurements with other devices or access points in the network. The IEEE presently has a working group specifying these protocols but essentially this is a complex measurement process.

This means that the devices will not be spectrally efficient and more important from a users perspective will not be power efficient irrespective of the network topology adopted. The use of mesh-network topologies for example would result in particularly poor spectral and power efficiency. White-space devices would therefore be likely to have much poorer duty cycles than equivalent cellular devices.

15.13   Power Efficiency Loss as a Result of a Need for Wide Dynamic Range

The need for a wide dynamic range in the receiver front end would further reduce session duty cycles by increasing DC power drain. This would effectively invalidate any business plans predicated on the use of mobile or portable equipment.

15.14   Uneconomic Network Density as a Function of Transceiver TX and RX Inefficiency

Unless white-space devices somehow escape the fundamental laws of physics, it will be hard to realise transmission or receive efficiency across the required operational bandwidth.

The lack of consistent channel pairing would make an FDD band plan problematic, so the assumption has to be that this would be a TDD interface. This will result in additional loss of receive sensitivity particularly when devices need to transmit and receive at the same time.

This will compromise the measurement capabilities of the devices (see spectrum sensing above) but additionally implies a network density that would be unlikely to be cost economic. Mesh networks are not a solution. (See the section on signalling efficiency above.)

TDD also implies a loss of capacity over extended distances, a function of the time-domain guard band overhead needed to accommodate round-trip delay. This implies a relatively dense network topology.

Note that TV receivers are inherently insensitive due to their need to tune over an extended frequency range. This does not matter if signals are being received through a high-gain roof-mounted antenna but it does matter for devices with internal antennas and/or for devices used at ground level. White-space devices will be used typically at ground level and may or may not have internal antennas. They will need to tune across similar bandwidths but additionally have to generate transmit power for the uplink, not something a television has to do. This will result in further receive desensitisation that will translate into lower downlink data rates and/or denser more costly networks.

15.15   Cognitive Radios Already Exist – Why Not Extend Them into White-Space Spectrum?

Cognitive radios already exist in mass market applications. DECT cordless phones are one example and it could be argued that cellular handsets are cognitive in that they measure channel quality. The admission control algorithms in LTE networks for example are based on the channel quality indicator measurements received from LTE handsets.

The difference here is that the search and measurement parameters are relatively modest. DECT handsets work over just ten channels within a 20-MHz channel allocation. WiFi measurement and access algorithms at 2.4 and 5 GHz are similarly comparatively straightforward. Most 2.4-GHz systems only use three 22-MHz channels across the 80-MHz bandwidth allocation.

In cellular networks, the measurement options are theoretically broad but practically narrow. Channel access will usually be in the same band and often on the same channel (a simple shift to another time slot).

In DECT and WiFi and cellular networks, the devices are measuring identical signal waveforms to the waveforms they need to demodulate. White-space devices have to detect different wave forms that may be similar or different to the waveforms they need to demodulate. This is a significantly more complex task.

15.16   An Implied Need to Rethink the White-Space Space

This suggests that white-space devices as presently conceived will have poor transmission efficiency that will translate into low uplink data rates and/or limited uplink range and a limited RF uplink power budget. The limited power budget will be compounded by a relatively high measurement and signalling overhead that will reduce session duty cycles. The devices will not work as well as cellular devices and/or WiFi and WiMax devices and will cost more due to smaller market volumes. This would tend to suggest that present white-space business models are fragile at best.

However, as white-space proponents point out, hundreds of MHz of spectrum are unused or under used at certain times at certain places. This white-space spectrum has a social and economic value. The problem is that white-space devices as presently conceived are not an efficient mechanism for realising that value. The answer may be to encourage the broadcasters (the NAB in the US the EBU in Europe) to develop a white-space device specification that could be integrated with existing (DVB) and planned (ATSC) portable TV specifications.

This would have several benefits.

The present dispute over potential interference problems would be resolved. A new business model could be developed based on a closer coupling between terrestrial TV and two-way wireless internet access. It would provide a broader industrial base over which RF development costs could be more efficiently amortised.

It would be even better if the cellular community could be encouraged to work with the broadcast community on a common white-space standard to provide compatibility with ATSC and DVB and LTE 700- and 800-MHz devices.

Similarly, the future of terrestrial TV will be dependent on building closer relationships with cellular service providers. TV transmissions from cellular infill sites are for example the only credible way forward if ATSC-based portable TV is ever going to work.

15.17   White-Space White House

So in theory, white-space devices are a great idea – an opportunity to realise value from presently unused or underused spectrum.

As such, the 4th November decision to make a swathe of new unlicensed spectrum available for innovative services must still look superficially attractive to the new US administration – a populist policy with tangible social political and economic benefits. In practice, these practical benefits can only be realised if devices can be developed to work in most, if not all, global markets. This is politically, commercially and technically challenging and implies a need to work with rather than against the broadcast and cellular community and to work across national and international boundaries.

In common with licensed spectrum, unlicensed spectrum incurs access costs that have to be recovered from access charges or from other sources such as tax revenue or market subsidy. To quote President Roosevelt 'We have never realised before our interdependence on each other'.

Seventy five years on and faced with similar recessionary pressures we should recognise that interdependency implies a need for to explore and exploit collaborative rather than competitive market opportunities and avoid the unnecessary and presently unsustainable costs introduced by conflicting market interests.

15.18   LTE TV

This brings us to LTE TV.

As documented earlier in this chapter, considerable time, money and effort has been spent on mobile TV standardisation and network and handset development with to date no commensurate fiscal return.

It has been difficult to realise a return on investment partly due to competing standards, (DVB H, Media FLO, ISDBT, DMB DAB and MBMS, mobile broadcast and multicast service) and partly due to network and handset development limitations and performance constraints compounded by regional and national differences in mobile-TV frequency-band allocation and incompatible standards.

Although mobile TV has been successful in some markets, for example Japan, it has failed to achieve global scale and in most markets there has been a marked reluctance by consumers to pay for mobile TV content.

The longer-term economics and effectiveness of hybrid satellite and cellular networks such as DVB SH in Europe or ATC (ancillary terrestrial component) hybrid satellite and terrestrial networks in the US also remain unproven, a topic addressed in the following chapter.

This might lead one to question the merit of introducing yet another mobile TV standard to service a global market that is presently nonexistent. However, the inclusion of multicast broadcast single-frequency networks, MBSFN, as a work item in present and future LTE releases suggests that there is still a measure of vendor and operator confidence that mobile TV remains a worthwhile investment opportunity.

Remarkably, it may be and MBSFN could be used to realise additional revenue and improve returns on existing spectral and network investment both for the cellular and broadcast community.

15.18.1   The Difference Between Standard LTE and MBSFN – The Technical Detail

MBSFN is based on 7.5-kHz subcarrier spacing rather than the 15-kHz used in standard LTE. This doubles the symbol length from 66.7 microseconds to 133.4 microseconds and allows the cyclic prefix to be increased from 4.69 microseconds to 33.33 microseconds.

The cyclic prefix provides a time-domain guard band between symbols to compensate for delay spread in the radio channel. A 4.69-microsecond cyclic prefix allows a delay spread

Table 15.4 Broadcast technology comparisons

System	Broadcast Wide Area			Cellular	Cellular broadcast
	DRM	DAB	DVB-T	LTE	LTE MBSFN
Frequency	<30 MHz	100 MHz, 220 and 1400 MHz	700 MHz	Between 700 and 2.6 GHz	
Range	500 km	74 km	67 km	5, 30, (100 km)	
Channel spacing	9 kHz	1.536 MHz	7.6 MHz	1.4, 3.0, 5, 10, 15, 20 MHz	5 MHz
Max gross data rates	25 kbps	2.304 Mbps	5 to 31.7 Mbps	10 to 30 Mbps or above depending on cell size	
Modulation	QAM	QPSK	QPSK 16 QAM 64 QAM	QPSK 16 QAM 64 QAM	
Number of subcarriers	204	1536	1705 (2 k) 6817 (4 k)	300 across 4.5 MHz	600 across 4.5 MHz
Subcarrier spacing	41.66 Hz	1 kHz	4.46 kHz 1.116 kHz	15 kHz	7.5 kHz
Symbol duration in microseconds	26 660	1246	1120	66.7	133.4
Guard interval in microseconds	2660	246	7 to 224	4.69	33.33

of 1.5 km; a 33.33 microsecond cyclic prefix allows a delay spread of 10 kilometres. The delay spread is the difference in path length caused by multipath in cellular networks. In SFN broadcast networks, the delay spread is also a function of receiving the same signal from more than one transmitter.

The 33-microsecond cyclic prefix used in MBSFN means that TV transmissions can be broadcast simultaneously from multiple node B cellular base stations without causing inter-carrier symbol interference in the receiver.

Table 15.4 compares DRM, DAB, DVB broadcast networks with LTE and LTE MBSFN.

The number of subcarriers defined for DVB T and LTE excludes edge-of-band subcarriers used to provide a frequency domain guard band. For example the 8 K DVB- T multiplex has 6817 one-kHz subcarriers nestled within 8 MHz (8 k times one kHz) channel spacing.

In practice, a number of DVB-T networks were implemented as 2-k rather than 8-k networks due to receiver chip set limitations, for example in the UK where initial implementation was realised in 1997. This prevented the implementation of a (more spectrally efficient) single-frequency DVB-T network. Although the LTE MBSFN is described as a single frequency network, it could be deployed as a cluster of SFN enhanced node B base stations with frequency reuse from cluster to cluster.

Present standards work is focused on either FDD or TDD implementation, known as down-link optimised broadcasting within existing cellular allocations, for example the TDD bands within Band I at 2 GHz.

15.18.2   MBSFN Investment Incentive – Cost Saving and Revenue Opportunities

This, however, fails to capitalise on some potential major cost saving and enhanced revenue opportunities that together could provide the required incentives for MBSFN investment. These can be identified by analysing the problems that the broadcasting and cellular industry have to solve, some of which we have identified earlier.

As can be seen from the table, the 8-k subcarrier OFDM used in DVB-T allows large cells to be deployed. This means that DVB-T broadcast networks can be very cost economic. However, to provide adequate coverage, low-power infill sites have to be deployed that add capital cost and running cost and reduce broadcast spectral efficiency.

Although this provides adequate coverage at an adequate link budget for TVs (including high-definition TVs) with a roof-top aerial, there is generally not enough signal strength to support portable receivers. Portable TV receivers have been the Cinderella of the TV industry for many years but the availability of lap tops with DVB or ATSC modems will potentially result in a much broader market reach. The top of the range high-definition Sony Vaio[16] is one example. However, the signal levels in most countries are not sufficient to deliver a sufficiently consistent user experience to support mass-market adoption. Deployment of high-density SFNs based on existing cellular infrastructure would overcome this problem. The difficulty is that portable DVB devices do not work well unless connected to an outdoor aerial or a satellite dish or cable. These limit the portability of the device.

A roof-mounted antenna has about 12 dBi of directional gain. A built-in broadband TV antenna in a lap top will usually have negative gain, typically −7 dBi and will be looking for a signal severely attenuated by height loss (10 dB), additional building penetration loss (8 dB) and location variation due to standing waves (10 dB), resulting in a received signal up to 47 dB lower than the value planned for fixed outdoor reception. It would be prohibitively expensive to increase flux levels from the existing terrestrial broadcast networks to support acceptably ubiquitous portable TV reception, particularly portable HDTV reception.

Cellular base stations, particularly cellular base stations designed to support the 700- and 800-MHz cellular bands could, however, function as relay repeaters and could provide a cost-effective alternative to dedicated TV repeaters. Cellular handsets or at least some cellular handsets and mobile broadband devices including lap tops with embedded LTE modems or dongles will have 700- and 800-MHz LTE transceiver functionality. It is therefore not inconceivable to increase LTE 700- and 800-MHz receiver bandwidth to accommodate DVB T transmissions.

Operators might question the point of including DVB reception in mobile broadband devices particularly given that many of the services are available free to air. This, however, ignores the indirect revenue gain that is potentially achievable from coupling free to air TV reception with two-way mobile broadband and is pragmatically probably the only way in which cellular operators will ever achieve a return from 700- and 800-MHz cellular infrastructure and spectral investment.

The ability to receive broadcast TV off air would reduce traffic load on LTE data networks delivering a capacity gain. Hill-top TV transmitters could also be used to provide extended cellular coverage. An upper limit cell radius of 100 km is included in the LTE specification.

---

[16] http://www.sony.co.uk/product/vn-z-series.

15.19   Summary

The proximity of cellular receive and transmit bands at 700 and 800 MHz to UHF broadcast TV channels opens up the possibility of supporting digital TV transmission via cellular base stations and digital TV reception in handsets and mobile broadband devices without significant additional hardware cost. Rebroadcasting of national and local TV from cellular transmitters would allow HDTV to be delivered to portable receivers without the need to connect portable devices to a fixed antenna.

Cellular site and hardware costs including backhaul overheads could be amortised across cellular and broadcast services. There would be no need to ring fence or repurpose TDD bandwidth at 2 GHz for broadcasting. These channels could be used as originally intended for mobile broadband access.

The coupling of local and national linear TV broadcasting with mobile broadband would unlock new direct and indirect revenue streams both for the broadcasters and cellular operators and would provide the basis for developing innovative mass-market consumer electronic products that could be clearly differentiated from present product and service offerings. The combination of the additional revenues realisable from a more broadly based user experience combined with more broadly amortised costs could significantly improve profitability for all involved parties.

15.20   TV or not TV – That is the Question – What is the Answer?

The launch of Apple's iPad has generated countless column inches of industry comment but no one seems to have quite nailed the answer as to what users will do with the device. The consensus seems to be that iPads will change the way that media is consumed and information is accessed on the move.

However, the form factor of the device, smaller than a lap top but larger than a smart phone, suggests a sit-down or sit-back experience – very different from how mobile phones are used today. If this is true, then some form of TV connectivity would seem sensible. This could be IPTV over WiFi or cellular but equally could include ATSC and/or DVB T.

It has become fashionable to dismiss terrestrial television as an increasingly irrelevant delivery medium, but in practice terrestrial broadcasters still hold a number of trump cards in terms of linear and nonlinear content, delivery bandwidth and political, social and business collateral.

The generic argument is that terrestrial TV will continue to be one of the most economic and power-efficient delivery platforms for content and information and that the use of MPEG and JPEG encoding and compression schemes at the application layer and IP protocols at the transport layer, if combined with closer physical bearer integration, would result in a closer coupling of these traditionally separate delivery platforms.

The BBC are presently promoting a new standard to be known as DVBT2M for mobile phones and have a mobile iPlayer offer which at Summer 2011 was getting 8 million requests per month. Attention was also being given to developing dual-screen delivery formats with a tablet displaying supporting content to a program being shown on TV.[17] Commercially, a TV license fee could be a way of collecting money for digital content rights.

---

[17] Stephen Baily, General Manager BBC R and D, presenting at Cambridge Wireless conference in Cambridge 27 June 2011.

We suggested that LTE could potentially provide the basis for this closer integration and that this process of technology and engineering convergence would translate into new market and business opportunities. However, in order to maintain its ubiquity as a delivery medium, terrestrial digital TV would need to be received by portable terminals used indoors without an aerial. This implies a very substantial increase in signal strength and/or an improved link budget of the order of several tens of dBs. High-definition broadcasting would need another 10 to 12 dB, though whether this would ever be worthwhile for portable form factor devices is open to debate.

Either way, the only practical way in which acceptably consistent TV coverage could be realised would be to rebroadcast the DVB T or ATSC multiplex via local cellular transmitters or via inbuilding repeaters. Alternatively, TV signals could be piped down from a roof-top aerial and rebroadcast from an indoor femtocell. However, even with signal rebroadcasting, portable receivers would benefit from having improved sensitivity and selectivity. Several factors now make this a more plausible scenario.

Cellular base-station transmitters in the 700-MHz (US) and 800-MHz (European) UHF band need to be sufficiently linear to transmit the cellular OFDM signal with minimal distortion. It is therefore not unrealistic to consider adding ATSC or DVB T broadcast to the TX chain.

Cellular handsets and portable lap tops will have to add 700- and 800-MHz UHF LTE TX/RX functionality. It is therefore not unrealistic to consider adding DVB T and ATSC receive functionality to these devices. Indeed, it could be argued that extended RF connectivity might be a precondition of mass-market adoption for the emerging crossover form factor devices that combine traditional lap top, net book and smart-phone functionality, of which the iPad is one example.

However, there are some practical RF design and performance issues that have to be addressed before these products can be realised. Substantially these are determined by the characteristics of the UHF band and the need for these devices to provide cellular connectivity in the 850/ 900, 1800, 1900, 2100 and 2600 MHz bands.

So, for example, it might seem sensible to have an ATSC and DVB T receiver that worked across the whole UHF band from 470 to 862 MHz and a transceiver that could cover the cellular bands from the bottom of the US 700 MHz band at 698 MHz to the top of the proposed European UHF cellular band at 862 MHz.

In practice, such a device, even with adaptive matching and tracking filters, would have unacceptable sensitivity and selectivity on the RX path, poor transmission efficiency on the TX path and the dynamic range needed would incur an unacceptable power drain. Adding other bands introduces additional switch paths that introduce insertion loss and poor isolation.

The link budget gain of working at 700 MHz or 800 MHz is about 8 or 9 dB when compared to 1800 or 1900 MHz, and rather more when compared to 2600 MHz. This is a function of reduced propagation loss. Inbuilding penetration in particular should be notably better and should translate into improved coverage and/or higher average data throughput rates.

However, these gains are academic if offset by RX and TX efficiency losses in the user's device. The form factor of an iPad or net book or lap top provides some extra space but this advantage can be offset by higher noise floors generated by display drivers or other proximate processor activity.

Additionally, the decision has to be made as to whether the TV receiver should be expected to work *simultaneously* with broadband cellular. This would mean additional parallel processing but more importantly would also require careful implementation to avoid desensitisation of the TV receiver by the LTE transmit path.

Commercially, it would seem that this is a problem worth solving. Online connectivity viewed in parallel with broadcast content could provide the basis for a whole new generation of interactive two-way content coupled with real-time peer-to-peer or peer-to-multipeer communication.

Technically, this is a nontrivial challenge. The assumption is often made that this kind of dual functionality will become easier to deliver as software-defined radios and/or cognitive radios become more pervasive.

Unfortunately, this is only partly true. Cellular transceivers today still route band-specific signals through band-specific filters and band specific power and low-noise amplifier components. All of these devices have individual matching components.

Getting acceptable performance from a TV receiver integrated with a cellular multiband transceiver would probably also require discrete low-, mid- and high-band signal paths if the whole UHF band had to be covered. Even if digital TV is eradicated globally from the upper UHF bands then a single receive chain would still seem unlikely for a device capable of receiving DVB T and ATSC broadcast.

The alternative is to manufacture devices that are regionally or nationally or band specific, but this frustrates scale economy and prevents products being used universally. The holy grail is to produce high-Q tracking filters that can be channel rather than band specific combined with active components that can maintain their efficiency over extended operational bandwidths, but these devices remain an ambition rather than a present reality, at least in present transceiver designs.

But incremental progress is being made. The industry news flow tends to concentrate on baseband devices but in practice and as stated in earlier chapters the advances being made in new passive and active materials including RF MEMS, silicon on sapphire and BST-based devices are likely to be at least as important.

The future of mobile TV will therefore be dependent on materials and manufacturing innovation.

15.21   And Finally the Issue of Potential Spectral Litigation

Parts of this chapter have highlighted the coexistence issues implicit in the WRC 2007 decisions. In this final section we review the spectral litigation risks that are arising as terrestrial broadcast, cellular networks and white-space devices compete for UHF spectrum and why these risks may increase over time.

An important objective of any regulatory process is to avoid or minimise the impact of spectral ownership disputes. An increase in the number of disputes and related costs implies a failure of the regulatory process.

The US 700-MHz auction and its aftermath from 2007 to 2011 provides an example. The context is a block of spectrum called A block that a number of mobile broadband operators bid for when it became available postdigital switchover. A block is immediately adjacent to the top of the TV broadcast band. The bidding entities including Cellular South and US Cellular supported by Metro PCS and Cox Communications alleged that AT and T and Verizon Wireless misused their market dominance to prevent vendors from making user equipment and base stations available capable of accessing the spectrum. The litigation and advocacy work was done through a 700 MHz Block A Good Faith Purchasers Alliance.[18]

Although Verizon had invested $2.57 billion in A block there were considerable performance costs implicit in adding the band to either Verizon or AT and T base station and user equipment. AT and T had no technical or commercial incentive to cover A Block and were more concerned as to how they repurposed and integrated the TDD bandwidth at channel 55 and 56 acquired from Qualcomm.

For A block owners you could say this was either a case of caveat emptor or an example of the regulator bringing spectrum to market that was not fit for purpose. A regulatory environment that fails to take into account engineering reality inevitably results in the destruction of industry and end user value, trading short-term treasury gain against long-term loss to the national economy. In this case a $50 billion industry investment was compromised by the physical limitations of a few 50-cent components in the front end of a mobile broadband transceiver.

15.21.1   Other Prior and Present Examples

One would have thought the FCC would have learnt from prior examples. In February 2002[19] a decision was taken to make bandwidth available between 3.1 and 10.6 GHz for ultrawideband technologies. The operational bandwidth was not acceptable in other countries. Other regulators responded with their own variations, thereby invalidating any potential business case for UWB. A decision to let the market decide on technology produced two competing standards that created additional market fragmentation. As suggested earlier, similar mistakes are being made on white-space legislation presently being introduced with minimal consideration for global spectral implications or user equipment RF development economics.

15.21.2   The Cost of Technology Neutrality

If the US 700-MHz band was in some way unique in terms of regulatory approach then this would probably not matter that much. However, the disconnect between regulatory policy and physics is more generally pervasive. A doctrine of technology neutrality or at least an absence of technology direction compounds the problem and generally results in a loss of spectral efficiency. This in turn increases the risk of spectral litigation. Mixing CDMA, HSPA, WiMax and GSM in the 1900 or 850 bands is one example.

One of the reasons we have regulators is to arbitrate between competing spectral property interests. This extends both to interindustry and intraindustry disputes. The amount of litigation could thus be viewed as a proxy measure of the individual and combined effectiveness of the regulatory process. In the US the 700 MHz spectral disputes are presently confined to interoperator disputes, but in years to come could easily involve the broadcast industry challenging interference from A block mobile broadband TX and public protection agencies challenging interference from Verizon upper block mobile TX. Both are the direct result of a poorly implemented band plan. Similar disputes are likely in L band and S band as mobile operators realise the potentially adverse fiscal and operational impact implicit in the present

---

[18] http://ecfsdocs.fcc.gov/filings/2010/04/30/6015588771.html.
[19] http://www.fcc.gov/Bureaus/Engineering_Technology/News_Releases/2002/nret0203.html.

repurposing of mobile satellite spectrum for terrestrial use. In Europe, similar disputes are simmering over the allocation and auction processes being proposed or implemented for the 2.6-GHz extension band linked directly or indirectly to 800-MHz deployments. Operators generally are concerned that their spectral holdings should be balanced across all available bands or rebalanced to address legacy allocations that in retrospect can be regarded as being competitively unfair.

This increase in tension can be ascribed to a number of factors.

Additional bands are being allocated with terms and obligations that differ from previous allocations. Each additional band introduces additional insertion loss and reduced isolation through the switch paths and filters in user equipment. This will result in unforeseen and unexpected interference issues but also reduces spectral efficiency.

Multiple technologies are being introduced into legacy bandwidth with minimal consideration as to the likely impact on proximate bandwidth and other user communities. This will result in unforeseen and unexpected interference issues, but also reduces spectral efficiency.

Existing bands are being extended to accommodate these multiple technologies. The 10-MHz extension to the 850 band in the US is an example. This will result in unforeseen and unexpected interference issues but also reduces spectral efficiency.

Channel bonding is being introduced to meet an assumed market demand for peak data rates of up to 1 Gbps. This will result in unforeseen and unexpected interference issues but also reduces spectral efficiency.

MIMO is being introduced in parallel with channel bonding on similar assumptions. MIMO may result in unforeseen and unexpected interference issues and will certainly reduce average throughput rates, effectively a reduction in system efficiency.

Operational requirements are being imposed on operators without due consideration of the spectral implications. Mandatory E911 support for example fails to take into account that the second harmonic of 787.5 MHz, between C band and D band mobile transmit in the US 700 MHz band falls directly on the GPS receive frequency in L band at 1575 MHz. Terrestrial use of L band may also compromise GPS front-end receiver performance, a problem that Light Squared had to contend with through 2011.

All of the above increase performance uncertainty both in terms of band-to-band performance, within-band channel-to-channel performance, sensitivity to hand and head effects and changing operational conditions including temperature and battery charge state. This makes quality of user experience service level agreements harder to model and manage. All of the above are therefore likely to increase rather than decrease 'within industry' interoperator disputes over spectral ownership rights, compounded by the combination of a reduction in spectral efficiency coupled with an increase in operator-to-operator, user-to-user interference. All of the above are also likely to increase 'between industry' interindustry disputes over interference and spectral ownership including disputes between the mobile broadband industry, the broadcasting industry, the mobile satellite industry, public safety radio industry and cable industry (set-top box interference).

An adversarial approach to future spectral allocation, for example the repurposing of C band between 3400 and 3800 MHz for FDD/TDD bands 22/41 and 23/42 makes it less likely that these separate industries will be able to work together to resolve technical issues. Interindustry cooperation at the technical level is already complicated and frustrated by competition policy. TDD/FDD coexistence will also introduce additional system level complexity which will be technically and commercially hard to resolve.

The problem is compounded by a false belief by the regulatory community that all that is needed is to define spectral transmission masks. This ignores the growing need to define and enforce receiver selectivity and dynamic range.

It is also increasingly inappropriate for the FCC to assume that the world should follow US policy. In terms of mobile subscriber market size, China and India are both more than twice as large as the USA. Given that 48% of all connections are now in Asia it should be obvious that the US now lacks sufficient economy of scale to support a nationally specific band plan and/or a nationally specific technology mix. Operator-specific band plans and operator-specific technology solutions within the US market make even less economic sense.

China is arguably the only market with sufficient volume and local design and manufacturing capability to be able to support national- or operator-specific band plans and technologies, but even this does not mean that such policies make any kind of long-term economic sense. The problem is that in all markets, China, the US and Rest of the World, spectral allocation and auction policy is based on economic modelling that fails to capture relevant cost and value dynamics. So as a reminder of what is in effect the ongoing narrative of this book there are, we would argue, five separate but coupled domains that need to be analysed:

15.22   Technology Economics

This is the area most directly coupled to the standards process. Our contention here is that false market ambitions, specifically high peak data rates per user are introducing unnecessary complexity and compromising system performance to the point at which performance loss/economic loss litigation becomes likely. For example, it might be expected that at some stage operators will need to protect network performance by introducing pass/fail margins into conformance testing. This is valid and understandable but will highlight vendors who are presently shipping products that only meet conformance performance requirements under an unrealistically narrow range of operational conditions.

Note that some devices can now take well over 1000 hours to go through conformance test – performance testing potentially takes longer. This is a cost that no one wants to absorb. This will result in an increase in litigation cost.

15.23   Engineering Economics

This is the area most closely coupled to the spectrum allocation and auction process. Our contention here is that false policy objectives, specifically the maximisation of short-term gains for national treasuries over longer-term economic gains have resulted in spectrum sales that ignore the laws of physics. The problem is compounded by present standards policy that compromises spectral efficiency. Caveat emptor may provide a measure of protection but an increase in litigation seems to be an inevitable outcome of the present policy approach.

15.24   Market Economics

The problem here is that standards and regulatory policy has failed to adapt to the change in relative market importance between Asia and the rest of the world. If subscale markets, for example the US, continue to pursue a nationally specific standards and regulatory agenda then it is understandable that Asian markets, particularly China, will want to do the same. The result will be an increase in litigation and more aggressive protectionist legislation.

15.25   Business Economics

This is the problem of the 50 cent components compromising the $50 billion investment. The fact that most RF components don't scale has made it hard for the RF component industry to deliver products that can meet associated performance requirements. Newly dominant markets are understandably using their market leverage to extract significant performance promises from this underfunded and underresourced sector of the industry. The result will be an increase in Chapter 11 filing that will further inhibit RF innovation and investment.

15.26   Political Economics

Back to where we started. Regulators will become increasingly exposed to litigation from entities who have bid for spectrum that is not fit for purpose or compromised either by poorly executed standards policy or, equally damaging, by the absence of a standards policy. The problem is compounded by regulatory policies that fail to take account of basic engineering reality and present RF component limitations and R and D constraints.

15.27   Remedies

The standards process would benefit from being less focused on an assumed need for high peak data rates and more directly focused on delivery cost economics. Standards and spectral policy need to be more closely coupled and engineering cost needs to be directly factored into band allocation and auction policy. Some mechanism must also be devised to encourage more effective collaboration between different sectors of the industry, mobile broadband, public safety, broadcasting and the mobile satellite sector and closer economic integration of wireless, cable, copper and fibre delivery systems.

Regulators are always under pressure to promote country-specific solutions to benefit their national stakeholders. The aim is laudable but history shows that this approach can be disastrous. In the early 1990s, Japan, at that time the world's second largest national economy – went its own way with mobile cellular. As a result, Japanese manufacturers were left floundering, unable to service both a quirky domestic market and a brutally competitive global market. In the USA, Block A Purchasers rail against manufacturers who ignore their special needs. But manufacturers who wish to compete in the global market cannot afford to divert expensive engineering effort to rescue spectrum owners left stranded by poor regulation.

In this context it is important to differentiate between constructive tension, the mechanism by which economic progress is achieved through efficiently ordered market competition and destructive tension, a general outcome of a poorly ordered market and misapplied regulatory intent.

A similar differentiation can be made between constructive litigation, the valid arbitration of competing market interest and destructive litigation, the outcome of poorly implemented spectral and standards policy. Destructive litigation is unnecessary and wasteful but can be avoided provided potential causes are identified at an early stage – a challenge and opportunity for the regulatory community.

16

Satellite Networks

16.1 Potential Convergence

Getting coverage to deep rural areas or out at sea or on deserted mountain tops or in the middle of the desert where no one lives can be technically challenging and economically problematic. With the possible exception of coverage in the middle of the ocean, cellular networks can provide coverage more or less anywhere, but the cost can become prohibitive relative to income.

Cellular networks are generally speaking optimised for capacity and increasingly now high per user peak data rates. The wider channel spacing and wider operational bandwidths needed generally reduce the coverage available in any particular band.

This chapter looks at low-earth orbit and geostationary satellite systems but we start with two-way radio systems and specialist user requirements. As with terrestrial cellular networks, satellite networks and two-way radio networks are increasingly doing a lot more than just supporting voice traffic and are becoming increasingly information centric and, in the jargon of the sector, 'mission critical'.

At the application layer substantial standards commonality is beginning to emerge across all delivery systems, for instance the use of MPEG and JPEG encoding and compression schemes and MPEG-based content descriptors. At the transport layer IP protocols are becoming more pervasive and potentially allow much greater transparency across multiple radio bearers. However, the benefits of this transparency can only be realised if a closer commonality between physical radio bearers can be developed over time.

16.2 Traditional Specialist User Expectations

Specialist users include public safety agencies, fire, police, ambulance, defence users and nonemergency users in transport and distribution, energy and small to medium business users.

Traditional expectations in terms of user functionality include open channel working between multiple users based on push to talk functionality, back-to-back direct mode operation where users can talk to each other directly without a base station in between and ground-to-air/air-to-ground communication, particularly important for military users or event management where aerial surveillance is required.

Making Telecoms Work: From Technical Innovation to Commercial Success, First Edition. Geoff Varrall.
© 2012 John Wiley & Sons, Ltd. Published 2012 by John Wiley & Sons, Ltd.

Table 16.1   Specialist user needs

Wide-area coverage	Better than cellular particularly in rural areas and/or within buildings in urban environments
'All informed' user capability	The ability to hear other users in an 'open' channel
'Instant access' onto available channels	Call setup time has to be less than 250 milliseconds
Multigroup announcements	
Wide-area broadcast messages	
Dynamically changeable priority levels	
Secure authentication and encryption	
Voice clarity	Includes task optimised noise cancellation, specialist hands-free and whisper phones for covert surveillance
Press to talk	With option to work full duplex if needed
Talk groups	Geographical or functional
Interworking/interoperability	Problematic as we shall see later
Storm plans/special event plans	Preset response to particular events including disaster recovery contingency planning.
Sleeper phones	The ability to stun phones remotely and re-enable them as listening or more recently listening and watching and sensing devices.
Ruggedised hand sets	Waterproof, dustproof, shockproof fireproof and explosionproof handsets

Latency has always been important and the benchmark generally has been to deliver onto channel times of less than 250 milliseconds.

In terms of radio performance, the most important metric has traditionally been to deliver good rural coverage and good urban inbuilding penetration for two-way voice communication.

In Europe this has been achieved by working at VHF or UHF with both the handsets and base stations working at higher powers than cellular radio. The narrow channel spacing used, typically between 6.25 and 25 kHz and the relatively narrow operational bandwidths, typically 5 MHz or less, makes it possible to deliver good receive sensitivity.

In the USA there are substantial two-way radio system deployments in the 800-MHz band. In common with European systems these operate at higher power than cellular systems. APCO 25 portables, for example, are allowed to produce 5 watts and the base stations can produce up to 500 watts. APCO is unusual in that it is standard developed by a specific user community, the Association of Public Safety Communication Officials.

Table 16.1 summarises some typical traditional specialist user needs.

From the above, it can be seen that it is not always easy to move users from a two-way radio system onto a cellular system and provide them with sufficiently equivalent functional capabilities. The attempted migration of Sprint Nextel iDEN users to Sprint CDMA a few years ago is a legacy example of some of the challenges inherent in this process.

16.3   Impact of Cellular on Specialist User Expectations

Conversely, much of the functionality that we take for granted in cellular handsets, particularly multimedia functionality and the data rates needed to support the real-time exchange of multimedia content is either not available in a two-way radio or commands a substantial cost premium and/or weight and size and/or duty cycle penalty.

Not all specialist users need or want multimedia capabilities in their handsets, but some do and this present minority will likely increase over time. Multimedia functionality implies audio and video capture bandwidth using high resolution cameras and high resolution display bandwidth using high-resolution displays. These functions are only immediately useful if matched with efficient high data rate uplink and downlink radio bearers.

It is not impossible to deliver adequate data rates over two-way radio systems. There are wider-band implementations of two-way radio systems potentially capable of meeting multimedia specialist user expectations but substantial engineering effort and investment will be needed to make the handsets and networks work efficiently. The narrowband technologies are optimised for voice or low bit rate data so are inherently unsuitable for real-time multimedia. These narrowband systems are spectrally efficient but this is of marginal relevance to most specialist end users.

We can illustrate these practical data rate limitations by reviewing first the technologies used and secondly the spectrum into which those technologies are presently deployed.

16.4   DMR 446

This is a digital PMR standard developed through ETSI and called 446 as 446 MHz is where some license exempt spectrum has been made available in some countries, for example the UK.

It is neither intended nor suitable for multimedia transmission. DMR products are available from Motorola, Icom, Kenwood, Tait and Entel and could potentially be deployed into other UHF and particularly VHF PMR applications. As we shall see later there is a problem of scale economy with these products and their target markets.

16.5   TETRA and TETRA TEDS

TETRA is originally a European standard for trunked radio based on 25 kHz channel spacing with a 4 slot frame structure. A more recent extension of the standard, TETRA Enhanced Data Service (TEDS) scales the 25 kHz channel spacing to 150 kHz and uses 16-level QAM to support up to 500 kbit/s as a 'close to cell' data rate.

TETRA handsets are available from Motorola, Kenwood, Sepura and Selex. Nokia has not developed handsets for this market since 2005. Even assuming general availability of TEDS complaint handsets, present TETRA cell densities would mean these higher data rates could only be supported in limited geographic areas.

This is not presently a major problem. Many specialist user communities are rightly hesitant about adopting new technologies or new variants of existing technologies. The sale of the TETRA Airwave network in the UK in 2007 for close to £2 billion pounds proves that good revenues and margins are available based on the provision of voice combined with relatively low data rate services. However, it is reasonable to expect that this will change over time particularly if the performance gap between TETRA radios and cellular handsets continues to widen, which it will for reasons that we explain later.

16.6   TETRAPOL

The 'other' European trunked radio standard introduced by Matra, now EADS, based on narrowband 12.5 kHz channels using GMSK. Tetrapol delivers very good range when used for voice services but is bandwidth limited for multimedia.

16.7   WiDEN

WiDEN is an evolved version of iDEN, the Integrated Digital Enhanced Network standard developed by Motorola in the early 1990s. The frame/slot structure is either 3 slot or 6 slot similar to cellular US TDMA. When combined with 64-level QAM modulation this can deliver per user data rates in the order of 100 kbit/s, but no WiDEN products are presently available.

16.8   APCO 25

APCO25/Project 25 handsets have 12.5-kHz or 6.25-kHz channel spacing and a one- or two-slot air interface using QPSK. The handsets have good range performance but are bandwidth limited for multimedia. Manufacturers include Motorola, Tait, Maxon and Bosch.

As can be seen, none of the available two-way radio technologies summarised in Table 16.2 are ideal for multimedia. The standards that are capable of supporting the higher data rates needed do not yet have products available.

Table 16.2   Comparison of present two-way radio technologies

Digital PMR (DMR 446 ETSI)	TETRA TEDS	TETRAPOL	WiDEN	APCO25
6.25 kHz or 12.5 kHz channels or 12.5 kHz with two time slots, 2 level or 4-level FSK, supports peer-to-peer in unlicensed spectrum or through a repeater or trunked system, typically uses very low bit rate 2 kbit/s codecs and power output limited to 500 milliwatts	TETRA is a 4 slot 25 kHz channel spaced physical layer, TETRA TEDS is an evolved version with channel spacing scaleable up to 150 kHz and modulation options up to 64-level QAM yielding potential gross data rates of 500 kbits	Originally developed by MATRA now EADS, a narrowband 12.5 kHz air interface using GSMK. Very good for long-range voice, bandwidth limited for multimedia	WiDEN an evolved version of iDEN which traditionally uses a 25-kHz channel with 3 or 6 time slots very similar to TDMA cellular where a 30-kHz channel is divided into 3 or 6 time slots, potentially 100 kbits/s data rates	12.5 kHz one or two time slot channels or 6.25 kHz channels, QPSK modulation, 9.6 K/bit/s data rate

16.9   Why the Performance Gap Between Cellular and Two-Way Radio will Continue to Increase Over Time

This is not specifically a technology constraint but more a consequence of limited engineering resources. Substantial engineering resources are needed to develop and ratify standards.

Substantial engineering resources are then needed to interpret these standards and translate them into performance-competitive cost-competitive radio products, handsets and base stations.

This is one reason why the performance gap between cellular phones and two-way radios continues to widen. Silicon vendors and handset manufacturers producing cellular and mobile broadband quad and quintuple band chip sets are servicing a unified market that is now running at over one billion units per year.

Companies like Nokia and the silicon and component vendors supporting Nokia and other Tier 1 cellular handset manufacturers have to treat this dominant market as an absolute priority when deciding on the allocation of engineering development resources. This 'market pull' gravitational effect also helps to focus efforts on meeting other user requirements such as interoperability and roaming.

The opportunity costs of servicing markets that are several orders of magnitude smaller compounded by the 'divide down' effect of needing to support multiple standards for these minority markets makes it extremely difficult to justify investment in two-way radio handset or infrastructure radio hardware and software development.

The result is that products come to market more slowly, are more expensive, and are generally more limited in terms of their radio functionality, at least as far as overall data rates are concerned.

16.10   What This Means for Two-Way Radio Network Operators

So this tells us two things. Specialist user expectations are changing over time. These expectations are partly driven by personal exposure to rapidly evolving cellular handset form factor and functionality.

These expectations include the assumption that multimedia capabilities can and should be made available in small form factor portable devices. This expectation extends to include the assumption that multimedia capabilities should be available whenever and wherever voice service is available.

This is presently a major challenge for cellular service providers but is an even greater challenge for two-way radio network operators with networks designed and dimensioned for voice and standards that have been historically driven by voice performance metrics.

16.11   Lack of Frequency Harmonisation as a Compounding Factor

Two-way radio hardware development might be more attractive if there was at least some degree of global commonality in terms of band plan allocation.

Table 16.3 shows a rather over simplified representation of present European VHF two-way radio and radio/TV allocations

Table 16.3 European VHF two-way radio and radio/TV allocations

Low Band VHF	FM radio	High-Band VHF	Band 111 Subbands 1 and 2	Band 111 DAB
30–88 MHz	88–108 MHz Possibly being repurposed using DRM or DAB	108–174MHz	174–217 MHz Old black and white 405 line TV	217–230 MHz 7 × 1.55 MHz channels

Table 16.4 European UHF two-way radio and radio TV allocations

European public safety and security		European nonemergency services		European TV
Tetra and Tetrapol 380–385 MHz	390–395 MHz	Tetra and Tetrapol 410–430 MHz DMR446 NMT450 now CDMA 450	450–470 MHz	470–872 MHz

Table 16.4 shows a simplified representation of present European UHF two-way radio and radio/TV allocations.

Part of the debate is now focused on whether LTE can deliver the required functionality for this specialist market and somehow combine that with global-scale economy. This in turn requires some commonality to be achieved between US, European, Asian and rest of the world band allocations.

In the US, discussion has focused on whether or how to deploy LTE for a public-safety national wireless broadband network and regional networks in the 700-MHz band. The US debate provides a reference for the discussions that are taking place on LTE public-safety networks in Europe, Asia and the rest of the world and provides an insight into some of the spectral and standards issues that need to be resolved in order to translate present market ambition into practical reality. As can be seen from Tables 16.5 and 16.6 there is no commonality between the US band plan and the LTE 700 band plan in Asia.

Table 16.5 Upper band 700-MHz LTE including the public-safety bands

Band description	Frequency (MHz)	Mob/UE **TX** or **RX**	Bands Used by/Owned by
C (Band 13)	746–757	**RX**	Verizon LTE
D (Band 14)	758–763	**RX**	Regional public-safety LTE broadband
	763–768	**RX**	Nationwide public-safety LTE broadband
	768–776	**RX**	Narrowband public safety
C (Band 13)	776–787	**TX**	Verizon LTE
D (Band 14)	787–793	**TX**	Regional public-safety LTE broadband
	793–798	·**TX**	Nationwide public-safety LTE broadband
	798–806	**TX**	Narrowband public safety

Table 16.6 LTE 700 band in Asia (Region 3)

698	698–703	703–748	748–758	758–803	803–806	806–824
Top of the TV band	5 MHz guard band	45 MHz LTE Mob TX	10 MHz centre gap	45 MHz LTE Mob RX	3 MHz guard band	PPDR TX P25 or iDEN

16.12 The LTE 700 MHz Public-Safety-Band Plan

It has been just over ten years since the 9/11 attacks in the US. The attacks prompted a major reassessment of public-safety communication needs that have been at least partially reflected in spectral allocation and auction policy and to a lesser extent in technology policy.

The upper 700-MHz band plan adopted by the FCC on 1 July 2007 allocated a 2 by 5 MHz duplex band at 763 to 768 and 793 to 798 MHz for a single nationwide public-safety broadband network and a second duplex band known as D band/Band 14 at 758 to 763 and 788 to 793 MHz to be auctioned with public-safety requirements. Both channel pairs work as reverse duplex with mobile transmit in the upper duplex. This means that the two lower channel pairs supporting base transmit (the downlink) are adjacent to Verizon's ten MHz of C band LTE 700 downlink (Band 13).

There are also proximate 6-MHz allocations for regional and national narrowband networks, dangerously proximate in terms of coexistence between Verizon upper band LTE transmit and the narrowband public-safety receive channels. These narrowband allocations are technology neutral, but theoretically could support either 2 by 3 MHz or 4 by 1.4 MHz LTE channels, not exactly narrowband but relatively narrowband when compared to other LTE networks.

The US LTE 700-MHz band plan between 698 and 806 MHz is unlikely to be adopted in Asia (with very good reason). The most likely band plan in these markets (basically most Asian markets excluding Japan) is a 2 by (rather ambitious) 45 MHz duplex pair implemented as a standard duplex as shown in Table 16.6.

Putting this US regulatory setback to one side, the promise of LTE public safety spectrum in the US market and associated federal funding has prompted infrastructure vendor interest.

In July 2010, Motorola announced a pilot LTE public-safety network in San Francisco as a broadband overlay to the existing APCO 25/Project 25 narrowband digital radio system. APCO25 is a user requirement body overseen by the Association of Public Safety Communications Officials. Project 25 (P25)[1] fulfils a similar role to the TETRA Forum[2] and ETSI TETRA standards process[3] in European and ROW markets. P25 radios have overlapping spectrum with analogue and iDEn digital two-way radios. The networks and band allocations are also described functionally as public protection and disaster relief (PPDR).

At the APCO Conference in August 2010 there were vendor announcements from Alcatel on public-safety LTE integration with EADS supply public-safety 911 despatch data systems. EADS also supplies TETRA systems outside the US. Nokia Siemens announced joint projects

---

[1] http://www.tiaonline.org/standards/technology/project_25/.
[2] http://www.tetramou.com/.
[3] http://www.etsi.org/website/Technologies/TETRA.aspx.

with Harris, a P25 supplier. On 7 September Motorola and Ericsson announced they would be working together on LTE public-safety solutions

Public-safety vendors and regulators have a chequered history when it comes to delivering interoperability. TETRA and P25 radios for example are incompatible both in terms of technology (the air interface) and band allocation. TETRAPOL does not interwork with TETRA. Within the US and Europe and Asia, legacy analogue networks are still operational. Interoperability can often therefore only be realised by supporting multiple digital and analogue radio systems making scale economy hard to achieve.

In the US, the white-space initiative (see Chapter 15) sometimes also described as super WiFi will introduce complexity and uncertainty when managing multiple radio system coexistence particularly if public-safety networks are designated as safety critical.

The LTE 700 MHz public-safety networks could work completely separately from all other radio systems but from a specialist user perspective it would be sensible to have at least some interoperability with other networks. For example, public-safety rural data coverage requirements and user expectations would suggest additional infrastructure will be needed in rural areas. Conversely, public-safety bodies are unlikely to have the finance to build dense urban networks. Public safety users will expect and probably insist on broadband connectivity with the same coverage footprint as existing narrowband voice networks. Economically, this means that public-safety LTE 700 networks and commercial cellular LTE 700 networks will need to find some way of working together technically and commercially.

Table 16.5 suggests the obvious starting point for the US market at least would be a dualband radio capable of accessing Verizon LTE upper C band (Band 13). However, this would increase the operational pass bandwidth of the Verizon Band 13 radios by 18 MHz. This will compromise performance and add cost and weight, an unattractive option for Verizon and Verizon subscribers.

16.13   The US 800-MHz Public-Safety-Band Plan

An alternative option is to produce dualband P 25/LTE or iDEN/LTE radios.

In terms of band allocation, legacy APCO 25/P25 radios are implemented with mobile TX at 821 to 824 MHz (immediately adjacent to the US 850 cellular mobile transmit band between 824 and 849 MHz) and mobile RX at 866 to 869 MHz (immediately adjacent to the US 850 mobile receive at 869 to 894 MHz).

However, Release 9 of the 3GPP specifications proposes three extensions/variations to the US 850 band including adding 10 MHz at the lower end of the present Band 5 allocation that includes all of the legacy P25 800 MHz channels and some of the proposed new channels. P25 radios and iDEN radios are now coming to market capable of supporting the frequency bands shown in Table 16.7. This extends P25 and iDEN into channels previously used and still used for analogue two-way radio. The table also shows the impact of the proposed extension of the US 850 cellular band.

As stated earlier iDEN radios have 25-kHz channel spacing with a transmit power of up to 1 watt and a six-slot QPSK modulated TDMA physical layer.

P25 radios can have 12.5, 20 or 25 kHz channel spacing and an output power of between one and three watts (also QPSK modulated). In-car mobiles can have an output power anywhere between 10 and 35 watts. Base stations can be up to 100 watts. This is serious radio, optimised for rural coverage and deep urban inbuilding penetration.

Table 16.7 US 800-MHz public safety – old and new band plan

806 to 810	**MOB/UE TX**	Conventional PMR	**New Project 25**
810 to 816	**TX**	Trunked PMR	**TX or iDEN**
816 to 821	**TX**	iDEN/wIDEN	**806 to 824 MHz**
821 to 824	**TX**	APCO 25	**18 MHz or 8 MHz with no US 850 overlap**
824 to 849	**TX**	US 850 cellular	**Proposed extension down to 814 MHz**
851 to 855	**MOB/UE RX**	Conventional PMR	**New Project 25**
855 to 861	**RX**	Trunked PMR	**RX or iDEN**
861 to 866	**RX**	iDEN/wIDEN	**851 to 869 MHz**
866 to 869	**RX**	APCO 25	**18 MHz or 8 MHz with no US 850 overlap**
869 to 894	**RX**	US 850 cellular	**Proposed extension down to 859 MHz**
896 to 902	**TX**	Trunked PMR	**iDEN SMR 896 to 902 MHz**
934 to 940	**RX**	Trunked PMR	**iDEN SMR 934 to 940 MHz**

Multiband versions of P25 radios also have to be capable of working at VHF between 136 and 174 MHz and UHF between 380 and 470 MHz. The theory is that LTE will provide a broadband bridge between these legacy networks and will allow the public-safety sector to benefit from global-scale economies with an opportunity to amortise research, development and manufacturing investment across billions of user devices per year. From a specialist user-experience perspective this should provide an opportunity to source standard form factor phones, smart phones, tablet/slates and lap tops at consumer prices.

Ideally, this user equipment would also be able to access commercial LTE networks.

It would seem initially attractive to consider integration with the US 850 band rather than LTE 700. Future iterations of P25 could include a 5-MHz LTE channel adjacent to the US 850 commercial networks as or when or if they transition to LTE.

The problem with this from an RF performance perspective is that even extending the US 850 band by 10 MHz to 35 MHz decreases user equipment sensitivity by 1 dB. Here, we are suggesting adding 18 MHz to create a passband of 43 MHz. This would involve at least as much loss again.

Coupling P25 radios with the LTE channel allocations in the 700-MHz band would require a dual-mode radio to support P25 TX at 806 to 824 MHz and theoretically at least low-band LTE 700 TX down at 698 MHz. This is 126 MHz of operational bandwidth equivalent to a 16% ratio of the centre of the band. The stretch on the RX path would be similar. This is technically challenging and commercially unlikely. In practice, the user equipment would be several radios in one box.

A P25 dual-mode radio that interoperated just with the LTE national public safety 700-MHz network would be easier to implement and the narrowband public-safety bandwidth could be included as well, but the devices would still need to stretch on the receive path from 758 MHz to 869 MHz (111 MHz) or 746 MHz to include Verizon Band 13 (123 MHz). Either option would still need a dual or triple front end to be sufficiently efficient.

The problem could be solved technically by throwing away the P25 and iDEN radios, reallocating the bandwidth to US 850 operators and just having LTE 700 police and SMR radios. However, this would be hard to sell to public-safety and specialist mobile radio users. They would quite rightly point to some of the difficulties encountered when Sprint Nextel iDEN users were forcefully migrated on to the Sprint CDMA network.

Alternatively, the LTE mobile broadband network could be regarded as a separate network function with legacy networks and new P25 and iDEN radios providing voice coverage and specialist user requirements such as all informed user capability, <250 millisecond press to connect onto channel access times, multigroup and wide-area broadcast messages, dynamically changeable priority levels, extra secure authentication and encryption, task-optimised noise cancellation, storm plans and special-event plans, back-to-back working, ground-to-air links, surveillance functions and ruggedised user equipment – all that special stuff. This, however, implies parallel standards and parallel development, manufacturing and test processes[4] and significant interworking and interoperability challenges.

An alternative would be to use LTE for broadband and narrowband, including voice but voice coverage both in rural areas and in terms of urban inbuilding penetration would need to be absolutely as good as existing narrowband specialist radio.

LTE achieves high peak data rates by using wide channels (up to 20 MHz) in a wide variety of widebands (190 MHz at 2.6 GHz for example) with longer-term support of eight or more bands. The 'cost' of delivering these high peak data rates is that multiband LTE user equipment will be inherently less sensitive than single-band narrowband radios.

Specialist mobile radio networks have to support fewer bands, often in the past only one, and have much narrower operational bandwidths and channel spacing. Narrow channel spacing allows for a narrowband IF. A narrow operational band allows for high-Q filters and a highly efficient transmit and receive chain. This means that sensitivity and selectivity and transmit and receive efficiency are generally better than wider-band devices. As a result, voice coverage in specialist radio networks is likely to be better, particularly in rural areas or for deep inbuilding penetration.

The IP voice used in LTE networks introduces additional packet header overheads that have to be taken out with a compression algorithm. Extracting a 12-kbps voice stream from a wideband 20-MHz channel is also computationally expensive. The computational overheads and compression clock cycles will also introduce delay overhead that would be better avoided if press to talk/press to connect needs to be supported. Talk groups and back-to-back working will also be needed, which will require additional standards work.

For applications where high peak data rates are needed, LTE will provide peak data rates that specialist radio networks will find it impossible or at least difficult to match.

For voice and text applications, or where average data rates are more important than peak data rates, for example in larger cells, narrowband networks will provide better voice quality and coverage and probably a longer data duty cycle. LTE could compete on coverage and building penetration if the voice or data traffic was heavily error protected, but this will compromise spectral efficiency. If the bandwidth is inexpensive this does not matter. If the bandwidth is expensive, it matters a lot. Heavy error protection will also reduce TX and RX efficiency, which will compromise the user data duty cycle. This does not matter if the user is attached to a vehicle power supply. It does matter if the user is out on an eight-hour shift on foot or bike or horseback (not an uncommon requirement in crowd control situations).

---

[4] http://www.p25.com/.

16.14 Policy Issues and Technology Economics

So the suitability of LTE for public safety in the US depends partly on spectral allocation and auction policy, partly on technology policy, partly on the application mix and partly on available budgets.

It seems unlikely that LTE 700-MHz networks on their own could meet all the needs of the public safety and/or SMR sector. This means that it will be impossible to discontinue either P 25 or iDEN radio and network system development and deployment. This loads parallel product development manufacturing and test cost onto the sector.

Some of this cost could be reduced if future iterations of P25 and iDEN could be dovetailed into the Release 10 (LTE Advanced) standards process. A 5 MHz channel could be used for LTE leaving sufficient channel bandwidth for narrowband Project 25 voice connectivity.

This would of course be entirely possible but presently improbable due to understandable vested interest both in the specialist radio vendor and specialist user community. Sector procurement policy is also understandably cautious about what might be perceived as relatively radical change.

While it is possible to design military specification radios that can access multiple bands using multiple technologies, these devices are neither low cost nor particularly energy efficient. They are also heavy. Public safety and emergency-response user requirements could be best met by flexible radios capable of accessing commercial LTE radio bands and the public-safety bands with RF performance at least as good as existing specialist radio products at a price point close to consumer price expectations. At present it seems unclear as to how this will be achieved.

The US 700-MHz band plan is a salutary example of what happens when spectral allocation and auction policy is decided by economists who ignore engineering advice and abrogate regulatory responsibility by allowing the market to decide technology policy. Out of necessity, operators and vendors have to respond to short-term market expectations. These are incompatible with the long-term decisions that need to be taken on future technology options.

The US may well end up with well-executed and well-integrated broadband and narrowband public-safety networks but they will be neither technically nor commercially efficient unless a much greater degree of user equipment RF front-end flexibility can be realised. This in turn requires a level of investment to which the industry presently has limited visibility.

In Europe and the Rest of the World there is an obvious opportunity to use at least part of the LTE 800 band (between 790 and 862 MHz) to support public-safety broadband connectivity particularly as these bands overlap the Project 25/iDEN bands in the US.

This would mean that standard LTE 800 user equipment and network hardware and software could be used in the US public-safety market in the APCO band, providing the global-scale economy needed by the sector.

The (rather major) snag with this is that LTE 800 may be deployed as a reverse duplex with mobile RX at 790 to 821 MHz and mobile TX at 831 to 862 MHz. This is because there is concern in the broadcasting community that transmissions from LTE mobile users will interfere with terrestrial TV reception, a process known as 'hole punching'. Alternatively, the band could be deployed as a standard duplex with TV signals retransmitted from cellular base stations. Apart from opening up opportunities for integrated LTE 800/Project 25 public-safety networks we would also get reliable digital TV on portable receivers – always handy in a national emergency and particularly relevant in countries presently attempting to reinvigorate

Table 16.8   US and European/ROW 800-MHz band plan implemented as standard duplex

	US	EUROPE/ROW
	MOB TX P25/iDEN/US850	MOB TX LTE 800 790
806 to 809	Narrowband P25 or iDEN Could also be two by 1.4 or one by 3 MHz LTE	
809 to 814	Broadband P25 LTE or iDEN LTE (5 MHz)	
814 to 849	Extended US 850 LTE	821 MOB RX LTE 800 831
851 to 854	Narrowband P25 or iDEN Could also be two by 1.4 or one by 3 MHz LTE	
854 to 859	Broadband P25 LTE or iDEN LTE (5 MHz)	
859 to 894	Extended US 850 LTE	862
		MOB TX LTE 900 880
896 to 902	iDEN SMR TX	
		915 MOB RX LTE 900 925
934 to 940	iDEN SMR RX	
		960

local TV. It would also make TV more resistant to interference from white-space transmissions. The band plan if implemented with LTE 800 as a standard duplex (mobile TX in the lower paired band) would look something like Table 16.8.

As an added bonus it can be seen that LTE 900 overlaps with the upper iDEN bands at 900 MHz, which could be implemented as four by 1.4 MHz LTE bands, two by 3 MHz LTE or one by 5 MHZ LTE with some guard band.

Getting a band plan like this underway would, however, require a regulatory environment that creates incentives to encourage cooperation between the cellular and broadcasting industry, the public safety radio user, standards and vendor community and white-space vendors and investors on both sides of the Atlantic, preferably bridging private-sector and public-sector interest.

An adversarial spectral allocation process designed to maximise income from spectral auctions makes this cooperation harder to achieve. This is compounded by a failure to realise any substantive global harmonisation, particularly in the LTE 700 band. A lack of cross-sector and/or transatlantic let alone global thinking and direction in standards setting also does not help.

International spectral and standards policy together have a direct impact on user equipment cost and performance and user experience value. Regional standards now make very little economic sense. Nationally specific standards make even less sense. Vendor-specific standards only make sense to the vendor.

If a local or regional band allocation or adopted standard does not scale on a global basis, this needs to be factored into bid valuation and network return on investment expectations. This seldom seems to happen.

16.15   Satellites for Emergency-Service Provision

On 10 July 1962 Telstar was launched into an elliptical orbit going round the earth every two hours and 37 minutes at a height of between 1000 and 6000 kilometres. The satellite had been built at Bell Telephone Laboratories, was roughly spherical, about 34 inches (880 mm) in length and weighed 170 pounds (77 kilograms). Working on 14 watts of power from a solar cell array the satellite received signals at 6 GHz and transmitted at 4 GHz and successfully relayed television pictures, telephone calls and faxes.

The launch was privately sponsored, which in itself was an innovation. The consortium included AT and T, the British Post Office and the French National PTT. The US ground station was in Andover in Maine, the UK ground station was on Goonhilly Downs and the French Ground Station was at Pleumeur-Bodou in North Western France. The BBC managed the conversion between the 525 line (US) and 405 line (UK and European) television standards. The ground-station antennas were huge, weighing 380 tons, with an aperture of 3600 square feet (330 square metres). The antennas were housed in radomes the size of a 14-storey building and had to track the satellite with a pointing error of less than 0.06 degrees.

Sixty years later satellites continue to play a fundamental role in telecommunications, but probably most impressively are powerful enough to transmit signals to hand-held devices. It is this part of the story that we particularly want to tell.

Just over thirteen years ago two companies, Iridium and Globalstar started providing a service to mobile users with hand-held phones from two low-earth orbit constellations.

The Iridium[5] project, championed, engineered and financed by Motorola, involved launching sixty six satellites into low-earth orbit to provide cellular-type services at a time when cellular networks were becoming increasingly ubiquitous and cellular service increasingly competitively priced. The system was and still is a spectacular engineering success; a tribute to largely US-based engineering resource, but at the time was a fiscal failure. Figure 16.1 shows the constellation and north south orbital paths.

The business model was predicated on the existence of a user community who would prefer not to use their cellular phone or two-way or short-wave radio as a preferred communications device. This user community would instead choose a system where the phones and phone service were made available at a substantial premium with poor indoor coverage, packaged in a form factor similar to a Motorola World War walkie-talkie radio.

Globalstar launched a competing constellation of 48 higher-altitude satellites. Like Iridium these were an engineering triumph but at the time a fiscal failure. Both Iridium and Globalstar went into Chapter 11 administration.

But life moves on and moves in mysterious ways. A retrospectively prescient decision was taken not to deorbit the satellites but to maintain both constellations and continue to service and develop a loyal group of specialist users. And then came 9/11, and the second Gulf War and

---

[5] Iridium is the 77th element and/originally there were to be 77 satellites in the constellation. Most Iridium found on the earth's surface is from meteorites.

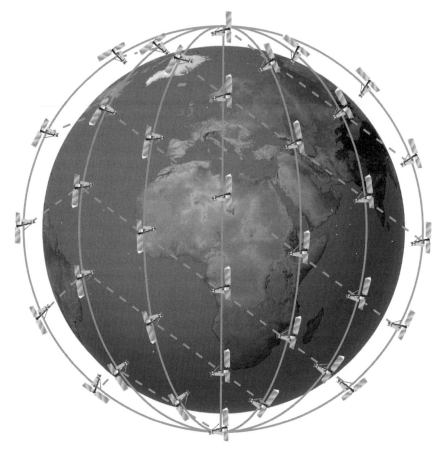

Figure 16.1   The Iridium Constellation. Reproduced with the permission of Iridium.

Afghanistan and Hurricanes Katrina and Rita and the Asian Tsunami, the Madrid bombings and the 7/7 bus and tube bombings in London and more recently the forest fires in California plus earthquakes in Turkey, Haiti, China, New Zealand and Japan, a few famines and floods and other natural and unnatural disasters around the world.

16.16   Satellites and Cellular Networks

Cellular networks were, and are, not always ideal to provide first-responder support in these often hazardous and naturally or unnaturally chaotic unwanted and unpredictable events. For example, towers and/or terrestrial telephone links can be blown up or blown down.

Iridium has always had a number of inherent resiliency advantages both over terrestrial only networks and other satellite networks. It was, and is, the first and only civilian low-earth orbit satellite system to implement intersatellite switching, reducing dependency on any single ground facility.

As a LEO (low-earth orbit) constellation, round-trip latency is 20 milliseconds, substantially lower than the 133 milliseconds of a MEO (medium earth) or the 500 milliseconds of a GEO (geostationary) satellite system. This makes speech and latency-sensitive data exchanges easier to support.

Additionally, the satellites have proved to be significantly more robust than expected and continue to provide service across the US, Alaska, Hawaii, the Pacific Ocean (as an integral part of the now updated tsunami warning system) and other hard-to-reach parts of the world.

Iridium therefore had a perhaps unexpected opportunity both politically and financially to justify new investment in a replacement constellation and updated service platforms, to negotiate innovative collaborative deals with other traditional and nontraditional service providers and possibly in the longer term to justify preferential access to new spectral allocations at L band between 1518 and 1675 MHz or S band between 1.97 and 2.69 GHz. Iridium and a number of other entities have similar plans. Iridium have executed on that opportunity with a fully funded proposal to launch a new constellation.

Whatever we write at this point will be out of date by the time you read it but essentially the new constellation (called NEXT) will do everything the old constellation did but better and with enhanced data and environmental sensing functionality. The best thing is to go to the Iridium NEXT[6] web site to see what has happened.

16.17 The Impact of Changing Technology and a Changed and Changing Economic and Regulatory Climate – Common Interest Opportunities

This opportunity has to be seen within the context of a substantially changed and changing economic and regulatory climate. Satellites are attractive again as investment opportunities.

Partly, this shift is technologically driven.

Satellites can now pack more processing power into a much smaller space. Advances in RF and baseband hardware have delivered a steady year-on-year increase in functionality per kilogram of orbital weight. Solar panel arrays are more efficient and can deliver more onboard power to support wider-bandwidth two-way communication.

Smart antenna technologies have improved over the past ten years so available power can be more accurately and adaptively deployed. Improvements in station-keeping efficiency and hardware reliability have helped to increase the life span of satellites. An operational life of 15 years is now a realistic expectation even for the traditionally shorter-lived low-earth orbit platforms.

A reasonably broad choice of launch options and some innovative mission-insurance solutions have helped trim launch costs. All these factors together have contributed positively to the overall economics of providing or updating and upgrading satellite-based services.

Iridium has the advantage of an existing constellation, an established and loyal user base and a track record of providing emergency-service support.

It has to be said that cellular operators have not been as conspicuously successful at nurturing and serving specialist user communities. The lack of service immediately post Hurricane Katrina for example was understandable but resulted in politically costly censure.

---

[6] http://www.iridium.com/about/IridiumNEXT.aspx.

Cellular operators would do well to review their service offerings for the public-safety sector and ensure that these sectors are at least adequately represented in their overall customer mix. Just focusing on consumer and corporate users and/or consumer and corporate applications is probably not wise either financially or politically in the present unstable world climate. This suggests an opportunity for cellular operators to work with satellite service providers on integrated services for specialist users.

Conversely, Iridium have to manage their user base across a transitional period where the existing constellations are past their nominal end-of-life expectancy. Globalstar, a competitor constellation had a number of RF hardware failures that reduced availability of their S band voice services and both Globalstar and Iridium have to finance and launch new satellites over the next five years. At the time of writing Iridium appear to be successfully achieving this.

This suggests a need to work with rather than against cellular operators. Each party has something the other party needs, always a good basis for a collaborative venture. There may be additional opportunities to work with other satellite operators with medium-earth or geo-stationary satellite systems, two-way radio service providers and the broadcasting community. Most broadcasters for example have a public-service remit that extends to providing emergency broadcasting services in response to local, national or international emergencies.

Satellites have always been a politically sensitive sector and so has satellite spectrum, particularly the allocations in L band and S band which are shown in Table 16.9. Some of that spectrum has been acquired or allocated on advantageous terms and can potentially be repurposed beyond an original remit to provide specialist broadcasting and/or emergency-service provision. For example, spectrum originally intended for broadcast TV could be extended to embrace a much broader multiplex of essential and nonessential service propositions.

There is substantial scope here for special pleading on the basis of social need, for example the provisioning of broadcasting and/or emergency service communications in emerging countries. Special pleading can, however, sometimes be a smoke screen for more prosaic economic ambition.

Cellular operators are right to be wary of new competitors who could be potentially successful at leveraging political influence into preferred access to new or existing spectrum. To an extent the very specific public service obligations imposed on the public-safety bands in the upper band UHF US auction are an early indication of similar battles that will be fought internationally over the next three to five years in the ongoing L band and S band allocation and auction process.

Such an adversarial approach to spectral allocation is inappropriate in a world where radio communications, particularly integrated radio communications, have an increasingly important role to play in emergency-service provision.

16.18   And Finally – Satellite and Terrestrial Hybrid Networks

Here, unashamedly, I returned to a 2007 technology topic on satellite and terrestrial hybrid networks, a sector that I have always been sceptical about.

In 2007 this was a gung ho investment opportunity. One of the big players who emerged in the 2007 to 2011 'window of opportunity' is a company called LightSquared who inked an agreement with Inmarsat.

I cannot resist including this particular news item by Ian Scales dated 21 June 2011 from Telecom TV.com. Table 16.10 puts the GPS issue into perspective.

Table 16.9 Terrestrial broadcast and satellite coexistence in L Band

L Band Terrestrial Broadcast		MHz	MHz	MHz	MHz	MHz	MHz	MHz	MHz
DAB	Broadcast audio	1452.96 1478.64 25 MHz							
DVB	Broadcast video							1670 1675 5 MHz	
L Band Satellite broadcast									
S DAB	Broadcast audio		1480.352 1490.624 10 MHz						
World Space	Broadcast audio	1453.384 1490.644 35 MHz							
L Band Satellite Two-Way Mobile									
Inmarsat Thuraya ACES				1525 1559 34 MHz			1626.5 1660.5 34 MHz		
Mobile Satellite Ventures				1525 1559 34 MHz			1625.6 1660.6 34 MHz		
Iridium						1610 1626.25 16 MHz			
L Band /S Band Two Way Mobile									
Globalstar						1615.15 1626.15 10 MHz			2483.5 2500 15MHz
L Band GPS Satellite Positioning									
GPS				1575					
Galileo				1575					
Glonass					1602 1615				

Table 16.10 GPS satellite constellations

Name	Orbit	Number of satellites	L band spectrum MHz	Status
GPS	MEO at 20 200 km	24 operational satellites, six nearly circular orbits	1575.42	Fully operational since 1994
Galileo	MEO at 23 222 km	30 satellites, in three Walker orbits	1575.42	Planned to be fully operational by 2012
Glonass	MEO at 19 100 km	24 satellites	1602–1615 MHz	Fully operational

LightSquared management and its investors are either on a roller-coaster ride of "ducking, diving, dodging and weaving," as those with South London entrepreneurial spirit might describe it, or they are following a master plan of great cunning and complexity.

In case you've not been following, LightSquared is a US start-up, backed by a prominent investor fund, which has developed plans for the building of a US-wide wholesale 'hybrid' LTE network, using both conventional LTE deployment (base stations, femtos, etc.) and satellite-delivered LTE for those hard to get-at places. With this carrot (rural coverage) it has secured regulatory clearance, investor and vendor backing and, apparently, contingent AND concrete wholesale customers (operators who will buy capacity from LightSquared to sell on to their customers). LightSquared also seemed to move quickly from PowerPoint to BuildPoint, with a satellite launched and deals with other carriers and equipment vendors signed.

But the 'Radio Network God' (whoever she is) is not easily placated by a slick presentation and few partner announcements. LightSquared ran into trouble over the interference problems it would allegedly visit on the Global Positioning System which operates in adjacent spectrum.

A campaign to stop LightSquared was promptly mounted which turned into a ginger group (SaveOurGPS.org, described as a cross-industry collection of manufacturers) and the company agreed to independent testing to make sure there was no interference.

Recently, it was supposed to report back to the FCC on the results, but has been delayed on the grounds that LightSquared needed to consider and analyse, leading to obvious speculation that the satellite part of the plan had been gotcha'd, interference was indeed a real problem, and that someone, somewhere had some explaining to do to LightSquared investors on why the technical due diligence failed.

To bolster confidence, details were then leaked on a complex network sharing deal with Sprint, which seemed to indicate that the terrestrial part of the enterprise was still on track and viable, no matter what happened with the satellite part (which will now, no doubt, be presented as a peripheral aspect of the build plan).

A cynical observer might conclude that leaked details are often preferable to properly presented details because crucial flaws can remain obscured behind a wall of "no comment, we are in a quiet period" (a stance that has become more company moto than PR tactic, and deployed by many companies, not just LightSquared).

Now, the very latest twist in tail circuit is that LightSquared has a produced satellite spectrum 'Plan B' that nobody appears to have guessed existed. The line is that instead of using the spectrum it was initially going to use for the satellite part of its LTE service, it has produced another chunk of spectrum which will serve almost as well and is parked further away from the spectrum used by the Global Positioning System so won't interfere with it. This spectrum was to be used later on in the system's development plan, the company has indicated.

To bring this 10-MHz block of spectrum into operation, however, it will have to work with satellite operator, Inmarsat, to roll out services.

So problem solved? Of course not.

SaveourGPS.org says the new plan is a figleaf. Even at the lower end of the MSS band (which it's using) the LightSquared service would still interfere with many GPS receivers as well as the precision receivers that LightSquared itself has conceded will be affected. The results of the interference study, it says, already confirm this – the only way forward is for LightSquared is to move out of the MSS band completely.

This story has some way to run yet.

The moral of this tale is of course it is really important to do technology due diligence before you throw millions and sometimes billions of dollars at a business plan.

But now let's go back to that July 2007 technology topic.

In July 2007 dual-mode networks that combine satellite and GSM cellular services already existed. Thuraya, a network operator servicing the United Arab Emirates and ACEs servicing Asia were two examples.

Hybrid satellite/terrestrial systems are different in that they use terrestrial repeaters to combine the wide-area coverage capabilities of satellites with the urban coverage and in building capabilities provided from terrestrial networks. These may or may not be associated with existing cellular or terrestrial broadcast networks.

The terrestrial repeaters are described in the US as ancillary terrestrial components (ATC) and in Europe and Asia as ground-based components. In China the networks are described as satellite and terrestrial interactive multiservice infrastructure (STiMI).

Hybrid networks were already used in the US to deliver audio broadcasting for in-car and in-flight entertainment. Widespread deployments of these systems were planned both for audio and TV broadcasting and for two way cellular service provision.

These deployments exploited already allocated L Band and S Band spectrum in US and rest of the world markets. Some of this spectrum has been gifted to the operators. Typically these allocations are not owned by traditional terrestrial cellular or terrestrial broadcast service providers. As such, they represented a competitive threat and by implication a collaborative opportunity for the cellular network operator and traditional terrestrial broadcast community. Present S band satellite spectral allocations are shown in Table 16.11. Their relationship to cellular allocations above and below 2 GHZ are shown in Table 16.12.

Hybrid networks were, and possibly are, an integral part of the convergence presently taking place between satellite and terrestrial cellular, TV broadcast and broadband and narrowband data-delivery systems. This convergence offers positive crossover value opportunities but these opportunities need to be qualified within the present and future context of the satellite industry. They also need to be technically efficient. If a network is not technically efficient it will not be commercially efficient.

16.19   Satellite Spectrum and Orbit Options

The satellite systems of interest to us are deployed either into L Band between 1518 and 1675 MHz or S Band between 1.97 and 2.69 GHz and are either in geostationary geosynchronous (GSO) orbits at 35 000 km, in medium-earth (MEO) orbits between 10 000 and 20 000 km or in low-earth (LEO) orbits or 'high' low-earth orbits (HLEO) between 700 and 1400 km. For the sake of comparison, the US Space Shuttle orbits at 350 km.

Some highly elliptical orbits such as the Molniya or low polar orbits provide optimised coverage for countries at extremely northern or extremely southern latitudes. 'Seeing' geostationary satellites from these latitudes can be problematic. The choice of orbit determines the number of satellites needed to provide a particular coverage footprint, the size and position of the satellites determines their functionality. The spectrum into which the satellites are deployed and the proximity of this spectrum to other users also determines their functionality, particularly in terms of system interoperability with other terrestrial or satellite networks.

LEOS have the advantage of low round-trip latency, about 20 milliseconds compared with 133 milliseconds for a MEO and 500 milliseconds for a GSO satellite. Geostationary satellites have the advantage that they stay in the same place when viewed from the earth. This simplifies handover and radio planning algorithms when servicing mobile users.

Table 16.11 Present and proposed S band satellite and terrestrial (ATC) broadcast and two-way radio networks

	MHz	MHz	MHz	MHz	MHz	MHz	MHz
XM Sirius	Broadcast 192 audio channels		2332.5 2345.00 12.5 MHz				
Mobaho Japan	Broadcast 11 video channels including HDTV 25 audio channels 3 data channels					2630 2635	
S-DMB Korea	Similar to Mobaho					2630 2635 25 MHz	
Terrestar and ICO US	Broadcast and two way	2000 2020 20 MHz	2180 2020 20 MHz				
ICO Europe	Broadcast and two way	1997.5 2010 12.5MHz	2187.5 2200 12.5 MHz				
Eutelsat SES Astra	Broadcast but potentially broadcast and two way	1980 2010 TBD	2170 2200 TBD				
WiFi	Potential integration with satellite receivers						2400–2480 80 MHz

Table 16.12 US S Band cellular allocations including satellite/terrestrial hybrids

Band	Frequency (MHz)	Operational Bandwidth (MHz)
AWS	1710–1755	45
PCS 1900	1850–1910	60
Sprint/ Nextel	1910–1915	5
PCS1900	1930–1990	60
Sprint/ Nextel	1990–1995	5
ICO/Terrestar	2000–2020	20
AWS	2110–2155	45
ICO/Terrestar	2180–2200	20
WCS(Sprint/Nextel)	2305–2320	15
SDARS (XM/Sirius)	2332.5–2345	12.5
WCS (Sprint/Nextel)	2345–2360	15
WiFi	2400–2480	80

The size of the satellite determines the size of the antenna array, sometimes upwards of 20 metres in large geostationary satellites. The size of the array determines the uplink and downlink gain available, particularly if adaptive spot beam-forming techniques are used.

Similarly the size of the solar panel array, sometimes upwards of 40 metres in large geostationary satellites, dictates the amount of power that can be generated, which determines both the coverage and capacity.

Advances in launch technologies have made it possible to launch satellites weighing over 5000 kg into geostationary orbit. Advances in RF and baseband hardware deliver a steady year-on-year increase in functionality per kg of orbital weight.

Conversely, microminiaturisation techniques have made possible new generations of super small satellites though these tend to be used for more specialist low-orbit or deep-space exploration applications. However, terrestrial network hardware costs have also reduced over the past twenty years by roughly 15% year-by-year and functionality has greatly increased.

In particular the rapid growth in subscriber numbers served by terrestrial cellular networks and terrestrial broadcast networks has attracted engineering investment that in turn has improved the delivery cost efficiency of these networks. Present upgrading of the DAB terrestrial networks in the UK to provide improved coverage and higher data rates provides an example.

In terms of user devices, the economies of scale available to the cellular industry effectively dictate the radio functionality included in mobile handsets. In the past, these factors have invalidated a number of apparently promising satellite based business models. New satellite ventures therefore have to be approached with caution and the relative merits and demerits of each option need to be carefully considered. This is particularly true when significant amounts of spectrum are being allocated by regulators either for satellite-based services or for new hybrid satellite terrestrial network propositions.

Satellites have delivered telecommunications, TV and data for over 50 years, the original triple-play proposition. This interdependency has determined the economics of the industry.

Over the past twenty years improved power output (downlink capability) and sensitivity (uplink capability) has allowed satellites to play an increasing role in delivering communications to mobile devices including mobile handheld devices. Hybrid satellite terrestrial networks are a logical next step, but have to be qualified in terms of the additional value that they deliver to existing terrestrial networks.

16.20   Terrestrial Broadcast and Satellite Coexistence in L Band

Table 16.9 referred to earlier shows the allocations to terrestrial broadcast, satellite broadcast and satellite two-way services in L Band.

16.21   Terrestrial DAB Satellite DAB and DVB H

The L Band allocations for DAB and DVB H were identified, although these allocations were not universally available in all countries. DAB was intended primarily for radio but could carry a TV multiplex of up to 7 video channels on a 1.7-MHz OFDM channel.

DVB H was implemented as a trial in New York delivering 75 TV and music stations transmitted from 74 terrestrial sites covering 475 square miles. The network used a 5-MHz channel rather than the standard 8 MHz used for DVB H at UHF or S band. DVB H could have been implemented in L band as a DVB SH network with terrestrial repeaters but lacked global scale and therefore could not be progressed.

16.22   World Space Satellite Broadcast L Band GSO Plus Proposed ATC

World Space provided radio services to Africa and Asia from two geostationary satellites, Afristar launched in 1998 and Asia star launched in 2000. These satellites use the same air interface as XM radio. Plans to launch a third satellite, Ameristar, to serve South America were not implemented as these L band frequencies were used by the US Air Force.

This satellite was then intended to be repurposed to provide European coverage with a particular emphasis on Italy. The satellite would transmit and receive with an air interface theoretically compliant with the ETSI satellite digital radio standard that was being extended to include S UMTS and DVB SH and a legacy set of standard documents known as GMR, the geostationary mobile radio standard. World Space had terrestrial repeater licenses for Bahrain and the United Arab Emirates and lobbied for similar license concessions in its other addressable L band markets.

16.23   Inmarsat – L Band GSO Two-Way Mobile Communications

Inmarsat have traditionally provided mobile and fixed communication services to the maritime, aeronautical, land mobile and remote-area markets. Recent investments have focused on increasing data rates as part of their Broadband Global Access (B GAN) network.

Space to earth links are at 1525 to 1559 MHz and earth-to-space are at 1626.5 to 1660.5 MHz. The network supports user data rates up to 256 kbps. Terminals are typically 1.3 kg or above and power consumption is in the region of 14 watts. There are ten satellites in a geostationary (GSO) orbit. The latest satellites launched have a planned 13-year operational life and a 9 metre antenna array.

16.24   Thuraya 2 L Band GSO Plus Triband GSM and GPS

Thuraya provided satellite service to mobile hand-held devices that also supported triband GSM and GPS with coverage optimised for the United Arab Emirates and the Middle East. The original geostationary satellite was launched in 2000 with a second in 2003 and a third in 2004. Uplinks and downlinks were the same as Inmarsat. The Thoraya 2 satellite had/has a 12-metre antenna array. The GPS frequency at 1575.42 sits reasonably conveniently between the L band uplink and downlink frequencies used in the handset. There were no stated plans for a hybrid ATC network.

16.25   ACeS L Band GSO Plus Triband GSM and GPS

More or less equivalent to Thuraya combining service from one geostationary satellite launched in 2000 and positioned to provide optimal coverage over Asia rather than the Middle East. Inmarsat and Aces proposed to combine their network proposition to provide global coverage. There were no stated plans for a hybrid ATC network.

16.26   Mobile Satellite Ventures L Band GSO Plus ATC

Mobile Satellite Ventures planned to launch two geostationary satellites in 2009/2010. These were large satellites weighing 5500 kg with a 22-metre antenna array and 500 spot beams. Coverage was to be optimised for the US, Southern Canada and Latin America.

The company had a license to deploy a 30-MHz paired band with the lower duplex between 1525 and 1559 paired with the upper duplex between 1625.5 and 1660.6 MHz based on proposals first tabled to the FCC in 2001. This became the basis for the LightSquared business case.

The mobile uplink was to be processed by both satellites with the signal being diversity combined, a technique also used by Globalstar. Terrestrial transmitters were to be added to existing cellular sites to provide additional coverage and capacity. Essentially this was an extension of the terrestrial repeater principle to two-way communication based on a combination of terrestrial cells with a radius of between one and five kilometres nesting within a satellite cell with a radius of about 100 km. This brought Mobile Satellite Ventures into direct competition not only with terrestrial broadcasters but also with cellular service providers. Mobile Satellite Ventures claimed to have substantial patent-based intellectual property regarding space/terrestrial frequency reuse, beam forming with larger arrays and handover algorithms. There was some crossover of patent value and personnel with XM radio.

16.27   Global Positioning MEOS at L Band GPS, Galileo and Glonass

Table 16.10 referenced earlier summarized the global positioning MEOs.

The US-managed GPS satellites have a 7.5-year design life, but satellites are lasting 12 years.

Galileo is a European initiative with coverage, positioning accuracy and satellite lock times optimised for Europe. The satellites will have a 12- to 15-year design life. Galileo has the same downlink frequency as GPS and could be expected to be implemented as standard in future handset designs.

It is theoretically convenient to build an L band transceiver with GPS capability, though care has to be taken to avoid desensitisation of the GPS or Galileo signal within the handset. This appears at the time of writing to be the fatal flaw in the LightSquared business plan.

GPS is of course now in a wide range of devices including the latest Sony 'Vita' PlayStation (Chapter 9). Most smart phones and of course all navigation devices, usually integrated with a GPRS receiver to capture traffic jam alerts to support optimised routing. It seems like we may be the last generation to be able to read a map.

16.28   Terrestrial Broadcast and Satellite Coexistence in S Band

Table 16.11 summarised present and proposed S band hybrid satellite and terrestrial broadcast and two-way radio networks

16.29   XM and Sirius in the US – S Band GEO Plus S Band ATC

XM[7] and Sirius[8] are two operators in the US providing MP3-quality audio radio for in-car and more recently in-flight entertainment. The two companies are presently in merger discussions.

XM has four 15-kW geostationary satellites XM1 (Rock), XM2 (Roll), XM3 Rhythm and XM4 'Blues'. XM1 and XM2 have suffered some fogging on their solar panels.

The satellites work with 1500 terrestrial repeaters that are each technically capable of delivering an ERP of 25 kW. Sirius has an additional three satellites also in geostationary orbit.

The air interface is specified in a standard known as SDARS[9] (Satellite Digital Audio Radio Service). The networks operate in a 12.5-MHz band between 2332.5 and 2345 MHz with four of the six radio carriers dedicated to the satellites and the other two channels dedicated to the terrestrial repeaters.

Satellites are QPSK to maximise power efficiency. The terrestrial repeaters use COFDM.

User terminals have two separate antennas, one for the satellite signals and one for terrestrial signals. The signals are combined in the receiver at baseband using maximal ratio combining. Alternatively, simple voting is used to choose the strongest signal.

The receivers tend to be traditional superhets with a 75-MHz IF commonly also used in TV receivers. Antenna systems for these devices are complex[10] though the design requirements for receiving satellite and terrestrial signals simultaneously are now well understood.

There is a present proposal to deliver more localised services from the terrestrial repeaters. The National Association of Broadcasters in the US opposes this.

16.30   Mobaho in Japan and S DMB in South Korea – S Band GSO Plus ATC

Mobaho in Japan and S DMB had a similar network configuration but implemented at 2.6 GHz. A single geostationary satellite covers Japan with a left-polarised beam and S

---

[7] https://www.siriusxm.com/player/.
[8] http://www.siriusxm.com/servlet/ContentServer?pagename=Sirius/CachedPage&c=Page&cid=1018209032790.
[9] http://www.tvtower.com/xm-radio.html.
[10] http://www.mwrf.com/Articles/Index.cfm?Ad=1&ArticleID=5892.

Korea with a right-polarised beam. A 12.226-GHz transponder on the satellite provides a downlink to terrestrial transponders to provided services in subways and tunnels.

16.31   Terrestar S Band in the US – GSO with ATC

Terrestar[11] were granted access rights to two by 10 MHz allocations at 2000 to 2020 and 2180 to 2020 MHz. The band is shared with ICO. The launch of the initial Terrestar 500 spot beam geostationary satellite was originally planned for November 2007, but has been rescheduled for September 2008.

Terrestar have a joint venture with Orbcomm[12] who have a legacy 30-satellite LEO constellation of micro-160-watt satellites with an uplink at 137–138 MHz and downlink at 400–401 MHz. Orbcomm went into Chapter 12 with Iridium and Globalstar but re-emerged and refinanced and at the time of writing provides services to M2M markets including telematics and asset tracking.

16.32   ICO S Band GSO with ATC

The FCC granted ICO[13] a license for the other two by 10 MHz allocations at S Band with a mobile uplink between 2000 and 2020 MHz and a downlink at 2180 to 2020 MHz. An initial satellite launch is planned for the end of 1997 optimised for US coverage.

The ground-based components receive the satellite signal in K band, down convert to 2 GHz then transmit in synchronisation with the satellite signal. Additional satellites would provide improved inbuilding penetration.

The suggestion is that this network could provide up to 50 TV channels to mobile handsets and is therefore a potential competitor to Media FLO though the network would have other functionality including two-way voice and data, interactive multimedia and disaster-relief capabilities.

16.33   ICO S Band MEO at S Band with ATC

ICO planned a MEO constellation of 12 satellites in two inclined orbits at 10 400 km. The launch of the first satellite failed, but the second satellite went into orbit in June 2001 and is operational and capable of providing services to Africa, Europe and Asia. The ITU/CEPT S Band allocation to ICO specified a 12.5-MHz mobile uplink between 1997.5 and 2010 MHz and a mobile downlink from the satellites between 2187.5 and 2200 MHz.

There has been regulatory discussion as the continuing validity of ICOs claim to this spectrum. ICO base their claim to continued access rights on the basis of their legacy investments and stated plans to repurpose present and future satellites to support a DVB SH ATC network as a joint venture with Alcatel Lucent. Alcatel Lucent are leading a consortium with Sagem,

---

[11] http://www.terrestar.com/.
[12] http://www.orbcomm.com/.
[13] http://www.ico.com/.

Philips and DiBcom known as Television Mobile Sans Limite[14] (Mobile Television without Limits) to promote wider adoption of the DVB SH standard.

China have several parallel standards initiatives using satellite and terrestrial interactive infrastructure and other localised reinterpretations of the T DMB and DMB T standards.

16.34   Eutelsat and SES ASTRA GSO – 'Free' S Band Payloads

Eutelsat[15] has twenty three geostationary satellites. SES Astra[16] has 36 satellites in 25 orbital locations. Together, they deliver over three thousand TV stations and over 2000 radio stations to 200 million cable and satellite homes. The transition to HDTV represents a challenge and opportunity to these providers and implies a possible future need to work more closely with the terrestrial broadcasting community.

There are no stated plans by either entity to implement a hybrid ATC network, however, an S Band payload wase included on the Eutelsat W2A satellite launched in 2008/9 and the companies have a joint venture working on delivering additional S band capacity.

The technology, engineering and launch costs of these platforms are already very adequately amortised across a substantial and largely captive subscriber base. S Band capacity can therefore be added at minimal incremental cost. It could be presumed that Intelsat would have similar economies of engineering, sourcing and subscriber scale that could be applied in a similar way. Other present and proposed S Band only providers need to consider the impact this could have on their future operational margins.

16.35   Intelsat C Band Ku Band and Ka Band GSO

Intelsat[17] provides bidirectional transponder services to corporate and national governmental markets predominantly at C Band (3.4 to 7.025 GHz), Ku Band (10.7 to 14.5 GHz and Ka band (17.3 to 30 GHz). Intelsat has 51 geostationary satellites and uses spot beams to provide global, hemi, zone or spot beam coverage predominantly to fixed users. There are no stated plans for a hybrid ATC network. In common with Eutelsat and SES Astra, Intelsat has the benefit of multiple satellite constellation economies of scale and a large corporate subscriber base over which to amortise future S band engineering investments.

16.36   Implications for Terrestrial Broadcasters

As can be seen from the above, a number of these hybrid satellite terrestrial networks represent new competition for traditional terrestrial broadcasters. The ability to deliver TV to mobile users either from terrestrial repeaters or a satellite and/or simultaneously from both provides a measure of additional flexibility not available to terrestrial-only broadcasters. The addition of an uplink to support interactivity provides additional differentiation.

---

[14] http://www.generation-nt.com/alliance-alcatel-archos-pour-la-tv-mobile-dvb-h-en-bande-s-actualite-40220
.html.

[15] http://www.eutelsat.com/home/index.html.

[16] http://www.ses-astra.com/business/.

[17] http://www.intelsat.com/.

Conversely, most of the hybrid satellite terrestrial operators require access to terrestrial sites and terrestrial subscribers, many of which are largely under the control or influence of the incumbent traditional broadcast community. This suggests that collaborative rather than competitive ventures would be likely to yield better shareholder returns for all parties.

16.37 Implications for Terrestrial Cellular Service Providers

A number of these hybrid satellite terrestrial networks represent new competition for traditional cellular service providers. The ability to deliver full duplex voice, video and broadband data services to mobile users either from terrestrial repeaters or a satellite and/or simultaneously from both provides a measure of additional flexibility not available to terrestrial cellular operators.

16.38 The Impact of Satellite Terrestrial ATC Hybrids on Cellular Spectral and Corporate Value

Mobile satellite spectrum was largely allocated in 1997 and was either gifted or acquired on advantageous terms particularly when compared with the sums subsequently spent by cellular operators on PCS, WCS and most recently AWS spectrum. This incongruity has allowed the satellite operators to refinance and revalue. The FCC ruling to allow ATC terrestrial repeaters has substantially helped in this revaluation process.

The regulatory intention is to encourage investment in satellite spectrum that has remained fallow for over ten years. Some US cellular operators might be minded to question whether this is in retrospect or prospect an equitable arrangement.

Table 16.13 shows the IMT S Band cellular and satellite allocations in Europe.

16.39 L Band, S Band, C Band, K Band and V Band Hybrids

The available radio frequency spectrum carries on beyond Ka band into V band, the millimetric band between 30 and 300 GHz. These are shown in Table 16.14 with a summary of allocations at lower frequencies.

Table 16.13 IMT S Band allocations

Standard	TDD	T-UMTS	S-UMTS	TDD	T-UMTS	S-UMTS
Frequency	1900	1920	1980	2010	2110	2170
MHz	1920	1980	2010	2125	2170	2200
Operational bandwidth, MHz	20	60	30	15	60	30
			ICO MEO			ICO MEO
Frequency MHz			1997.5			2187.5
			2010			2200
Operational bandwidth MHz			12.5			12.5

Table 16.14   ITU frequency bands including V Band

UHF	S DAB	L Band	GPS and Galileo	S Band	C Band	Ku Band	Ka Band	V Band (Millimetric)
235 400 MHz	1452 1492 MHz	1518 1675 MHz	1575.42 MHz	1.97 to 2.69 GHz	3.4 to 7.025 GHz	10.7 to 14.5 GHz	17.3 to 30 GHz	30 to 300 GHz
Military mobile	TV and radio	Civilian mobile	Positioning	Cellular and broadcasting	TV, radio and data broadcast			Military mobile

These V band networks are ambitious satellite-based bidirectional broadband communication projects and include an advanced global EHF satellite hybrid MEO GSO network[18] providing broadband connectivity for US Stealth Bomber aircraft.

This network is being implemented between 36.1 and 51.4 GHz and is a dual constellation hybrid combining a MEO constellation for low latency exchanges with a constellation of geostationary satellites for less latency-sensitive uploading/downloading.

Participation in these major military projects allows US vendors in particular to amortise engineering investments that can be translated to civilian applications. The EHF project suggests that future commercial networks may combine terrestrial repeater coverage with hybrid MEO and GSO satellite coverage.

16.40   Summary

There have been rumblings of discontent in the cellular and terrestrial broadcast community that new generation hybrid satellite terrestrial networks are really terrestrial networks with an ancillary satellite component rather than satellite networks with an ancillary terrestrial component.

These networks are being deployed into spectrum that was acquired at a fraction of the cost of recent (past 7 year) cellular spectral investments. A number of the operators have also emerged from Chapter 11 administration with engineering and launch investments largely paid for by their creditors rather than consumers. This would seem to confer an unfair advantage to these companies.

This may of course be true but pragmatically there will be a high level of interdependency between satellite /terrestrial ATC networks, terrestrial broadcast and terrestrial cellular networks both at infrastructure and subscriber level. Interdependency implies collaborative profit opportunity. More tellingly, a number of the propositions will have to be closely coupled with already-established terrestrial service providers in order to achieve long-term financial viability.

In particular, the significant economies of scale available to the cellular industry effectively dictate the radio functionality included in any handset. Satellite operators need to consider this factor with particular care. The positioning of Eutelsat and SES Astra is potentially

---

[18] http://www.satnews.com/stories2007/4128.htm.

advantageous given their substantial existing satellite and subscriber assets and ability to amortise engineering and marketing investments over a large and secure existing revenue base. Intelsat has potentially similar advantages. S Band payloads on any of these satellites are essentially free.

This factor combined with other positive satellite technology and cost trends including lower launch costs, larger more efficient pay loads and improved uplink and downlink performance suggest that satellites will play an increasingly positive role in future terrestrial mobile broadband service provision.

At the very least, satellites need to be more actively factored into present and future cellular and terrestrial broadcast business planning. Telstar marked the start of a new technical and commercial era in telecoms.[19] Sixty years on it feels that the story has only just begun.

---

[19] Incidentally, July 2011 was the 50th anniversary of Yuri Gagarin's flight almost coinciding with the last planned mission of the American Shuttle.

Part Four

Network Software

17

Network Software – The User Experience

Back to earth again, the next five chapters address various aspects of network software. Chapter 18 of 3G Handset and Network Design looked at traffic-shaping protocols in some detail. This chapter provides a short reprise of traffic shaping and tracks the changes that have taken place over the past eight years.

In 2002 the discussion revolved around the scale of the efficiency gain potential of IP networks and specifically IP networks with a radio network attached. At the time, Nortel were claiming that the 3 G air interface combined with an IP network would reduce delivered cost per bit by a factor of twenty. This was absurd. Nortel is no longer in business, a salutary warning to companies that allow marketing to become disconnected from engineering reality.

The issue was that while it is very possible to achieve performance gain there is normally an associated performance cost in some other area that is often not factored into the marketing message. In this case, the performance cost of the theoretical efficiency gain was a loss of network determinism. In January 2003 after the book was published we wrote a technology topic on IP network processor performance limitations. The summary of this is as follows.

17.1 Definition of a Real-Time Network

The IEEE defines a real-time operating system as 'A system that responds to external asynchronous events in a predictable amount of time'. Handset hardware and software engineers, for example, need to make sure that application layer and physical layer functions are executed within known and predictable time scales in response to known and predictable external events and requests – this is often described as a deterministic process.

We can apply a similar definition to a real-time network as 'A network that can provide throughput to asynchronous traffic in a predictable amount of time'. Network hardware and software engineers need to make sure that router processors can process offered traffic within known and predictable time scales. This in turn requires an understanding of the offered traffic

Making Telecoms Work: From Technical Innovation to Commercial Success, First Edition. Geoff Varrall.
© 2012 John Wiley & Sons, Ltd. Published 2012 by John Wiley & Sons, Ltd.

mix and offered traffic properties, particularly any time interdependencies present between different channel streams.

17.2 Switching or Routing

The general assumption was, and is, that there will be more routing and less switching over time and less use of hardware, greater use of software to reduce cost and improve flexibility. However, software-based routing introduces delay and delay variability – the time taken to capture a packet, the time taken to check the header and routing table, the time packets spend in buffers waiting for the router to deal with other packets and the time packets spend in buffers waiting for egress bandwidth to become available (queuing delay).

Traffic-shaping protocols such as Diffserv and MPLS can help manage queuing delay (i.e. reduce delay for some packets by increasing delay for other packets) but queuing delay can still be 20 or 30 milliseconds and varies depending on traffic load. Transmission retries (using TCP/IP) introduce additional variable delay.

17.3 IP Switching as an Option

Rather than treating packets individually, a sequence of packets can be processed as a complete entity – the first packet header contains the properties and bandwidth requirements of the whole packet stream – this is described as flow switching or IP switching – 'a sequence of packets treated identically by a possibly complex routing function'. This would seem like a good idea, particularly as session persistency increases over time. However, session properties and session characteristics tend to change as a session progresses, so more flexibility is needed. It is difficult to deliver flexibility and deterministic routing or switch performance.

17.4 Significance of the IPv6 Transition

An additional challenge for an IPv4 router is that it never quite knows what type of packet it will have to deal with next. IPv6 tries to simplify things by using a fixed-length rather than variable-length header and by reducing the 14 fields used in IPv4 to 8 fields – the protocol version number, traffic class (similar to type of service in IPv4), a flow label to manage special priority requests, payload length, next header, hop limit, source address and destination address. This is one of the reasons Japanese vendors are keen on mandatory IPv6 in routers – it makes deterministic performance easier to achieve.

17.5 Router Hardware/Software Partitioning

Performance can also be made more predictable by adding in a hardware coprocessor (or parallel coprocessors) to the router. Figure 17.1 shows the packet-flow sequence in a packet processor.

To minimise delay, the router divides packet processing into a number of tasks.

The data comes in from the physical layer and is demultiplexed in accordance with the MAC (medium access control) layer rule set, packets within frames within multiframes. The packet

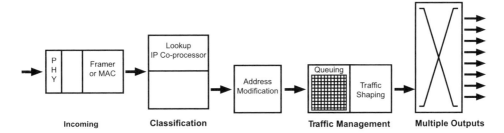

Figure 17.1   Router packet processing.[1]

is then sent for classification. At this point a hardware coprocessor may be used to improve look-up performance. Performance is defined by the number of searches per second, the number of entries in the table, and whether or not multiprotocol tables are used. Multiprotocol tables are needed if differentiated classes of service are supported. A software-based standard processor-based solution can take several hundred instruction cycles to classify a packet with QoS and/or security attributes. A coprocessor can perform the task in a single clock cycle but lacks flexibility. It can only be used when the decisions to be taken are largely predefined and repetitive. This is hardware-based switching rather than software-based routing.

17.6   The Impact of Increasing Policy Complexity

This looks like a good solution but there's a drawback. Session complexity increases over time. For example, we might need to support an increasing number of multiuser to multiuser multimedia exchanges. In these exchanges, policy management can become complex. Policy rights might be ascribed to the user such as a priority right of access to delivery or storage bandwidth. The user, the user's device or the user's application may be (likely will be) authenticated and the user's traffic may have end-to-end encryption added for additional security. In a multiuser session, the security context may change/will change as users join, leave or rejoin the session. This implies substantial flexibility in the way in which packets are handled, hard to realise in a hardware coprocessor.

In an ideal world we would like to combine consistent deterministic throughput and flexibility. In practice this is hard to achieve. One (simple) way to achieve consistent performance and flexibility is to overdimension transmission bandwidth in the IP network but this has a cost implication. Consistent network performance therefore comes with a price tag attached. Having to handle highly asynchronous traffic increases that price tag. Bursty bandwidth is expensive bandwidth.

The theoretical understanding of this can be traced back to 1909 when A K Erlang published his 'Theory of Probabilities and Telephone Conversations' – we still use the Erlang as a unit of measurement to dimension Node B, RNC and core network capacity and (rather dangerously) use Erlang traffic-arrival formula to design network processor components.

---

[1] Reproduced from 3G Handset and Network, Design Varrall and Belcher, Page 406, Figure 17.1, John Wiley, ISBN 0 471 22936-9.

The good thing about voice traffic is that it is reasonably predictable. Multimedia traffic is also quite predictable though we have less experience and knowledge of 'mixed' media/'rich' media traffic-arrival patterns (and less experience of how we should treat the traffic). This means that networks don't always behave as expected – their behaviour is not consistent with traditional queuing theory. The anecdotal experience is that as we move from voice to a mix of voice, data and multimedia, networks become progressively more badly behaved.

One reason for this bad behaviour is the increasing burstiness of the offered traffic that effectively puts network components (network processing and network router buffers) into compression. This results in packet loss and (if using TCP) transmission retries.

These 'bandwidth effects' can be due to transmission bandwidth constraints, buffer bandwidth constraints and/or signalling bandwidth constraints. Traffic-shaping protocols (TCP, RSVP, MPLS, etc.) may help to modulate these effects or may make matters worse! It is thus quite difficult sometimes to know whether we are measuring cause or effect when we try and match traffic-arrival and traffic-throughput patterns.

Benoit Mandelbrot (1982),[2] and subsequently Kihong Park, Willinger[3] and others have characterised network traffic in terms of a series of multiplicative processes and cascades – the end result is a turbulent network.

Whether using TCP or UDP, the network effectively behaves as a filter, including bulk delay and group delay effects and creates backpressure when presented with too much traffic or bursty traffic (which puts the network into compression). The effect is similar to the reflection coefficient in matching networks (VSWR).

The problem of turbulent networks is that it is difficult to predict the onset of turbulence – the point of instability, the point at which a 'fluid' flow becomes laminar. The reason that it is difficult to predict the onset of turbulence is that turbulence is turbulent. Lewis Fry Richardson (the uncle of Sir Ralph Richardson) produced some seminal work on turbulence just prior to the First World War when there was a particular interest in knowing how well aeroplanes could fly. Richardson was so inspired by the mathematical complexity (and apparent unpredictability) of turbulence that he wrote a poem about 'little eddy' behaviour – the behaviour of the little eddies or whorls that you notice in dust storms or snow storms.

Big whorls have little whorls which feed on their velocity And little whorls have lesser whorls and so on to viscosity. Figure 17.2 illustrates this concept of the network as a filter, attempting to smooth out offered traffic turbulence.

17.7 So What Do Whorls Have to Do with Telecom Networks?

Well, in order to understand network instability, we need to understand the geometry of turbulence. At what point does the flow of a liquid or gas or packet stream go from smooth to laminar (an unsmooth flow) and what are the cumulative causes.

---

[2] The Fractal Geometry of Nature, Benoit B Mandelbrot, 1983 Edition, ISBN 0-7167-1186-9.
[3] Self-similar Network Traffic and Performance Evaluation, Edited by Kihong Park and Walter Willinger, Wiley, New York, 2000 Edition, ISBN 0-471-31974-0.

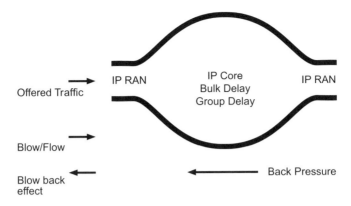

Figure 17.2   The network as a filter (the turbulent network).

In a wireless wide-area network, the root cause is the increasingly wide dynamic range of the traffic being fired into the network from cellular handsets. Release 99 3GPP1 specified that data rates could vary on the uplink between 15 and 960 kbps from frame to frame (from one 10 millisecond frame to the next 10 millisecond frame). Ten Releases later the bandwidth excursion has increased significantly and is mapped to two millisecond and in the longer term one-millisecond and half-millisecond frame lengths. This is bursty bandwidth coming from multiple similar sources. Unfortunately, merging these traffic streams together does not necessarily result in traffic smoothing – bursty data streams aggregated together may produce even burstier traffic streams. This burstiness exercises (effectively compresses) RF and DSP components in the radio layer and router and switch components in the network. RF and DSP components in compression go nonlinear and create intermodulation and distortion. Router and switch components in compression produce packet loss and packet delay (first-order effects) and loss and delay variability (second-order effects). These effects can be particularly damaging for nonelastic/inelastic traffic.

Kihong Park adds to the anxiety by suggesting that this offered traffic (in particular the burstiness of the offered traffic) can become strongly 'self-similar'. Self-similar traffic is traffic that shows noticeable (and when scaled, remarkably similar) burst patterns at a wide range of time scales – typically four or five orders of magnitude, milliseconds, seconds, tens of seconds, hundreds of seconds, thousands of seconds.

It's rather like qualifying the effect of using multiple OVSF codes at the radio layer, we can predict the dynamic range of the burstiness but not when the burstiness will occur, through we can predict how often it will occur.

Kihong Park's work is based on multiple-source variable bit rate video. The traffic coming from mobile broadband devices is not dissimilar, therefore we need to consider the impact of this traffic on network resource provisioning.

Self-similar traffic has long-range dependence – this means that it has a cumulative 'fill effect' on the buffer, in other words we need more buffer bandwidth than traditional telecom traffic theory would suggest. However, if we add additional buffer bandwidth we not only increase end-to-end delay (first-order effect) but end-to-end delay variability (second-order effect). Given that as we shift towards a richer media multiplex with an increasing percentage of conversational traffic exchanges our traffic by definition will become less elastic, this will be problematic.

17.7.1   So What's the Answer?

Philippe Jacquet, one of the contributors to Kihong Park's book paints a gloomy picture 'Actual router capacitors are dangerously underestimated with regard to traffic conditions'. While this is almost certainly true it is also reasonable to say that provided transmission bandwidth is adequately provisioned then next-generation networks will work quite well but will cost rather more than expected.

It also implies a significant shift away from present network processor design trends and a shift in the functional partitioning between software switching (flexible but slow) or hardware switching (fast but inflexible).

You can of course have flexible *and* fast but end up with overcomplex and expensive parallel processing engines heavily dependent on overcomplex and expensive memory (S-RAM and CAM).

Decisions on software/hardware partitioning self-evidently have to be based on the requirements and characteristics of the traffic being processed. As traffic becomes increasingly asynchronous it becomes progressively harder to control buffer delay and buffer-delay variability. However, it's not just 'burstiness' that needs to be considered when deciding on network processor performance.

17.7.2   Longer Sessions

First, let's consider the impact of longer sessions. Sessions are getting longer partly because file sizes are getting bigger but also because sessions are becoming more complex. Multimedia sessions tend to last longer than voice calls, multiuser sessions tend to last longer than one-to-one conversations and conversational exchanges last longer than best-effort exchanges.

As session lengths increase it becomes more economic to switch in hardware or, put another way, you don't need lots of software instructions to switch a long session – it's a relatively simple and deterministic transaction.

17.7.3   Shorter Packets with a Fixed Rather than Variable Length

Packet size is reducing over time. This is because multimedia components are source coded using periodic sampling – typically 20 milliseconds for voice (the syllabic rate). Video sampling rates depend on the frame rate and source coding used but generally you will end up with a MAC layer that produces a stream of fixed-length packets of typically 40 bytes. The decision to use fixed rather than variable-length packets is driven by the need to get multiplexing efficiency even when the data rates coming from multiple sources can be widely different (video encoding is moving increasingly towards variable rate).

It is obviously absurd to take a 40-byte payload and add a 40-byte IPv6 header to each packet, so the idea of treating each packet as an individual entity in a multimedia network is no longer valid. As we move towards short, fixed-length packets it makes sense to hardware switch. Figure 17.3 illustrates this shift.

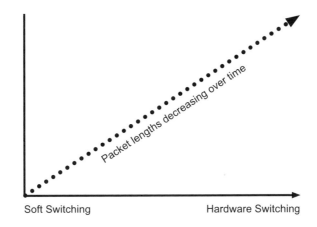

Figure 17.3   Hardware switching increases as packet length reduces.

17.7.4   Fixed Routing

As session length increases, we move progressively towards providing **less** rather than more routing flexibility.

Traditionally in an IP network, routing flexibility has been promoted due to its ability to improve transmission bandwidth utilisation and increase network resilience. However, the cost is delay (first-order effect) and delay variability (second-order effect). As traffic becomes more inelastic over time we become less tolerant to any change in routing trajectory. Routing flexibility is antithetical to network determinism.

Conversely, if we have fixed routing we don't need to have routing tables, thus saving on CAM memory. If we don't need packet-by-packet classification we don't need classifiers (saving on S-RAM). If we also decide we don't need or cannot tolerate buffering because delay and delay variability compromises our inelastic traffic then we can also get rid of all or at least some of that expensive S-DRAM and DDR DRAM.

Also, as we reduce the buffering used in network components we reduce the need to manage memory – the software processes implicit in memory management introduce delay and delay variability in the end-to-end communications channel. So it may be that the need for flexibility and the benefits of flexibility in next-generation networks are overstated. What is really needed is predictable and stable end-to-end performance that can be mapped to clearly defined quality of experience metrics. The easiest and lowest-cost way to achieve this may be to overdimension network delivery bandwidth.

So the thesis is that as session lengths increase and as we move towards fixed-length packets and fixed routing trajectories it makes more sense to hardware switch and less sense to software switch.

This goes against the present trend in which a number of vendors are promoting general-purpose network processors for IP RAN and IP core network applications (including Node B and RNC routers). The selling point of these network processors is that they offer flexibility (which we are saying will not be needed). We also have to consider that this flexibility has a cost in terms of additional processor, memory and bus bandwidth overhead.

Apart from cost implications, these overheads imply higher power consumption and some performance cost in terms of additional end-to-end delay and delay variability.

One option is to introduce hardware coprocessors to speed up time-sensitive tasks but this can result in quite complex (and difficult to manage) hardware/software configuration. Somehow, we need to work out what would be an optimum trade off point.

17.8   Packet Arrival Rates

First, we need to decide how much time we have to process a packet. A 40-byte packet has an arrival rate of 160 nanoseconds at OC48 (2.5 Gbps), 35 nanoseconds at OC192 (10 Gbps) and 8 nanoseconds at OC768 (40 Gbps). When you consider that memory access even in high-performance S-RAM is around 6 nanoseconds then you can see that you don't have much chance for multiple memory searches at higher throughput rates.

Consider also that classification software can take several hundred clock cycles and a router update can typically take well over 100 cycles. If you want line-rate or near-line-rate processing you would end up with some interesting clock speeds.

17.9   Multilayer Classification

We also need to consider the level of classification used on the packet. If this includes deep packet classification then a data payload search is needed. The data may be several hundred bits long and may be located at random within the payload.

As we move down the protocol stack we find we may have to 'parse' out MPLS headers (4 bytes) or RSVP/Diffserv in addition to the 20-byte IPv4 or 40-byte IPv6 header. This is shown in Figure 17.4.

Figure 17.5 shows that reading a header is relatively straightforward and deterministic but becomes much harder if we need to look into the payload. We introduce delay and delay variability into the classification process.

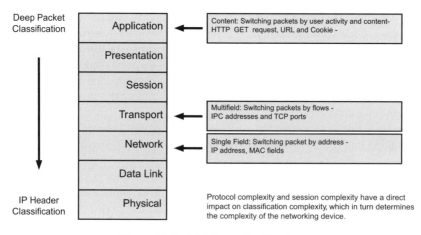

Figure 17.4   Multilayer classification.

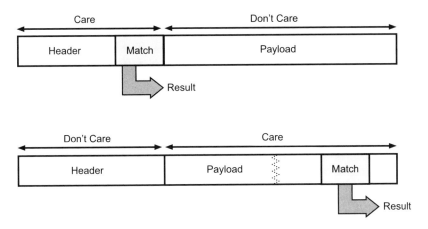

Figure 17.5   Header versus data payload searches.

The whole process of classification means we have to put the packet somewhere while we examine it. As we move down the protocol stack we need better and better performance from our memory and we can find we end up spending large sums of money, particularly when provisioning S-RAM and CAM (content-addressable memory). Note that these memory overheads are in addition to the buffer memory needed for queue management (prior to accessing egress bandwidth from the router). Figure 17.6 illustrates these classification memory overheads.

Note also that both classification and queue memory need to be substantially overprovisioned to deal with asynchronous and (at times) highly asymmetric traffic.

The end result is that router hardware and router software will cost more than expected and perform less well than expected.

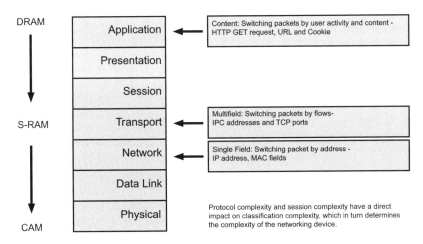

Figure 17.6   Classification memory overheads.

17.9.1   Summary – Telecom Cost Complexity

Telecom costing has always been a complex process involving amortisation of fixed facilities with a life span of 50 years or more (think telegraph poles and traditional telephone exchanges). Universal service obligations and access guarantees have always implied a substantial over-provisioning of access and network bandwidth.

17.9.2   Popular Myths – Packet is Cheaper

The transition from circuit-switched to packet-routed networks has been promoted as a mechanism for improving network and transmission bandwidth multiplexing efficiency that in turn is supposed to deliver a decrease in the cost of delivery measured in euroherz or dollarherz or eurobytes or dollarbytes of network transmission bandwidth.

Multiplexing efficiency is dependent on buffering. Buffering has two cost components – the cost of the memory and processor bandwidth needed to store, manage and prioritise buffered packet streams and the (largely hidden) cost of supporting differentiated classes of service.

This cost is dependent on the offered traffic mix – conversational traffic costs more to deliver than best effort because it cannot be buffered. This is of course only true if the multiplexing efficiency of buffering exceeds the cost of overprovisioning transmission bandwidth beyond the point at which buffering would no longer be required. If a significant percentage of offered traffic is real time/conversational, the assumption of a multiplexing gain from buffering is invalid.

Additional network efficiency is supposedly achievable by deploying flexible end-to-end routing. Again, any efficiency gain achievable is directly dependent on the offered traffic mix. If a significant percentage of offered traffic is real time/conversational, the assumption of a multiplexing gain from routing flexibility is invalid. The amount of bandwidth used increases as the number of hops increase and/or the routing trajectory becomes progressively more indirect.

17.9.3   IP Protocol and Signalling Overheads

At protocol level, the overheads of SS7 have been, or are being, replaced by the 'lighter signalling load' of the TCP/IP or UDP protocol stack working with traffic shaping and IP prioritisation protocols (MPLS, Diffserv, RSVP, etc.). Any efficiency gain achievable is directly dependent on the offered traffic mix. If a significant percentage of offered traffic is real time/conversational, the assumption of an efficiency gain from IP signalling and traffic shaping is invalid.

If average packet lengths are 1500 bytes or so, IPV4 or IPV6 address overheads are relatively trivial. However, if a significant percentage of traffic is multimedia and time-sensitive or time-interdependent, IP packet traffic (IP voice, IP audio, IP video) will be mapped on to fixed and short (typically 92 bytes or less) fixed-length packets and IP address overhead will be substantial. In radio systems, this can be equivalent to taking between 4 and 5 dB of link budget out of the radio system. Given that one dB of link budget (rather approximately) equates to a 10% increase in network density then it is clear that IP addressing and signalling overhead adds directly to the dollardensity costs of deploying a wireless network. Similarly, a multimedia multiplex will require fixed length packets in order to preserve the time-domain properties of the multiplex. The flexibility of variable-length packets (one of the claimed advantages of IP when compared to ATM) therefore no longer applies.

17.9.4   Connectionless is Better than Connection Orientated?

The always-on connectivity implicit in a packet network delivers efficiency benefits when handling small bursts of data. In a GSM network, the 1 to 2 seconds of setup delay in a circuit switched call (or 3 to 4 seconds in a Release 99 network) represents a major overhead when handling short bursts of data. However, these packet efficiency benefits are only valid with a short session duration. Most present evidence suggests that in a multimedia network, sessions are substantially longer than traditional telecom voice calls and are increasing over time. This suggests that any supposed efficiency benefits achievable from having a connectionless end-to-end channel are becoming increasingly invalid.

This also affects the overall efficiency of call-setup protocols such as SIP. A SIP message generates about 8 kilobytes of signalling load. A change in session property (a new user or additional content stream) requires a new SIP message exchange. As sessions become more complex over time, SIP becomes progressively less efficient and requires substantially higher signalling bandwidth than existing SS7 signalling.

17.9.5   The Hidden Costs of IP MMS

An IP MMS (IP multimedia subsystem) enabled multiservice network is rather like a postal or courier service but instead of two classes of service (first class or second class for postal services, priority or standard for couriers) there are 4 levels of service. Note that the postal service in the UK loses money on second-class mail and makes money on first-class mail. This is partly because first-class mail has a higher value but also because second-class mail costs more to deliver. The additional costs are incurred because the storage costs (and related storage administration) now exceed the benefits of holding back mail to fully fill increasingly inexpensive trucks (rather bizarrely the real costs of running a large delivery van up and down the motorway are going down rather than up over time). With international carriers, the same tipping points apply and as aviation costs have reduced (larger, cheaper more efficient aeroplanes) it becomes cheaper not to store and forward but just to forward. The same principle applies to indirect routing. Although the cost of indirect routing may decrease over time, the process still consumes additional bandwidth. This is also true of wireless mesh networks, which can be remarkably spectrally inefficient due to routing inconsistencies and associated user signalling overheads.

17.9.6   So is the Internet an Efficient Transport System?

Not really. It's robust because that's what it's designed for. It was, however, never designed to handle multimedia traffic and if you expect it to deliver the same end-to-end channel performance in terms of latency and jitter as an ATM or circuit-switched network, then any supposed efficiency benefits will rapidly disappear. All Internet protocols are inherently inefficient and either waste bandwidth through transmission retries (TCP/IP) or lost packets (UDP) or a mixture of both.

17.9.7   So Why is the Internet so Cheap and Why does it Work so Well?

Because it is presently grossly overprovisioned and has been financed by the pension funds that lost investors money in the dot com dot bust cycle.

17.9.8   So if IP Networks are Neither More Efficient nor Lower Cost than Circuit-Switched or ATM Networks, Why are Telecoms Operators Deploying Them?

Well that's a good question and one of the answers is that to an extent hardware costs are reducing as hardware (and to a lesser extent software) in the network becomes a commodity, but this is in reality a modest incremental process. Complex ATM hardware is being replaced with complex soft switch platforms with high clock speed processors, extravagantly provisioned high-performance buffer memory cards and unstable traffic-shaping protocols. Not really a big leap forward.

As the Internet becomes more aggressively loaded over time, the real costs of delivery will start to reappear and will need to be factored in to end-to-end delivery cost calculations. If the route to reduced cost compromises the user experience then this is a self-defeating process.

18

Network Software – Energy Management and Control

18.1 Will the Pot Call the Kettle Back?

In January 1999 we wrote a Technology Topic on Device Access Networks, 'Will the Pot call the Kettle Back? (1). We suggested that the protocols needed to allow devices to discover one another and then have a meaningful conversation were insufficiently standardised to support widespread market adoption. This was identified as a potential problem for Bluetooth enabled devices.

As with many prognostications we were partly right. Agreeing the Bluetooth procedures for device-to-device discovery and device-to-device communication proved to be a Herculean process and work is still ongoing in this area.

In November 2007 we thought it was time for a sequel – Will the Pot Call the Kettle Back? (2).

The original article focused on device-to-device communications in the home. Eight years later the focus had shifted to corporate, specialist and consumer machine-to-machine communication, the machine-to-machine (M2M) triple-play proposition. In 2007 it was estimated that there were ten microcontrollers for every human on the planet with the microcontroller population growing faster over time – a 60 billion unit target market. On that basis the machine-to-machine market represented a high-value high-growth revenue opportunity.

This was, and is, undoubtedly true but five years on our electricity, gas and water meters are generally still read by people, I buy my train ticket from a man in a ticket office (very helpful chap by the way), check the level of water in our water butt by lifting the lid and looking inside, check the oil level in the car by wiping the dipstick and open the curtains at home by hand. I use a conventional diary because the access times are faster and I use a map in the car because I hate being told what to do.

So the question is, are the big opportunities in the corporate, specialist or consumer domain, what are the market adoption barriers that are holding back genuinely ubiquitous machine-to-machine communication and how can these barriers be overcome and/or have they already been overcome, which must be the assumption behind some of the more bullish market forecasts.

Making Telecoms Work: From Technical Innovation to Commercial Success, First Edition. Geoff Varrall.
© 2012 John Wiley & Sons, Ltd. Published 2012 by John Wiley & Sons, Ltd.

18.2   Corporate M2M

There is a wide choice of technology for machine-to-machine communication; indeed the breadth of choice has proved more of a hindrance than an asset. If faced with too many options people may choose not to choose. For example, an M2M application connecting vending machines to a central hub can be realised using VHF or UHF two-way radio or a GPRS modem. More localised connectivity can be provided using WiFi and/or Bluetooth and/or ZigBee. White-space devices are also being propositioned for M2M applications.[1]

Machines have some advantages over people. They do not complain, though their owners might. Many corporate M2M applications are static, vending machines being one example. They are also often installed in basements where radio coverage is poor. Thus, the choice of radio technology may be dictated by propagation.

Many of the applications are low-bandwidth exchanges of a few kilobytes or less and may or may not be time critical. If power drain is not an issue, devices can be continuously connected. More often it makes more sense for devices to be polled or to send data at preassigned intervals or on an as-required basis – a 'running out of chocolate' message for example. One application that is apparently successful is a rodent trap that sends an SMS message when a visitor arrives. SMS can also be used to send control messages to remote devices.

A number of organisations provide bespoke solutions for corporate end-to-end customers. Examples include Vianet,[2] Arkessa[3] are an example in the UK as are Arkessa and Aeris.[4] Operators such as Orange have also been proactive in developing this market.

18.3   Specialist M2M

Specialist M2M applications are different from corporate M2M in that the application spend would typically be justified on the basis of public safety or security. The most pervasive example presently would be CCTV surveillance and/or speed cameras or traffic-monitoring devices. The bandwidth exchange for a PIR (passive infra red) detector would be a few tens of kilobytes. A picture from a surveillance camera would typically be 40 or 50 kilobytes.

As with corporate M2M these exchanges may be event driven or periodic. Images used as evidence in court need to be digitally water marked and audit trailed to minimise the risk of being challenged. Specialist M2M includes devices that are capable of working in extreme conditions including heat (-30 to $+80$ degrees), water and dust and capable of withstanding shock or continuous vibration.

Low-cost cellular modems are not always suitable for these applications. There are a number of vendors that service this market, for example KoreTelematics[5] in the US and TDC in the UK. There is obvious overlap between corporate and specialist user markets and commonalities in terms of need and requirement.

---

[1] http://www.neul.com/products.php.
[2] http://www.vianet.co.uk/.
[3] www.arkessa.com.
[4] http://www.aeris.com/.
[5] http://www.koretelematics.com/.

18.4   Consumer M2M

Consumer M2M applications are different from corporate and specialist M2M in that the application spend is justified on the basis of entertainment and/or personal convenience.

Getting a kettle to tell a teapot that is has reached boiling point might be entertaining but is not necessarily useful, though might be in special circumstances, for example making life easier and safer for home owners with sight disability.

Verizon in the US have a home automation portal called 'Control Point' that links WiFi with their LTE network to provide remote management of home security, lighting, heating, air conditioning and 'connected intelligent furnishings' (whatever they may be) from a smart phone, useful I suppose for people who own multiple homes and/or go on holiday a lot and/or don't get on with their neighbours.

Similarly, moving higher-bandwidth data around the home when you are in it offers undeniable potential both from a user-experience perspective and in terms of delivering new consumer product differentiation opportunities. Simple examples prevalent today would be the use of shared folders on a lap top and PC across a wireless router.

Several issues, however, need to be considered.

18.5   Device Discovery and Device Coupling in Consumer M2M Applications and the Role of Near-Field Communication

Our January 1999 Technology topic talked about the difficulties of device discovery and device coupling. Essentially, the question revolved around how to manage device discovery and device coupling policy. Eleven years later this problem still remains to be solved.

As you walk around a house, devices come into radio view and become available for service. Just because a device is available to use does not mean that you want to use it and the process of discovery and coupling takes time (bandwidth) and power. This is not a decision process that can be safely left to a machine, unless that machine has an understanding of the user and the context in which the device may or may not be enabled. The answer may be to use near-field communication.

NFC devices are in everyday use as travel cards. The 13-MHz transponder is touched on the turnstile at the tube station or pay button on the bus. Nokia have developed a parallel enthusiasm for NFC in hand-held cellular devices.

This would seem to make complete sense as a means for managing the device-coupling process. If you want two devices to talk or multiple devices to talk, you physically introduce them to one another. NFC has been briefly touched on (excuse the pun) in an earlier chapter, but obligingly the following press release just appeared in my in box and addresses payment applications – the logical extension from travel cards.

Danish Operators Join Forces on Mobile Wallet

By Mikael Ricknäs, IDG News Jun 23, 2011 2:40 pm

TDC, Telenor, TeliaSonera and Three are cooperating on the launch of a digital wallet service based on NFC (Near Field Communications) in Denmark, the operators said on Thursday.

The roll-out will start this fall, but it will take a few years before the technology reaches widespread adoption, according to the operators. The number of NFC-compatible phones and take-up among retailers will determine the pace, a spokesman at TDC said.

Consumers will be able to use their mobiles to pay for goods, services and travel, at a discount if digital coupons are available, and also open doors at hotels or borrow a book at the library, according to the operators' vision of the future.

For users, the operators working together will mean that they can switch from one to the other and still bring their wallet with them, the operators said.

For the project to become a success, the operators will need backing from all parts of the technology and merchant ecosystem, including phone makers, banks, card companies and last but not least retailers. The announcement didn't have any details on such partnerships.

Operators are increasingly getting together to push mobile payments using NFC. The announcement in Denmark follows the creation of similar projects in the U.S. with the Isis network.

The operators are hoping to gain greater control of the payments market by offering a single platform, according to market research company CCS Insight.

So possibly this marks the start of the big breakthrough, at least in the consumer part of the market.

18.6   Bandwidth Considerations

A payment transaction is at most a few kilobytes but some transactions could be significantly more bandwidth intensive.

This is not just an access-bandwidth issue but a storage-bandwidth issue. There is not much point in streaming a high-definition television programme to a device with a few kilobytes of buffer bandwidth and/or limited video and audio playback capability. For example, most of us would understand that there is not much point in coupling the VCR and a kettle unless of course it was a very exceptional kettle. There are also possibly at least four types of exchange that need to be supported, best effort, streamed, interactive and conversational:

Best effort is for non time critical exchanges, opening the curtains perhaps.

Streamed is for moving video from device to device.

Interactive is for gaming and other interactive pursuits (keep an eye on the teapot and the kettle they might starting placing bets against each other).

Conversational is for conversational devices – teddy bears that talk to each other and/or respond to a TV programme – hideous but here.

Within the WiFi 802.11 e QOS standards process these applications have been accommodated with the extended data channel access wireless media extension that supports background and best effort and video and voice bearers simultaneously on a 20-MHz time multiplexed channel aggregated into an eight-level queue at the access point.

This is a world away from the original contention-based medium access control that has been the basis for WiFi systems to date. In practice, it will probably work reasonably well most of the time, which remarkably enough seems to be acceptable to most consumers.

Similarly, Bluetooth 2.0 EDR now supports three simultaneous traffic streams for voice, data and device control. One suggestion is that Bluetooth 3.0 will extend this parallel approach by supporting WiFi at 2.4 and 5 GHz. Bluetooth would then become by default a local area radio access management protocol rather than a specific physical layer standard. Together with

ZigBee this would seem to address most of the foreseeable local area machine-to-machine requirements.

18.7   Femtocells as an M2M Hub?

An alternative is to consider the role that femtocells might play in the corporate, specialist and consumer M2M proposition. In corporate applications we have said that machines are sometimes positioned deep within a building. The loss from an outdoor cell into a building interior can be anything between a few dB to more than 40 dB.

In these situations a femtocell or femtocells can be installed to provide coverage. Similarly, specialist machine-to-machine applications may need to be supported underground and femtocells would again provide a coverage option. In both cases the economics of such a solution would probably need to be amortised over a range of uses including voice, broadband data and M2M. The exception may be safety-critical M2M where the application may have sufficient value to cover capital and running costs for the installation.

Operators are generally positive about femtocells because the real-estate cost of hosting a base station is transferred to the building owner or occupier. If the operator provides ADSL broadband as part of the package then the backhaul has been transformed from being a cost to a revenue item. Some infrastructure vendors are less positive – 'an excuse for a bad network' being one vendor's view.

The positioning of femtocells as a home hub is also less than clear. The same benefits to the operator apparently apply. The real-estate cost of hosting a base station is transferred to the consumer and if the operator supplies an ADSL line then backhaul becomes a revenue item.

However, in the home as indeed in many office and public spaces, femtocells are competing with WiFi, which although far from perfect does work and still has significant cost-reduction and performance-optimisation potential.

Rather than the single 5-MHz RF channel pair used in a femtocell, WiFi uses 20 MHz or in some cases 40 MHz (two bonded 20-MHz channels) to deliver 54 Mbits/s or 100 Mbits/s. The actual throughput rates are a fraction of this but still faster than a femtocell can manage. Although femtocell throughput rates will increase overtime so will WiFi.

Today, on a femtocell supporting four parallel voice channels the peak data may be enough for corporate or personal e mail but is not enough for broadband machine-to-machine communication in the home.

The range from a femtocell may be better as the allowed transmit power is higher and receive sensitivity is higher due to the frequency duplex. It must, however, be questionable as to how many houses are large enough for this benefit to count.

Thus, although femtocells may have a role to play in machine-to-machine connectivity in some corporate and specialist applications it is harder to see the rationale for home-hub applications. Arguably, all femtocells do is increase the user's energy bill.

And anyway, why use wireless given that most of these appliances are connected to a mains socket. The Home Plug Powerline Alliance[6] supports a set of standards for transferring IP

---

[6] www.homeplug.org.

packets between devices. This includes entertainment devices and entertainment hubs (back to those smart televisions we talked about in Chapter 9) and smart energy networks.[7]

18.8   Summary

M2M is a fast growth sector attracting the attention and support of a wide range of vendors and more recently a relatively broad cross section of operators offering bespoke solutions. Traditional VHF and UHF radio modems can still be cost and performance effective in many applications. In order to compete, cellular-based propositions need to be carefully costed, well deployed and well supported and for most applications have access to inexpensive (under utilized) bandwidth.

There are essentially three markets, corporate, specialist and consumer. There is substantial overlap between corporate and specialist applications and crossover opportunities between machine-to-machine and telemetry applications. Consumer machine-to-machine is substantially different. Entertainment applications are inherently more bandwidth hungry, energy management applications less so.

Traditional difficulties associated with device discovery can potentially be resolved by using NFC technologies integrated into cellular phones and hand-held communication devices. This would seem to suggest that consumer M2M could be potentially serviced by low-cost femtocells offering cellular compatibility.

However, the consumer M2M market is aggressively addressed by WiFi products that offer higher data rates than femtocells at lower unit costs. WiFi products still have substantial cost-optimisation and performance-gain potential. For these reasons it is hard to see how femtocells can score in this sector unless substantial subsidies are applied. Device-to-device data rates may be disappointing and dual cellular WiFi access may need to be provisioned, which would be unnecessarily complex and expensive. Outdoor to indoor coverage can also be expected to improve as network densities increase over time. Consumer M2M can be expected to include a wide variety of higher-bandwidth applications for example video sharing and lower-bandwidth exchanges including device to device telemetry and control.

Handset vendors promote the logic of using the cellular phone as a control point for these devices and applications. This begs the question as to whose phone controls the house in multiphone households. We will continue to open curtains by hand rather than by phone and talking teapots will remain as at most a niche product opportunity.

The case for cellular in consumer M2M remains unproven but corporate and specialist applications may provide a more than adequate market opportunity.

---

[7] www.homeplu.org/tech/smart_energy.

19

Network Software – Microdevices and Microdevice Networks – The Software of the Very Small

This is a chapter about very small things and the software that drives them. This includes very small things that when interconnected can become very large networks. In previous chapters we have discussed the evolving role that microelectrical mechanical systems are playing in cellular handset design and beyond that the role that nanoscale manufacturing may have in transforming material properties. The combination of silicon geometry scaling and microminiaturisation techniques together are transforming phone form factor and functionality.

There are, however, other devices that are acquiring new capabilities and shrinking both in size and the amount of energy they consume. These devices use manufacturing techniques at the scale of nanometres (one billionth of a metre) to build structures at atomic or molecular level that are combined in to devices that are either measured in micrometres (one millionth of a metre), or millimetres (one thousandth of a metre).

We review three classes of 'microdevice', 'Memory Spots' 'motes' (being developed in the US as a result of various 'smart dust' projects) and the Hitachi Mu chip (an ultraminiaturised RF ID device).

Two of these devices have failed to gain market traction for reasons that we need to analyse. All three device classes are, or were, intended to store information and/or collect information and/or provide identification. All of them are, or were, communication devices. A particular interest is to consider how we communicate with these communication devices and whether the cellular phone has a valid role in this communication process.

We compare these microdevice applications with a present larger form factor application (contactless smart cards) and suggest some technical and commercial commonalities. We discuss the present status of 'phone to device' communication systems, particularly RF systems and protocols and question whether there are plausible cellular phone and mobile broadband business models to justify development investment in this intriguing but challenging application sector. We highlight certain factors that suggest adoption time scales may be longer than commonly supposed.

Making Telecoms Work: From Technical Innovation to Commercial Success, First Edition. Geoff Varrall.
© 2012 John Wiley & Sons, Ltd. Published 2012 by John Wiley & Sons, Ltd.

19.1   Microdevices – How Small is Small?

MEMS and microminiaturisation techniques in combination with silicon geometry scaling are allowing us to deliver storage functionality, information gathering functionality, identification and communication capability in increasingly small devices.

Table 19.1 lists three form factors ranging from a grain of rice (small) to a grain of sand (very small) to a grain of dust (very very small). A grain of dust typically has a diameter of a few tens up to 200 or 300 micrometres, hence the term 'microdevice'.

We review products developed for all three form factors. All three product sectors present distinct communication challenges and (hence) potential communication value opportunities.

On 14 March 2003 Hitachi announced a product called a Mu chip. It was very small (or else was photographed against a very big finger). This device is illustrated in Figure 19.1.

Actually it was 0.3 millimetres square, half the size of the RF tags on the market at the time and was manufactured using a 0.18-micrometre process as opposed to the 0.35-micrometre process generally in use then. The chip operated at 2.4 GHz and stored a 128-bit ID and could be read from a foot away (30 cm). The antenna, which cannot have been very big, was attached to two electrodes at the top and bottom of the chip, which apparently was very clever.

The Mu chip was intended as an alternative to bar codes or conventional RF tags for use in supply-chain management, product traceability and security and was packaged as a complete system including readers, software and networking infrastructure. RF tags at the time cost about 40 cents each.

In practice, the device never achieved traction either in the business and asset tracking market or in the consumer space and the web site is now closed.

Table 19.1   Microdevice form factors and functionality – three examples

Small	Very small	Very very small
Grain of Rice < 4 millimetres	Grain of Sand <2 millimetres	Grain of Dust <500 micrometres
Example Product Memory Spots[1] (HP Labs)	Motes (Intel web site)[2]	Mu Chip[3] Hitachi
Information delivery device	Information collection device	Identification device (RFID)
2 mm by 4 mm including integrated antenna	1 mm by 1 mm (mote sized)	0.4 mm by 0.4 mm by 60 micrometre
256 kilobit to 4 megabit storage	Typically with sensing capabilities – temperature, light, vibration, acceleration, air pressure	0.15 mm by 0.15 mm by 7.5 micrometre
Data transfer rate up to 10 Mbits/s	Data transfer rate Device and application dependent	Data transfer rate 12.5 kbits/s
Distance – close coupled	Distance – 20–30 metres	Up to 400 mm

---

[1] http://www.hp.com/hpinfo/newsroom/press/2006/060717a.html.
[2] http://embedded.seattle.intel-research.net/wiki/index.php?title=Intel_Mote_2.
[3] http://www.hitachi.co.jp/Prod/mu-chip/.

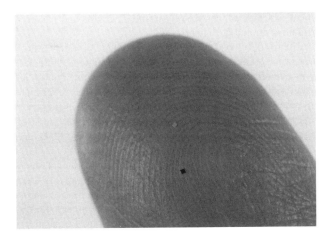

Figure 19.1   Hitachi Mu chip. Reproduced with permission from Hitachi.

On 17 July 2006 HP announced 'a grain sized chip that could be attached to almost any object, making information more ubiquitous'. The device was called the Memory Spot and is illustrated in Figure 19.2.

Here is the launch press release:

HP Unveils Revolutionary Wireless Chip that Links the Digital and Physical Worlds
   PALO ALTO, Calif., Jul 17, 2006
   HP today announced that its researchers have developed a miniature wireless data chip that could provide broad access to digital content in the physical world.
   With no equal in terms of its combination of size, memory capacity and data access speed, the tiny chip could be stuck on or embedded in almost any object and make available information and content now found mostly on electronic devices or the Internet.
   Some of the potential applications include storing medical records on a hospital patient's wristband; providing audio-visual supplements to postcards and photos; helping fight counterfeiting

Figure 19.2   HP Memory Spot. Reproduced with permission from HP Research Labs Bristol.

in the pharmaceutical industry; adding security to identity cards and passports; and supplying additional information for printed documents.

The experimental chip, developed by the "Memory Spot" research team at HP Labs, is a memory device based on CMOS (a widely used, low-power integrated circuit design) and about the size of a grain of rice or smaller (2 mm to 4 mm square), with a built-in antenna. The chips could be embedded in a sheet of paper or stuck to any surface, and could eventually be available in a booklet as self-adhesive dots.

"The Memory Spot chip frees digital content from the electronic world of the PC and the Internet and arranges it all around us in our physical world," said Ed McDonnell, Memory Spot project manager, HP Labs.

The chip has a 10 megabits-per-second data transfer rate – 10 times faster than Bluetooth™ wireless technology and comparable to Wi-Fi speeds – effectively giving users instant retrieval of information in audio, video, photo or document form. With a storage capacity ranging from 256 kilobits to 4 megabits in working prototypes, it could store a very short video clip, several images or dozens of pages of text. Future versions could have larger capacities.

Information can be accessed by a read-write device that could be incorporated into a cell phone, PDA, camera, printer or other implement. To access information, the read-write device is positioned closely over the chip, which is then powered so that the stored data is transferred instantly to the display of the phone, camera or PDA or printed out by the printer. Users could also add information to the chip using the various devices.

"We are actively exploring a range of exciting new applications for Memory Spot chips and believe the technology could have a significant impact on our consumer businesses, from printing to imaging, as well as providing solutions in a number of vertical markets," said Howard Taub, HP vice president and associate director, HP Labs.

The chip incorporates a built-in antenna and is completely self-contained, with no need for a battery or external electronics. It receives power through inductive coupling from a special read-write device, which can then extract content from the memory on the chip. Inductive coupling is the transfer of energy from one circuit component to another through a shared electromagnetic field. A change in current flow through one device induces current flow in the other device.

Memory Spot chips have numerous possible consumer and business-based applications. Some examples are:

- Medical records: Embed a Memory Spot chip into a hospital patient's wrist band and full medical and drug records can be kept securely available.
- Audio photo: Attach a chip to the prints of photographs and add music, commentary or ambient sound to enhance the enjoyment of viewing photos.
- Digital postcards: Send a traditional holiday postcard to family and friends with a chip containing digital pictures of a vacation, plus sounds and even video clips.
- Document notes: A Memory Spot chip attached to a paper document can include a history of all the corrections and additions made to the text, as well as voice notes and graphical images.
- Perfect photocopies: A Memory Spot chip attached to a cover sheet eliminates the need to copy the original document. Just read the perfect digital version into the photocopier and the result will be sharp output every time, no matter how many copies are needed, and avoiding any possibility of the originals jamming in the feeder.
- Security passes: Add a chip to an identity card or security pass for the best of both worlds – a handy card with secure, relevant digital information included.
- Anticounterfeit tags: Counterfeit drugs are a significant problem globally. Memory Spot chips can contain secure information about the manufacture and quality of pharmaceuticals. When added to a drug container, this can prove their authenticity. A similar process could be used to verify high-value engineering and aviation components.

Like the Hitachi Mu chip, the Memory Spot never gained mass-market traction, but both products were precursors of things to come. Sometimes ideas just get to market too early but that does not mean that these devices were not important or no longer have relevance.

Both the Memory Spot (HP Labs) and Mu chip (Hitachi) products were passive, generating power inductively from the interrogating device.

As stated in the press release the Memory Spot was a CMOS device with an intended storage capacity of between 256 kilobits and 4 megabits and a claimed transfer rate of 10 megabits per second. The device had a built-in antenna and could be embedded on a sheet of paper or any suitably friendly surface. Suggested consumer applications included adding a memory chip to photographs or postcards or books to provide an audio (voice or music) or imaging (still and video imaging) or extended text download capability (microdot applications). As such, it could be described as an 'information storage and delivery device'. The public announcement of the device created substantial interest.

In the middle 'grain of sand' category we have the various 'mote devices', also sometimes described as 'smart dust devices' with their genesis in US projects aimed at realising an autonomous sensing computing and communication system within a cubic millimetre (a 'mote'), though present devices are significantly larger.

The idea of these devices is that they can be scattered across an area and form a self-organising network of interconnected interactive objects that are either battery powered, solar powered or vibration powered. Their purpose is primarily to capture information on the physical world, for example temperature, ambient light, vibration, acceleration and air pressure – they are information-collection devices.

However, the question we are trying to answer is whether we want or need cellular phones to be able to talk to these microdevices? If we do, what are the technology, engineering and business challenges and opportunities? Why did the MU chip and Memory Spot fail to make market headway and do they have contemporary equivalents?

19.2   Contactless Smart Cards at 13.56 MHz – A Technology, Engineering and Business Model?

A starting point is to look at a (larger) device that we already talk to, the contactless smart card, and see what lessons we might learn. Contactless smart cards are similar to the Memory Spot in that they are typically close-coupled applications in which the two devices (the reader and the smart card) are either touching or within a few millimetres of each other. This is 'near-field communication'.

There is an international standard ISO14443 for contactless smart cards, also known as RF tags, operating at 13.56 MHz, close to the GSM clock reference at 13 MHz. The devices are ASK or BPSK modulated, either passive (load modulation) or active with a range from 0 to 200 millimetres. There are 4 types of contactless smart card (tag) categories determined by memory footprint and transfer speed detailed in Table 19.2.

In the UK, every time you use a London Train and Underground Oyster Card[4] the 13.56-MHz transmitter in the oyster terminal device at the turnstile is irradiating your 'smart' oyster card with RF energy, your oyster card then uses this energy to transmit your ID back to the device.

---

[4] http://www.tfl.gov.uk/tickets/14836.aspx.

Table 19.2   Contactless smart cards

Tag type	1	2	3	4
Memory	1 kb	2 kb	1 Mb	64 kb
Data rate	106 kbps	106 kbps	212 kbps	106–424 kbps

In 2006, when we first researched this topic there was no compelling technical reason why this function could not be included as a mobile phone function. It could be passive or active and unidirectional or bidirectional. If bidirectional, the phone could read information stored at the turnstile (time tables, delays, special travel offers) and of course the cost of any transaction would just be added to the monthly phone bill.

Five years on, NFC-enabled handsets are available and applications are finally beginning to emerge. The lesson here is that fast innovation adoption only happens when there are tangible benefits to all parties in the value chain. In this example, there is clearly user value in terms of convenience but the mobile-phone operator has become an intermediary in a transaction that has cost and risk with income going to a third party (the train company).

In the end, if there is a genuinely compelling user experience benefit then the innovation will happen, but much more slowly than people generally appreciate. The same lag effect meant that the Memory Spot and Mu products were never a commercial success. The devices potentially created user experience value but needed to be networked to realise that value and the incentive for third parties to invest in those networks was insufficient. The device business model worked. The network business model failed.

Anyway, back to Oyster cards. In practice, we still use dedicated smart cards not cellular phones to access these systems. This is because we do not have NFC transponders as a standard item in phones. The reason we do not have NFC transponders as standard items in phones is that it adds cost and the associated value model depends on having commercial agreements in place, which require crossindustry consensus and therefore take time to negotiate.

19.3   Contactless Smart Cards and Memory Spots – Unidirectional and Bidirectional Value

The Memory Spot in some ways represented a development of the contactless smart card business model but with the focus on information transfer rather than transaction facilitation. This implied a need for higher data rates (at 2.4 GHz) than those available using NFC (at 13 MHz). Although the Memory Spot was a contactless smart memory information dispensing device, the provision of information and transactional value are closely related. Reading about things prompts us to buy things.

This comes down to the simple principle that a unidirectional exchange has a certain value. Uploading our ID to a turnstile has convenience value, reading a Memory Spot would have had interest and information value. If the exchange can be made bidirectional then the value increases.

19.4   Contactless Smart Cards, RF ID and Memory Spots

The second microdevice example, the Hitachi Mu chip, was intended to function as an RF bar code and/or RF ID device. RF bar codes offer some advantages over conventional bar codes, for example nonline-of-sight and multiple read capability and additional address bandwidth to support electronic product codes rather than standard (universal) product codes. In 2006, they were also beginning to be used in passport systems though that proved complicated as well. For consumer and retail applications the problem remained one of cost. Even a few cents matters if it's a can of beans that you are trying to sell.

Traditional printed label bar codes, however, are effectively zero cost and have benefited from years of intensive investment in optical scanning techniques.

RF ID tags also have competition from other offerings such as long-wave magnetic systems[5] operating at wavelengths below 450 kHz. These system options have their own standard (IEEE P1902 Standard for Long Wavelength Wireless Network Protocol) and some application advantages, for example the ability to work underwater and/or underground.

As with RF tags, magnetic tags can be active or passive. The active devices have a claimed life of 10 years or more from a (coin-sized) lithium battery. Read rates are typically between 300 and 9600 bits/s.

RF ID tags and long-wave magnetic tags both have sufficient address bandwidth to support IPV4 or IPV6 addresses. This suggests a shift in functionality beyond present 'visibility network' or 'visible asset' applications to a broader application base in which wireless interrogation of an RFID prompts the interrogating device to access a web page.

To place this in the context of our Oyster card and Memory Spot examples, we do not necessarily need to download information from an embedded device other than an IP address that is then used to access information from a supporting web site.

In 2006 and 2007 the IEEE coordinated substantial standards work[6] on these long-wavelength device systems but five years on ubiquitous applications have failed to emerge. Probably an example of a solution to a problem that didn't really exist.

19.5   Contactless Smart Cards, RF ID, Memory Spot and Mote (Smart Dust) Applications

Our third microdevice example, mote-sized smart dust devices, seem very different. These devices are intended to be deployed as self-configuring *ad hoc* networks that interact with each other. These devices can potentially all have IP addresses and can be a part of a sensing and surveillance network. They combine local storage capability, intelligence and communications capability.

The relevance of these networks to cellular phones may seem tenuous. However, the protocols for *ad hoc* networking are well established and already deployed in a number of two-way radio system solutions so the concept of cellular phones and mobile broadband devices interacting with these devices is at least plausible.

19.6   The Cellular Phone as a Bridge Between Multiple Devices and Other Network-Based Information

The logic of using a cellular phone in any or all of the above applications is that the cellular phone provides a bridge to the outside world. Advantageously, of course, this is a toll bridge

---

[5] http://www.rfidjournal.com/article/articleview/2436/.
[6] http://www.rfida.com/apps/passive.htm.

with an efficient and robust revenue capture (billing) capability including pre- and postpayment collection.

The technical challenge of using a cellular phone is that we need to support additional wireless systems and protocols that add cost and complexity to an already overloaded product platform.

19.7 Multiple RF Options

Frequencies used presently include long wave (below 500 kHz), 1.95, 3.25, 4.75 and 8.2 MHz (typically used for antishop-lifting tags in retail stores), near-field communication in the 13 MHz ISM (industrial scientific medical) band, the 27-MHz ISM band, the 430–460-MHz ISM band available in Region 1 (Europe and Africa), the 902–916-MHz band available in Region 2 (North and South America), the 918–926-MHz band used for RF ID in Australia, the 2.35–2.45-GHz ISM band, a possible band at 5.4–6.8-GHz and/or the use of ultrawideband systems between 3 and 10 GHz.

UWB was potentially appealing in that there was a very scalable relationship between bit range and range (low bit rate/long range, high bit rate/short range). The frequency band (up to 10 GHz) is also useful for microdevice form factors. However, there was a massive standards battle combined with a failure to establish a consistent spectral policy interregionally, so UWB never made it to the market.

There does now seem to be a regulatory consensus that there should be three internationally agreed universal allocations for device-to-device communication, a low-frequency allocation at 125 kHz, the NFC allocation at 13.56 MHz and the 2.45 GHz allocation. These devices need to coexist with the other radio systems within the phone including wide-area cellular and local area (Bluetooth and WiFi) and receive only functions such as GPS.

19.8 Multiple Protocol Stacks

Standardised radio protocol stacks include Bluetooth[7] (optimised for throughput) and ZigBee[8] (optimised for low power consumption), both of which have well-developed vendor support. There are additionally operating systems optimised for low power consumption such as Tiny OS[9] and a range of specialised proprietary offerings.

19.9 Adoption Time Scales – Bar Codes as an Example

Contactless smart cards and each of the microdevice options referenced in this chapter (HP Memory Spot, smart dust motes and micro-RFID) all offer or offered intriguing and potentially compelling cellular phone device and system-level integration opportunities.

They are, or were, technically feasible and supported by radio standards and protocol stacks that are relatively mature. The devices do not necessarily need to reduce in size but will increase in functionality and reduce in cost, thus broadening their application profile. However, business

---

[7] http://www.bluetooth.com/Pages/Bluetooth-Home.aspx.
[8] http://www.ZigBee.org/.
[9] http://www.tinyos.net/.

models may take longer to evolve than might be expected. Traditional bar codes provide an example.

The beginning of bar codes as we know them today can be traced back to a patent for 'a Classifying Apparatus and Method' filed in 1949 by Bernard Silver and Norman Woodland. Silver and Woodland were graduates at the Drexel Institute of Technology and were responding to a local food store's request for an automated method of reading product information.

It took twenty five years for the idea to evolve and for standards to be agreed, mostly getting agreement on the Universal Grocery Products Identification Code that evolved into the Universal Product Code. The first UPC scanner was installed in a supermarket in Ohio in 1974. The first product to have a bar code was a packet of Wrigley's gum.

Bar codes are now ubiquitous, but it has taken another 25 years for this to happen. Bar codes have taken over 50 years to become universally adopted. Hitachi and HP could not afford to wait that long.

19.10   Summary

Optical bar codes provide an example of a product that is now an integral part of everyday life. The enabling idea was simple but required an enabling technology (optical laser scanning) to be available together with universal application standards in order to support ubiquitous deployment.

Today we have a new generation of MEMS-based enabling technologies that are allowing us to build supersmall devices that can collect and store information and/or perform identification or labelling tasks that help us to interact more efficiently with the physical world around us.

It is tempting to position the cellular phone as the 'device of choice' for communicating with this new generation of microdevices. The technical logic is that it is relatively easy to extend the present communications systems within the phone to include phone-to-device and device-to-phone communication applications. However, commercial logic suggests the business models to support these mass-market application sectors will take some years to emerge.

Consumer applications at the time of writing look as though they may take off, provided there is wide enough support from the handset vendor community. Examples on the RFID Journal web site[10] at June 2011 included contactless wrist bands for the Isle of Wight Festival, a ticketless check in system for Scandinavian airlines, a mental-health facility application and a smart-city system.[11]

The smart-city application is based on a sensor board that measures noise pollution (real-time noise maps), dust quantities, (real-time dust maps) the structural health of buildings, bridges and road, garbage levels (only empty bins when they are full), gas monitoring, environmental monitoring (CO_2 and NO_2) radiation detection and smart parking.

The sensor nodes communicate with the network using ZigBee[12] radio links mainly to ensure that sensor devices have a sufficiently low energy requirement to be able to be powered from solar power or a battery lasting several years or more.

And actually this is quite a nice example of radio hardware being developed with an application appropriate protocol stack with a radio range up to 12 kilometres using ZigBee,

---

[10] http://www.rfidjournal.com/article/articleview/8553.
[11] http://www.libelium.com/.
[12] http://www.ZigBee.org/.

albeit at a low data rate or Bluetooth or GPRS with some extended hibernation capabilities that reduce power drain to the sort of levels needed.

Note that the ZigBee standards process is reasonably closely linked with the Home Plug Alliance (see previous chapter) and the WiFi alliance. The ZigBee physical layer consists of two different options in three license-free frequency bands with sixteen channels at 2.4 GHz (the same as Bluetooth and WiFi) with a maximum data rate of 250 kbps, ten channels at 902 to 928 MHz (The US) with a maximum data rate of 40 kbps and one channel at 868 to 870 MHz with a maximum data rate of 20 kbps (Europe, Australia and New Zealand).

Both physical layers use direct sequence spread spectrum (DSSS) modulated O-QPSK with a 32 PN-code length and an RF bandwidth of 2 MHz in the 2.4-GHz band and BPSK modulation with a 15 PN-code length and an RF bandwidth of 600 kHz in Europe and 1200 kHz in North America.

And in the consumer space we may now finally be on the cusp of seeing broader market adoption. On 16 June 2011 Sony Ericsson announced a range of Android NFC phones[13] with NXP, the semiconductor business formerly owned by Philips.

So using our phones to pay for things finally looks set to happen. Using our phone to talk to other devices via Bluetooth already happens, hands-free Bluetooth headsets being the most ubiquitous example. Using our phone to interrogate and communicate with microdevice networks, for example sensor networks is also beginning to happen. We can go online to look at video feed from CCTV cameras, we can go online to turn appliances on and off at home, we can go online to turn lights on and off at home via ZigBee or WiFi-enabled wireless switches.

But just because we can do these things does not mean we will. Many of us are quite happy opening and closing our own curtains. A suitable point on which to end this chapter.

---

[13] http://www.nearfieldcommunicationsworld.com/2011/06/16/38069/sony-ericsson-to-launch-range-of-android-nfc-phones/.

20

Server Software

On 5 January 1941, while flying from Blackpool to RAF Kidlington near Oxford, Amy Johnson[1] went off course in poor weather and drowned after bailing out into the Thames Estuary. She was run over by the ship that was sent to her rescue.

It is assumed she lost her sense of direction, although an alternative theory is that she was shot down by the RAF after failing to provide correct identification. Glen Miller met a similar end (probably)[2] when hit by a jettisoned bomb. Either way it's a telling illustration that it is easy to get lost in a cloud and not a great idea to rely on less than foolproof identification procedures.

January 2010 marked the 25th anniversary of the cellular industry in the UK and I helped to organise a great big party[3] to help raise some funds for a new communications gallery at the Science Museum. The thing is that in an age of austerity it is not deemed politically correct to have a party unless you can prove the event was also informationally useful, which it was and to prove it I am going to crossreference some of the presentations from the predinner conference that have direct relevance to the topic of this chapter, server software. Two data points sourced from the Ericsson presentation[4] highlighted some of the issues.

Every two years the number of voice and data sessions doubles. In 2009, more than four exabytes (four billion gigabytes) of unique information were created. This information has to be stored somewhere and only has value if it can be accessed efficiently and effectively. It has been suggested that information and the applications that interact with that information are best accessed when needed rather than locally stored. This is commonly referred to as 'the cloud'. Information and applications and our relationships with other people and other things in the physical world around us will be mediated through what is in effect a virtual network.

At the time of writing there are somewhere between 300 000 and 400 000 applications (Apps) available for the iPhone. In fact, there are so many to choose from doing so many

---

[1] http://www.sciencemuseum.org.uk/onlinestuff/stories/amy_johnson.aspx.

[2] http://www.bbc.co.uk/dna/h2g2/A2654822.

[3] http://www.cambridgewireless.co.uk/cellular25/.

[4] http://www.cambridgewireless.co.uk/docs/John%20Cunliffe%20-%20Ericsson%20Ltd%20You%20Tube%20Link.pdf.

Making Telecoms Work: From Technical Innovation to Commercial Success, First Edition. Geoff Varrall.
© 2012 John Wiley & Sons, Ltd. Published 2012 by John Wiley & Sons, Ltd.

things that there are apps about apps rather like metadata tags that offer to help you choose what apps to download.[5]

20.1   The Wisdom of the Cloud?

The argument for or against centralised or distributed information is of course not new – the cloud is just a way of describing well-understood mechanisms for sharing and remotely accessing common resources. This means that 'the cloud' combines a mix of familiar benefits and risks.

Realisable benefits apart from the assumed cost savings of shared tenancy could theoretically include an improvement in social efficiency, social mobility and social inclusion, a part of the digital inclusion debate, an improvement in economic efficiency and economic equality, an improvement in intellectual efficiency and an improvement in environmental efficiency. The Ericsson presentation pointed out that using a mobile phone for a year has an equivalent CO_2 footprint to driving a car for an hour. If we assume that owning a mobile phone helps us to travel less then this should result in a net environmental gain. Also, apparently 17.4% of subscriptions in Sweden are machine-to-machine compared to 2.4% in Europe and 4.4% in the US. Some of these subscriptions will be for energy monitoring and energy control and environmental monitoring and environmental control, suggesting that additional environmental gains are achievable if the Swedish model were to be more widely applied.

Possible risks include compromised identification procedures, a proxy for safety and security and nonaltruistic political interference and control, of which more later. There are also some practical issues. One of the Ericsson predictions was that by 2013, 80% of all users will access the internet using a mobile phone. This is not the same as saying that 80% of all users will access the cloud by mobile phone but does suggest an expectation that wireless access will be relatively dominant.

However, mobile applications remain constrained by battery limitations. At times it can be more power efficient, faster, more convenient and more secure to store information locally rather than remotely.

And it is certainly true to say that at the time of writing there is no implicit incentive for application developers to make applications either power efficient, bandwidth efficient or code efficient.

Additionally, from a network perspective it is plausible to assume that wireline/guided media average data speeds may continue to increase faster than wireless data speeds, partly due to physics and partly due to available power. The only factor that would prevent this happening is a lack of investment in fibre, copper and cable connectivity that in turn would be a function of the returns achievable from the investment.

20.2   A Profitable Cloud?

On this basis, cellular network investment has historically looked attractive – a base station installed at the right place at the right time could pay for itself in a few months.

---

[5] www.magicsolver.com is one example.

It could be argued that markets deliberately structured to be overcompetitive combined with spectral auctions manipulated to maximise short-term returns to national treasuries have destroyed this advantage, though this may be about to change.

20.3   A Rural Cloud?

For example, there is some dependency on population density, which in turn determines the application mix. Rural communities in India as an example typically consume content (earlier we mentioned how popular FM receivers were in mobile phones sold in India) whereas urban areas tend to consume gaming applications.[6]

20.4   A Locally Economically Relevant Cloud?

Self-evidently, the impact of the cloud will be different in different countries. The weather is dramatically different in different places and so are the clouds.

There may also be regulatory inducement to ensure rural areas are covered either by imposing a levy or tax on urban access (as done in India) or through license conditions (Germany in the 800-MHz band, for example).

Poor rural economies, for instance in India, can be transformed by relatively small increases in connectivity, for example the SMS-based transaction applications discussed earlier in the book.

In April 2010 when we originally researched this topic Nokia had recently introduced their C Series phones targeted at emerging markets such as Indonesia. The product offering was coupled with Nokia's OVI store. OVI, Finnish for open door, who says the Finns don't have a sense of humour, was and at the time of writing still is intended to be Nokia's answer to the Apple applications offer that as of summer 2011 developed into the Apple iCloud.[7]

The OVI-compatible product range includes low-cost phones with quadband GSM, Bluetooth and WiFi connectivity. It is therefore economically relevant to emerging markets. In June 2011 a similar announcement was made that Series 40 products (market entry level) would support OVI. Note that locally relevant also means supporting local dialects, for example at least 26 dialects in India and 23 in Europe. Local economic relevance can also include applications such as Groupon integrated into a location-aware smart-phone application.[8]

20.5   A Locally Socially Relevant Cloud?

Thus, while the pundits have been generally dismissive of Nokia's applications offer and while there may be a perception of the offer being subscale in the US and some European markets it should also be noted that this may be part of an ongoing strategy that has seen the Ovi store[9] pulling ahead of Apple in emerging markets partly as a result of established distribution, partly

---

[6] R Swaminathan of Reliance Communications presenting at Future of Wireless International Conference in Cambridge, 27 June 2011.

[7] http://www.apple.com/icloud/.

[8] http://www.groupon.co.uk/.

[9] http://www.ovi.com/services/.

by having a wide range of low-cost and relatively simple access products and partly by adding applications to the store that are socially relevant to the local market.

This may reflect a growing recognition that the future of our industry is no longer being shaped by European or US consumer needs. The vendors best placed to take advantage of this shift may be most advantaged.

A hint of this can be read into this June 2011 press briefing from Nokia which highlighted the development of apps for their entry level Series 40 phones:

> "Consumers are downloading 6 million mobile apps and content each day from the Nokia Ovi Store, which now has a catalogue of more than 48,000 products available. While much of the emphasis has been on smartphone applications, Series 40 devices, which will enable apps for the next billion mobile phone users, have experienced more than 35 percent growth in download volumes in the last two months, making up about a quarter of the total downloads. More than 170 developers have each exceeded 1 million downloads from the Ovi Store."

Nokia also stated that they had OVI application and content download billing systems deployed and supported in 121 operator networks across 42 countries including Claro in Chile; Vodafone in Portugal; CSL, 3, SmartTone-Vodafone, PCCW and China Mobile in Hong Kong; and Etisalat in the United Arab Emirates and that the OVI store was available in more than 190 countries and 30 languages, with more than 90% of all store customers downloading apps in their own language. Applications made up about 70% of the total download with the rest being content with a general trend towards applications becoming more dominant in this mix. This probably puts them ahead of RIM and Microsoft but behind Google's Android Market and Apple's App store. Nokia's alliance with Microsoft[10] announced in 2009 and consolidated in 2011[11] is presumably a move to change this relative position over time. One developer, a company called Off Screen Technologies, claimed to have fulfilled more than 72 million downloads with 1000 applications being added per week. On this basis prognostications of the OVI door being closed would seem premature. It's more a case of building a new house with new windows and doors.

This application space is also opening up opportunities for specialist application developers, for example integrating trail guides into location mapping software.[12]

20.6   A Locally Politically Relevant Cloud – The China Cloud?

But life is not always simple. Politicians in developed countries like to talk about net neutrality and the recent experiences of Google in China suggest that political attitudes to a right of unfiltered access can be substantially different in other parts of the world.

China's ambassador to the UK, Madam Fu Ying, interviewed on the BBC (26/1/2/10)[13] made the point that the average income in China is equivalent to the average UK income in 1913 and suggested rather persuasively that it is inappropriate and arrogant to automatically

---

[10] http://www.microsoft.com/presspass/press/2009/aug09/08-12pixipr.mspx.

[11] http://conversations.nokia.com/2011/02/11/open-letter-from-ceo-stephen-elop-nokia-and-ceo-steve-ballmer-microsoft/.

[12] www.viewfinder.com.

[13] http://news.bbc.co.uk/1/hi/programmes/andrew_marr_show/7970581.stm.

assume that Western values and customs can or should be imposed on other cultures that may be socially and economically different.

Pragmatically, this means that compromises that may seem to be politically incorrect from a western perspective may at times be a necessary mechanism for increasing engagement between different cultures. The hope and aspiration has to be that this engagement results in the better parts of each culture being absorbed, shared and adopted over time.

20.7   The Cultural Cloud?

In this context, consider the following statistics from an Ericsson presentation.[14] In January 2010 there were 200 million active users on Facebook and 300 million on the Chinese community QC. 20% of all internet users were Chinese. Between the 25 and 31 January 2009, over the Chinese New Year, over 18 billion SMS messages were sent in China.

18 months later (June 2011) active Facebook users had grown to 500 million[15] and 21.4% of all internet users were Chinese.[16] This might seem like a small increase from 20% but the effect is massive. This means there are 420 million internet users in China, a population penetration of 31.6%, representing a growth of 1766.7% over the prior ten-year period.

The US comes in a distant second with 240 million users, population penetration of 77.3% and a growth of 151% over ten years. The US makes up 12% of the world market. Hardly surprising then that Nokia remain keen on adding application and content value to lower-end handsets.

The mobile phone is one of the world's most transformative devices with a capability to improve cultural understanding between countries traditionally and tragically distant in the past. The mobile phone and the cloud are potential partners in this process. Nokia remains a player with a number of unique market-positioning advantages that investment analysts seem to be presently underestimating.

The cloud is presented as an enabler of transformative change across consumer, business and specialist user ICT markets and Apple are considered to have achieved a first-mover advantage in this space. The cloud may well prove to be transformative but much of the opportunity analysis is presently very US centric. Asian rather than American clouds and rural clouds rather than urban clouds could well be the global economic and socially transformative weather makers over the next few years.

---

[14] http://www.cambridgewireless.co.uk/docs/John%20Cunliffe%20-%20Ericsson%20Ltd%20You%20Tube%20 Link.pdf.

[15] http://www.facebook.com/press/info.php?statistics.

[16] http://www.internetworldstats.com/top20.htm.

21

Future Trends, Forecasting, the Age of Adaptation and More Transformative Transforms

21.1 Future Forecasts

The ongoing narrative of the past twenty chapters has been that forecasting is essentially a process of looking behind and side to side. Behind means analysing past technical and commercial success and failure, but also whether past projections and predictions were right or wrong, if yes why, if not why. Side-to-side means analysing and comparing all device options, all delivery options including guided and unguided media and all network options in terms of technical and commercial efficiency including energy and environmental efficiency.

Our thesis throughout is that if a process or product is technically inefficient it is less likely to be commercially efficient, although I have had some wry comments about Microsoft having broken this rule and yes this is the second time this morning that my computer has crashed. In network terms this means that bandwidth functionality and the value realisable from that functionality must always exceed network cost amortised over the life of the network.

We have drawn on our own past work in terms of prior published books and articles and research undertaken over the past twenty five years and the advice and guidance and suggestions of many friends and colleagues in the industry.

In this last chapter we take a final look at our research over the last five years and dangerously and presumptuously use this to make some predictions for the next five and 50 years forward.

In February 2006 we wrote a piece about competitive networks. We suggested that future networks would need to develop new mechanisms for capturing user bandwidth and user value and that the most successful networks would be those that most successfully competed for billable revenue – aggressively competitive networks.

Competitive networks were defined as intelligent networks with ambition. They could also be described as 'smart networks' using 'smart' in the contemporary sense of a network having

Making Telecoms Work: From Technical Innovation to Commercial Success, First Edition. Geoff Varrall.
© 2012 John Wiley & Sons, Ltd. Published 2012 by John Wiley & Sons, Ltd.

a developed ability to make money. Competitive networks may have other obligations, which may include social and political gain. Smart competitive networks are efficient at translating social and political gain into economic advantage. Economic gain is the product of a composite of cost and functional efficiency and added value.

Added value is delivered through service provision and billing, but is dependent on a broad mix of enabling technologies and enabling techniques that evolve over time. This process of technology evolution is managed within a standards-making process that in itself determines the rate at which new technologies are deployed.

The standards-making process also introduces a cyclical pattern of technology maturation that is not naturally present in the evolutionary process, which is largely linear. This produces market distortions that reduce rather than enhance added value.

An understanding of the evolutionary process is therefore a useful precondition for identifying technologies that deliver long-term competitive advantage and provides an insight into how standards making could become, and probably will become, more productive over time.

21.2   The Contribution of Charles Darwin to the Theory of Network Evolution

To study the evolutionary process, where better to start than Charles Darwin. Inspired by amongst others the famous botanist, geologist, geographer and 'scientific traveller' the Baron von Humboldt (the current man), Darwin spent five years (1831–1836) on the HMS Beagle observing the flora and fauna of South America.

From these observations Darwin developed his theories of natural selection.

In parallel, similar studies on similar expeditions to the Amazon were leading Alfred Russell Wallace to develop and promote similar theories that in turn prompted Darwin (in 1859) to publish 'On the Origin of Species by Means of Natural Selection, or the Preservation of Favoured Races in the Struggle for Life'.

A summation of Darwin's theory:

> Evolution exists.
>
> Evolution is gradual.
>
> The primary mechanism for evolution is natural selection.
>
> Evolution and natural selection occur through a process of ongoing specialisation.

So, what relevance does Darwin and his fellow botanist and naturalist colleagues have to network design and network evolution?

Well, to paraphrase,

> The survival (or success) of each organism (network) is determined by that organism's (network's) ability to adapt to its environment.
>
> The fittest (most power and bandwidth-efficient networks) win out at the expense of their rivals (other networks) because they succeed in adapting themselves best to their environment.

21.3   Famous Mostly Bearded Botanists and Their Role in Network Design – The Dynamics of Adaptation

Darwin suggested that the process of evolution and adaptation occurs over millions of years. One way or another humans seem to have used those millions of years to achieve a competitive advantage over (most) other species. This competitive advantage is based on energy efficiency (output available for a given calorific intake), observational abilities (sight, sound, smell) and the ability to use observed information intelligently to avoid danger or to exploit locally available opportunities.

To us, the process of long-term adaptation is unnoticeable due to the time scales involved. The process appears to us to be static. However, part of the process of successful adaptation over time is to become more adaptive, particularly in the way in which we conserve energy, the way in which we manage to have energy available to support short concentrated bursts of activity and the way in which we couple these processes to observed information. Short-term adaptation functions are noticeable. We can experience them and measure their extent and effect. We can recognise and observe the dynamic nature of the process.

In network and device design, adaptation is an already important mechanism for achieving power and bandwidth efficiency and is becoming more important over time. As such, it is useful to study how adaptation works in an engineering context but using several million years of biological and botanical experience and several hundred years of biological and botanical study and analysis as a reference point.

21.4   Adaptation, Scaling and Context

We can describe almost any process in an end-to-end delivery channel in terms of three interrelated functions – adaptation, scaling and context:

i Adaptation is the ability of a system or part of a system, to respond to a changed requirement or changed condition.
ii Scale (scaling) is the expression of the order of magnitude over which the adaptation takes place (the range) and may also comprehend the rate and resolution of the response.
iii Context is the amount of observable information available in order for the system itself or some external function to take a decision on the adaptation process. The accuracy of the observation process and the ability of the system to interpret and act on the observed information will have a direct impact on overall system efficiency.

The function of the human heart provides a biological example. Our heart rate and blood pressure increases when we run. Our heart rate and blood pressure also increases when we use our observation system (sight, sound, smell, touch and sense of vibration) to perceive potential danger and/or local opportunity. As such, the process of adaptation is both reactive and proactive.

The scale in this example ranges from the lowest to highest heart rate. The context is either actual (we have started to run) or predictive (we know we might have to run). Human observation systems in themselves are adaptive. For example, our eyes can adapt to light intensities ranging from small fractions of a lux to over 100 000 lux. This is the scale of the adaptation process. The context, in this example, is the ambient light level.

21.5 Examples of Adaptation in Existing Semiconductor Solutions

Many present semiconductor solutions used in cellular phones, consumer and professional electronics products and network devices use adaptation to decrease power drain. In this context, adaptation is based on the ability to change the clock speed (heart rate) and/or voltage (blood pressure) of the system when presented with specific processing tasks (the context).

The scale is the highest to lowest clock rate and highest to lowest voltage. The rate of response (milliseconds or microseconds) and accuracy of response will directly impact overall system efficiency. They may also be additional 'policy' issues to consider, for example the charge state of the battery and external or internal operating temperatures.

Peripheral devices such as displays (one of the largest power-consuming items in most present appliances), also use adaptation by responding to changes in ambient light level or by decreasing frame rates, resolution and colour depth in response to a changed operational requirement.

21.6 Examples of Adaptation in Present Mobile Broadband Systems

Adaptation has always been an inherent part of cellular network design. First-generation analogue cellular networks used receive signal strength measurements to determine power and edge of cell handover thresholds (an intelligent use of observed information).

Second-generation networks made the handset work harder in that the handset compiles a measurement report that is a composite of received signal level and received signal quality (bit error rate and bit error probability). This is sent to the radio network controller that in turn makes power control and handover decisions. The handset is instructed to look at up to 6 base stations, its own present serving base station and five 'neighbour' handover candidates though there is support for more extended measurement reporting (up to ten base stations) in more recent releases. Power control is implemented within a relatively relaxed duty cycle of 480 milliseconds, though optional enhanced (120 millisecond) and fast (20 millisecond) power control is now supported.

So in these examples, the adaptation process is the combined function of power control and handover. The scale of power control is typically about 25 dB (first generation) or 30 to 35 dB in second-generation systems, the rate is either 480 milliseconds, 120 milliseconds or 20 milliseconds and the resolution is typically half a dB. The context is provided by the measurement report. Note that overall system efficiency is determined directly by the accuracy of the measurement report and the ability of the power control and handover algorithms to interpret and act on the observed information – this is algorithmic value.

Release 99 WCDMA introduced more aggressive power control (an outer loop power control every 10 milliseconds and an inner power control loop running at 1500 Hz), a substantially greater dynamic range of power control (80 dB) and the ability to change data rate and channel coding at ten-millisecond intervals.

HSDPA simplifies power control but allows data rates and coding and modulation to be changed initially at 2-millisecond intervals and in the longer term at half-millisecond intervals. LTE takes this process a step further and can opportunistically map symbols to OFDM subcarriers enjoying favourable propagation conditions.

The context in which the data rate and channel coding and symbol mapping decisions are taken is based on a set of channel quality indicators (CQIs). This is a composite value (one of 30 possible values) which indicates the maximum amount of data the handset thinks it should be able to receive and send, taking into consideration current channel conditions and it's designed capabilities (how many uplink and downlink multicodes and/or OFDM subcarriers and modulation schemes it supports).

There are, however, many additional contextual conditions that determine admission policy. These include fairly obviously the level of service to which the user has subscribed but also the local loading on the network. Admission policy may also be determined by whether sufficient storage, buffer or server bandwidth is available to support the application.

So, we have a 30-year example of evolution in cellular network design in which step-function changes have been made that deliver more adaptability over a wide dynamic range (scale) based on increasingly complex contextual information. It is this process of increasingly aggressive adaptation that has realised a progressive increase in power and bandwidth efficiency that in turn has translated into lower costs that in turn have supported lower tariffs that in turn have driven traffic volume and value.

21.7   Examples of Adaptation in Future Semiconductor Solutions

There is still substantial optimisation potential in voltage and frequency scaling. Present frequency scaling has a latency of a few microseconds and voltage scaling a latency of a few tens of microseconds but the extraction of efficiency benefits from scaling algorithms is dependent on policy management.

Scaling policy has to take into account instantaneous processor load and preferably be capable of predicting future load (analogous to our heart rate increasing in anticipation of a potential need for more energy). Proactive rather than reactive algorithms can and will deliver significant gains particularly when used with multiprocessor cores that can perform load balancing. These are sometimes known as 'balanced bandwidth' processor platforms. Note this is an example of algorithmic value.

21.8   Examples of Adaptation in Future Cellular Networks

Developments for near-future deployment include adaptive source coding, 'blended bandwidth' radio schemes that use multiple simultaneous radio bearers to multiplex multimedia traffic, and 'balanced bandwidth' IP RAN and IP core network architectures that adaptively manage buffer bandwidth, storage bandwidth and server bandwidth to support bandwidth efficient multiservice network platforms.

Adaptive source coding includes adaptive multirate vocoders, AAC/AAC Plus- and MP3Pro-based audio encoders, JPEG image encoders and MPEG video encoders, wavelet-based JPEG encoders and MPEG2000 Part 10 scalable video coders. The realisation of the power and bandwidth efficiency potential of these encoding techniques is dependent on an aggressively accurate analysis of complex contextual information.

'Blended bandwidth' includes the potential efficiency benefits achievable from actively sharing bandwidth between wide-area, local-area and personal-area radio systems. Admission control algorithms can be based on channel availability, channel quality and channel 'cost' that in an ideal world could be reflected in a 'blended tariff' structure to maximise revenues and margin in combination with a more consistent (and hence higher value) user proposition.

When combined with other downlink delivery options such as DAB, DVB, these 'blended bandwidth' offerings are sometimes described in the technical literature as 'cooperative networks'.

The technical and potential cost benefits of such schemes are compelling but adoption is dependent on the resolution of conflicting commercial objectives. The fiscal aims of cooperation and competition rarely coincide unless a certain amount of coercion is used.

'Balanced bandwidth' essentially involves achieving a network balance between radio access bandwidth, buffer bandwidth in the end-to-end channel, (including IP RAN and IP core memory bandwidth distribution), persistent storage bandwidth and server bandwidth and being able to manage this bandwidth mix proactively to match rapidly changing traffic loading and user-application requirements.

21.8.1   How Super Phones Contribute to Competitive Network Value

We have already stated that existing phones and mobile broadband devices play a key role in collecting the contextual information needed to support power control and handover and admission control algorithms.

Power control, handover and admission control algorithms are needed to deliver power efficiency (which translates into more offered traffic per battery charge) and spectral efficiency (which translates into a higher return on investment per MHz of allocated or auctioned spectrum).

21.8.2   Phones are the Eyes and Ears of the Network

However, we also suggested that there are three categories of phone, standard phones, smart phones and super phones. Standard phones are voice and text dominant and change the way we relate to one another. Smart phones aspire to change the way we organise our work and social lives. Super phones change the way we relate to the physical world around us.

We studied the extended image and audio capture capabilities of super phones and the ways in which these capabilities could be combined with ever more accurate macro- and micropositioning information.

Macropositioning information is available from existing satellite systems, for example GPS with its recently upgraded higher-power L2C signal and, at some stage, Galileo with its optimised European coverage footprint. Macropositioning is also available from terrestrial systems (observed time difference) and from hybrid satellite and terrestrial systems.

Micropositioning is available from a new generation of multiaxis low-G accelerometers and digital compass devices that together can be used to identify how a phone is being held and the direction in which it is pointed.

This moves competitive networks onto new territory. The eyes and ears of the network see more and hear more than ever before. The network has a precise knowledge of where users are, what they are doing and where they are going (direction and speed).

In simple terms, this knowledge can be used to optimise handover and admission control algorithms. More fundamentally, the additional contextual information becoming available potentially transforms both the adaptability and scalability of the network and service proposition. This has to be the basis of future mobility value in which handsets help us to relate to the physical world around us and networks help us to move through the physical world around us.

21.8.3   And Finally Back to Botany and Biology

Biological evolution may from time to time appear as nonlinear. The Cambrian Explosion 543 to 490 million years ago (when most of the major groups of animals first appear in the fossil record) is often cited as an example of nonlinear evolution.

Technology evolution may also appear from time to time to be nonlinear. The invention of the steam engine or transistor for example might be considered as inflection points that changed the rate of progress in specific areas of applied technology (the industrial revolution, the birth of modern electronics). This apparent nonlinearity is, however, a product of scaling and disappears if a longer time frame is used as a benchmark of continuing progress.

The essence of competition, however, remains relatively constant over time.

Have humans become more adaptive over time, have they adapted by becoming more adaptive? Possibly. Certainly we have become more adept at exploiting context partly due to our ability to accumulate and record and analyse the knowledge and experience of prior and present generations. And this is the basis upon which competitive networks will capture and deliver value to future users.

Network value is increasingly based on knowledge value but knowledge value can only be realised if access efficiency can be improved over time. Access efficiency in mobile broadband networks is a composite of power and bandwidth efficiency (the blended bandwidth proposition) and access efficiency. Access efficiency is dependent on the efficiency of access, admission and storage and server algorithms that anticipate rather than respond to our needs.

The asset value of cellular and mobile broadband networks has traditionally been denominated in terms of number of cell sites or MHz of allocated or auctioned bandwidth and number of subscribers. This still remains valid but these are not inherently competitive networks.

Competitive networks are networks that exploit accumulated knowledge and experience from present and past subscribers to build new value propositions that deliver future competitive advantage. This in turn increases network asset value. Efficient competitive networks are networks that combine access efficiency with power and bandwidth efficiency. An intelligent network with ambition and ability, a network that adapts over time by becoming more adaptive.

21.9   Specialisation

But in addition, networks need to find some way of differentiating themselves from one another, creating 'distance' in the service proposition. This is where Darwin's (and Wallace's) theories of specialisation begin to have relevance. The services required from a network in the Amazon and Malay archipelago and Galapagos Islands are different from the services required in Manhattan, Maidenhead and Mancuria. Radio and TV stations are increasingly specialised in terms of regionalised and localised content provision. Cellular and mobile broadband networks will need to develop similar techniques in terms of their approach to specialist regionalised and localised network service platforms. As we said earlier, smart competitive networks are networks that are efficient at translating social and political gain into economic advantage. This implies an ability to respond to extremely parochial geographic and demographic interests. A bee in Mongolia might be outwardly similar to a bee in Biggleswade but will have different local interests and requirements. A 'one size fits all' network proposition becomes increasingly less attractive over time. Chapter 19 used the Nokia OVI platform as an example of the competitive advantage achievable from localisation.

21.10   The Role of Standards Making

We argued the case that technology evolution is to all intents and purposes a linear process. It may appear from time to time to be nonlinear, but this is either due to an issue of scaling (not studying the evolutionary process over a sufficiently long time scale) or due to distortions introduced by the standards-making process.

The answer is of course to make the standards process more adaptive and to avoid artificially managed step-function generational changes. Of course the counterargument here is that artificially managed change can be exploited to deliver selective competitive advantage. This is, however, a manipulative process and manipulative processes tend to yield relatively short-term gains, the danger of a Pyrrhic victory. The destruction of the potential market for ultrawideband devices through a standards dispute is one example.

21.11   The Need for a Common Language

Value can also be destroyed through misunderstanding and poor advocacy. Poor advocacy can be particularly expensive if the result is decision making based on inaccurate or incomplete or poorly presented technical facts. As an antidote to this we recommend a short consideration of the work of Dr Ludwig Zamenhof, founder of the Esperanto movement.

Esperanto is often dismissed as a marginal curiosity with little relevance to the modern world. Dr Zamenhof had a minor planet named after him by Yro Vaisala, the Finnish astronomer and physicist, and is considered as a god by the Omoto religion. Aside from these distractions, his work on a universal language has real relevance to engineers and engineering. For example, the Chinese Academy of Science conducts a biannual international conference on Science and Technology in Esperanto and publishes a quarterly journal, 'Tutmondaj Sciencoj kaj Teknikoj'.

However, we are not on a mission to preach the virtues of Esperanto to engineers. More specifically, Esperanto has practical relevance to our own area of interest, user equipment and network design.

Esperanto is an example of a language that is both efficient and precise in the way that it describes the world around us. A language optimised for logical thought and analysis. For most of us, learning Esperanto is never going to be a high priority, even if it should be. We can, however, learn lessons from Esperanto in terms of how we approach decision making in network design.

Decision making is dependent on the accuracy with which a specific problem or set of choices is described. We suggest a number of **'descriptive domains'** that can be applied that help to clarify some of these choices. We show how 'descriptive domains' can be used to validate R and D resource allocation and partitioning and integration decisions.

21.11.1   Esperanto for Devices

The principles of Esperanto are applicable for human-to-human and device-to-device communication. New functionality in user equipment is often introduced on discrete components. Audio integrated circuits, voice and speech recognition, micropositioning MEMS or GPS integrated circuits are recent examples of new 'real estate' introduced as additional add-in components.

This places new demands on packaging and interconnect technology and creates a need for commonality (a common language) both in terms of physical (compatible pin count and connector) and logical (bus architecture) connectivity. The reason for the discrete approach is usually performance and/or risk related. The new function can initially be made to work more efficiently on a discrete IC and engineers responsible for other functional areas feel safer if the new functionality is ring fenced within its own dedicated physical and logical space.

Standards initiatives such as the Mobile Industry Processor Initiative[1] are then needed to ensure at least a basic level of compatibility between different vendor solutions. This process is neither particularly efficient nor effective and is generally a consequence of interdiscipline communication issues.

21.11.2   Esperanto for Engineers

This in turn is a language problem. Partly this has been solved by the use of English as a common language, though English is neither precise nor efficient. The problem is, however, not just one of spoken language but the descriptive language used.

Software engineers speak a different 'language' from hardware engineers, DSP engineers speak a different 'language' from RF engineers, imaging system engineers speak a different 'language' from audio engineers who speak a different 'language' from radio system engineers, who speak a different 'language' from micro- and macropositioning engineers, who speak a different 'language' from mechanical design engineers.

Silicon design engineers speak a different 'language' from handset design engineers who speak a different 'language' from network design engineers who speak a different 'language' from IT engineers. Engineers speak a different 'language' from product marketing and sales who speak a different 'language' from business modelling specialists who speak a different 'language' from lawyers and accountants who make the world go round. This is frustrating as, in reality; all these 'communities of interest and specialist expertise' have more in common than seems initially apparent. 'Descriptive domains' provide a route to resolving these interdisciplinary communication issues.

21.12   A Definition of Descriptive Domains

A descriptive domain is simply a mechanism for describing form and function The 'analogue domain' and the 'digital domain' provide two examples. The analogue domain can be widely understood in a modern context as a set of continuously variable values. In a linear system, the output should be directly proportional to the input.

---

[1] http://www.mipi.org/.

This is the *'form'* of the domain; its defining characteristic. The *'function'* of the domain is to provide a generic method for describing the physical world around us. Light, sound and gravity and of course radio waves are all in the analogue domain. The 'form' of the digital domain is defined as the process of describing data sets as a series of distinct and discrete values. The 'function' of the digital domain is to provide a generic method for describing analogue signals as discrete and distinct values.

Most people are comfortable with these descriptions but do not necessarily understand some of the practical implications of working in either domain. For example, an engineer might need to describe the fiscal merits of digital processing in terms that are accessible to an accounting discipline. This suggests the need to add in 'cost' and 'value' to the descriptive domain. Let's test the validity of this on some practical examples

21.12.1   Cost, Value and Descriptive Domains – A Description of Silicon Value

As with many technology-based industries, the telecommunications industry is built on a foundation of sand, also known as silicon. Silicon-based devices provide us with the capability to capture, process, filter, amplify, transmit, receive and reproduce complex analogue real-world signals. Silicon-based devices allow us to speak, send (data, audio, image and video), spend and store. Silicon value translates into software value that translates into system value that translates into spectral and network bandwidth value.

Decreasing device geometry delivers a bandwidth gain both in terms of volume and value. 'Bandwidth value' at the device level determines bandwidth value at the system and network level that, for cellular and mobile broadband networks, implies a return on spectral investment. By studying silicon value we determine how future software, system and spectral value will be realised.

As suggested in previous chapters, we can consider at least five value domains – Radio systems, Audio systems, Positioning systems, Imaging Systems and Data Systems. These are illustrated in Table 21.1. Silicon vendors and silicon design teams who successfully integrate these five domains will be at a competitive advantage.

Earlier in this chapter we described the concept of blended bandwidth and balanced bandwidth as interrelated concepts that could be used to qualify radio access and network transport functionality and radio access and network transport value. Integration is the process of blending and balancing.

The same principle applies at the silicon level. First, define the domains and then qualify how these domains add individual and/or overall value. Use this to qualify partitioning and integration decisions and R and D resource decisions.

'Blended bandwidth' (the horizontal axis in Table 21.1) implies a need to integrate each of the five domains and within each of the five domains, to integrate individual subsystem

Table 21.1   Value domains

R			A		P		I		D	
Radio			Audio		Positioning		Imaging		Data	
Wide area	Local area	Personal area	Voice	Audio	Micro	Macro	Image	Video	PIM	CIM

functions. The amount of crossintegration determines the 'breadth' of the blended bandwidth proposition.

For example, 'blended bandwidth' in the 'radio system domain' means getting wide-area systems (for example, HSPA or LTE or EVDO) to work with broadcast wide-area, local-area WiFi and personal-area Bluetooth and/or NFC local-area connectivity.

In the *'audio system domain'*, 'blended bandwidth' involves a successful integration of **voice** encoder/decoder functionality with *audio* encoder/decoder functionality (AAC AAC Plus, MP3Pro and Windows Audio). This includes functions such as voice and speech recognition.

In the *'positioning system domain'*, 'blended bandwidth' requires the integration of micropositioning systems (for example, MEMS-based low-G accelerometer devices) with macropositioning (GPS, A GPS or observed time difference systems). Micro- and macropositioning systems have the capability of adding value to all other domains and should be considered as critical to the overall silicon value proposition.

In the *'imaging system domain'*, 'blended bandwidth' means the integration of image and video and the integration of imaging systems with all other system domains. Imaging systems include sensor arrays, image processing and display subsystems. MMS is an imaging domain function.

In the *'data system domain'*, 'blended bandwidth' involves making personal and corporate information systems transparent to all other domains. SMS Text is inherently a data domain function.

'Balanced bandwidth' is how much functionality you support in any one domain and in any one function within each domain – the 'depth' of the domain.

For example, in the radio system domain, the choice for wide-area system functionality will depend on the mix of technologies deployed and the bands into which they are deployed.

For *local-area* radio system functionality, the choice would be between 802.11 a, b and g and related functional extensions (802.11n and MIMO).

In the *personal area*, the decision would be whether to support Bluetooth EDR and/or NFC.

In the *audio system domain*, the choice for voice would be which of the higher-rate voice codecs to support, the choice for audio would be which audio codec to support, other choices would be functional such as voice or speech recognition support, advanced noise cancellation, extended audio capture or advanced search and playback capabilities.

In the *positioning domain*, the choice is essentially rate, resolution and accuracy both in micro- and macropositioning. An improvement in any one of these metrics will imply additional processor loading.

In the *imaging domain*, the decision would be whether to support newer coding schemes such as JPEG2000 (imaging) or H264/MPEG Part 10 AVC/SVC or, in general, any of the emerging wavelet-based progressive rendering schemes.

In the *data domain*, decisions revolve around the amount of local and remote storage dedicated to personal and corporate data management systems and related data set management capabilities.

21.13   Testing the Model on Specific Applications

To be valid, we now need to show that the model has relevance when applied to specific applications. For example, a *gaming* application may be a composite of radio layer functionality, audio functionality, micro- and macropositioning functionality, imaging functionality and personal profiling (personal data set management). A *camera phone* application will already have wide-area radio access, may have enhanced audio, should probably have integrated positioning, will certainly have imaging and should have data functionality.

21.13.1   The Concept of a 'Dominant Domain'

This leads us towards defining future handsets in terms of their *'dominant domain'*. The 'dominant domain' of an audio phone is the audio system domain with potentially all other domains adding complementary domain value. The dominant domain of a location device is the positioning domain with potentially all other domains adding domain value. The 'dominant domain' of a camera phone is the imaging system domain, with potentially all other domains adding complementary domain value. Within the imaging domain, the dominant functionality is image capture (the optical subsystem and sensor array).

The 'dominant domain' of a games phone is also the imaging domain but the dominant functionality within the domain is, arguably, the display subsystem and associated 2D and 3D rendering engines. The 'dominant domain' of a 'business phone' is the data domain, with the emphasis within the data domain on corporate information management. All other domains are, however, potential value added contributors to the overall system value of the dominant domain.

21.13.2   Dual Dominant-Domain Phones

The above examples are reasonably clear cut, but let's take for example an ultralow-cost phone.

The dominant domain of an ultralow-cost phone is the wide-area part of the radio system domain and the voice part of the audio domain plus some parts of other domains (text from the data domain function, possibly basic camera functionality from the imaging domain). Being pragmatic, it is sensible to describe a ULC phone as a 'dual dominant domain device'.

In Chapter 19 we described how Nokia is working on introducing richer functionality to these entry-level devices. Note that they still retain a market volume advantage so can add functionality more aggressively without necessarily compromising product margins.

21.14   Domain Value

Each individual domain has a value and cost. The cost is functional and physical and is a composite of processing load and occupied silicon real estate. Value is a composite of the realised price of the product plus incremental through life revenue contribution as functionality is increased in any one domain.

Domain value is independent of partitioning or integration. Presently, individual domains may be on separate interconnected devices. For example, in the radio domain, Bluetooth has been historically separate from the mobile broadband RF and baseband and/or from embedded WiFi functionality. In the audio domain, speech codecs and audio codecs may be and often are, separate entities. Micropositioning is separate from macropositioning. Imaging is separate from video. For example, most camera phones have two cameras, one for still-image capture, one for lower-resolution video. The data domain may be distributed across several devices, for example, the application processor, the SIM card and so on.

This does not mean that domain value is not a valid approach for qualifying partitioning or integration decisions. Adding functionality to phones has increased the complexity of the decision process and implies a need for a level of interdomain understanding that is hard to achieve. Radio system specialists have not historically needed to know much about audio systems or positioning systems or imaging systems or data systems.

So the domain-value approach can be useful at engineering level either to work through the mechanics of partitioning or integration or to decide how much functionality to support in any one domain and/or to decide how to allocate R and D effort to achieve a maximum return. Similarly, the domain-value approach can help in developing future market models. Forecasting the future sales volume and value of voice phones, audio phones, location devices, camera phones and business phones and/or personal organisers and smart phones and tablets and dongles can become muddled by a lack of descriptive clarity.

Given that function determines form factor, domain-based functional descriptions provide a valid alternative approach allowing product families to be developed at the silicon and handset level with clearly differentiated functionally based cost and value metrics.

The same applies in market research. The telecoms industry, in our case, the wireless telecoms industry, is a classic example of a technology-driven rather than customer-driven industry.

This is altogether a good thing. However, in a technology-driven industry, 'listening to your customer' is probably the worst thing you can possibly do. Understanding your customer is, however, completely crucial. 'Understanding' in this context implies a quantitative understanding of the economic and emotional value of each of the domains and the perceived quality metrics of each domain.

21.15   Quantifying Domain-Specific Economic and Emotional Value

Such an approach is not particularly difficult and can be objectively based. For example, it is possible to quantify the economic and emotional value of wide-area radio system mobility and ubiquity and build models of how the value/cost metric changes as bandwidth and perceived quality increases over time.

It is possible to quantify audio system value on the basis of voice and audio value using well-defined and calibrated mean opinion score methodologies and to model how the value/cost metric changes as bandwidth and perceived quality increases over time. It is possible to quantify micro- and macropositioning value and to model how the value/cost metric changes as accuracy and fix speed increases over time.

It is possible to quantify imaging system value and to model how the value/cost metric changes as resolution and colour depth and perceived quality increase over time using well-defined and calibrated mean opinion score methodologies.

It is possible to quantify data system value both in terms of personal efficiency and corporate efficiency metrics. It might be argued that emotional value is hard to quantify. However, emotional value is part of the 'engagement cycle' that determines the 'soft value' (or 'fondness') that users feel towards particular service offerings. Intuitively, as emotional value increases, session length increases; a directly measurable metric.

Is domain value relevant to user-equipment design?

Yes. We have chosen silicon value as an example but it is equally applicable to handset research and design. It is a valid approach to developing handset technology policy over a forward three–five-year time scale.

Is domain value relevant to infrastructure and network design?

Certainly, all five value domains can be used to quantify cost metrics and value metrics in a radio network proposition. For example, radio-system costs and radio-system value, audio-system costs and audio-system value, positioning-system costs and positioning-system value, imaging-system costs and imaging-system value and data-system costs and data-system value can be and should be separately identified.

Is domain value relevant to content management?

Absolutely. Content has a direct impact on radio-system cost and radio-system value, audio value is an integral part of the content proposition but benefits from being separately identified as a cost and value component, positioning adds value to content, imaging is an integral part of the content proposition but benefits from being separately identified as a cost and value component, data systems are an integral part of the content cost and value proposition.

Note that audio, image and data costs are a composite of delivery and storage cost and delivery and storage value, both are usefully described in their individual domains.

Is domain value relevant to application value?

Yes. Chapter 20 provided examples of this.

Is it a universal model?

Yes. It is certainly universally useful as a mechanism or framework for getting engineering and marketing teams to work together on product definition projects. It provides an objective basis upon which R and D resource allocation can be judged and an objective basis for deciding on 'hard to call' partitioning and integration decisions.

The definition and development of user equipment with integrated radio system, audio system, positioning system, imaging and data system functionality demands particular inter-discipline design skills. Convergence increases rather than decreases the need for interprocess communication.

Making multimedia handsets work with multiservice networks introduces additional descriptive complexity and requires a closer coupling between traditionally separate engineering and marketing and business modelling disciplines. Developing consistent descriptive methodologies that capture engineering and business value metrics helps the interdiscipline communication process.

The challenge is to combine the language of engineering with the language of fiscal risk and opportunity. Descriptive domains provide a mechanism for promoting a more efficient and effective crossdiscipline dialogue – engineering Esperanto.

21.16   Differentiating Communications and Connectivity Value

Tom Standage in his book, The Victorian Internet,[2] explains how the electric telegraph delivered 'network value' to the rapidly industrialising nations of the nineteenth century. In other

---

[2] http://tomstandage.wordpress.com/books/the-victorian-internet/.

work[3] he identifies the role of coffee shops in the seventeenth and eighteenth century as centres for information distribution and social and information exchange. Customers were happy to pay for the coffee but expected information for free, free speech and free information as the basis for social, political and economic progress.

In the twentieth and twenty first century, the same principles apply but the delivery options have, literally, broadened. Present economic growth can be at least partially ascribed to the parallel growth of connectivity bandwidth. Connectivity implies the ability to access, contribute and exchange information. Connectivity is a combination of voice-, audio-, text- image-, video- and data-enabled communication (the internet) but crucially includes access to storage bandwidth (the World Wide Web and the cloud).

21.17 Defining Next-Generation Networks

So, noncontentiously we can say that next-generation networks are likely to be an evolution from present-generation networks and present-generation networks are, or will be, based on access to the internet and access via the internet to the World Wide Web.

This is straightforward. The problem is the user's expectation that access to this commodity (the sharing of human knowledge, experience, insight and opinion) is a basic human right analogous to the provision of sanitation and healthcare. There may, of course, be a valid argument that governments should provide internet access for free on the basis that the added tax income from the associated additional growth will provide a net economic, social and political gain.

Selective free access is of course already available, for example internet access in schools and internet access in libraries (free to pensioners). The provision of such services is analogous to the right of citizens to have access to free to air broadcast content and the related public service remit to 'inform, educate and entertain'. Interestingly, most of us in the UK are happy to pay a license fee (to the BBC though not perhaps to Mr Murdoch) on the basis that this makes it more likely that the content delivered in return is 'free' from political bias or at least overt political interference.

Of course, governments do not need to invest in communications networks because other people do it for them. However, the contention is that network access, broadly defined to include all forms of present connectivity, directly contributes to an increase in GDP. There are exceptions. Watching the Simpsons[4] or Neighbours[5] does not necessarily add significantly to the economic well being of the nation but the general principle prevails. Therefore, governments have a duty to ensure connectivity is provided at an affordable cost.

Most of us are willing, if not necessarily happy, to pay 40 or 50 dollars per month for the right of access to the various forms of communications connectivity available to us. Price perceptions are influenced by what we pay for other forms of connectivity – water, gas, electricity.

Emerging nations are, however, different. You cannot expect someone on a dollar a day to pay 50 dollars per month for a broadband connection. What's needed is an ultralow-cost network and ultralow-cost access to and from that network.

---

[3] http://www.economist.com/node/2281736?story_id=2281736.
[4] A popular US TV cartoon programme.
[5] A popular Australian soap.

21.18   Defining an Ultralow-Cost Network

21.18.1   Ultralow-Cost Storage

Storage costs are halving each year, a function of solid-state, hard-disk and optical-memory cost reduction and improvements in content compression methodologies.

The low error rates intrinsic to nonvolatile storage media suggest further improvements in compression techniques are possible, provided related standards issues can be addressed. A closer harmonisation of next-generation video- and audio-compression techniques would, for example, be advantageous. Other techniques such as load balancing across multiple storage and server resources promise additional scale and cost efficiencies.

21.18.2   Ultralow-Cost Access

This is dependent on the relative cost economics of fibre, copper and wireless access. All three of these delivery/access media options have enjoyed incremental but rapid improvements in throughput capability. Throughput rate increases have been achieved through a mix of frequency-multiplexing and time-multiplexing techniques.

Fibre and copper are both inherently stable and consistent in terms of their data rates and error rates. Error rates in wireless are intrinsically higher and more variable as a consequence of the effects of fast and slow fading. These effects are particularly pronounced in wide-area mobile wireless systems but are still present in fixed wireless systems particularly when implemented at frequencies above 3 GHz.

Higher error rates and uneven error rate distribution make it harder to realise the full benefits of higher-order content-compression techniques. Wide-area wireless mobility also implies a significant signalling overhead that adds directly to the overall cost of delivery.

All three access options (fibre, copper and wireless), have a related real-estate cost that is a composite of the site or right of access negotiation and acquisition cost and ongoing site or right of access administration costs.

Present fibre to the home (ftth) deployments, for example, imply significant street-level engineering and financial-resource allocation that has to be justified in terms of bidirectional data throughput value amortised against relatively ambitious return on investment expectations.

In cellular and mobile broadband networks, similar ROI criteria have to be applied to realise a return from spectral investment, site acquisition and ongoing site-management costs. Cellular and mobile broadband networks are, of course, in practice, a composite of radio access, copper and fibre. Wireless costs are reducing but whether they are reducing faster than other delivery options remains open to debate.

21.18.3   The Interrelationship of RF Access Efficiency and Network Cost

The escalation of spectral and site values has highlighted a perceived need for cellular and mobile broadband technologies to provide incremental but rapid improvements in capacity, the number of users per MHz of allocated or auctioned spectrum multiplied by typical user per data throughput requirements, and coverage (range). New technologies are therefore justified on the basis of their 'efficiency' in terms of user/data throughput per MHz, which we will call, for lack of a better term, 'RF access efficiency'. In practice, new technologies need to offer useful performance gains, lower real costs and increased margins for vendors.

21.18.4   The Engineering Effort Needed To Realise Efficiency Gains

However, new technologies, particularly new radio technologies, generally deliver initially disappointing performance gains. This is because the potential gains implicit in the technology can only be realised through the application of significant engineering effort.

21.18.5   The Impact of Standards and Spectral Policy on RF Network Economics

At this point it is worth considering the impact that standards and spectral policy have on RF network economics.

Standards policy has an impact on *technology costs. Spectral policy* has an impact on *engineering costs*.

Generally, technologies will have been specified and standardised to take effective advantage of available and anticipated device capabilities. Engineering is the necessary process whereby those capabilities are turned into cash and/or competitive advantage. Technology design and development has to be amortised over a sufficient market volume and value to achieve an acceptable return on investment. Engineering effort has to be amortised over a sufficient market volume and value to achieve an acceptable return on investment.

A poorly executed standards policy results in the duplication and dissipation of technology design and development effort. The lack of a mandated technology policy increases technology risk. This requires vendors to adopt more aggressive ROI policies. As a result, technology costs increase.

A poorly executed spectral allocation policy results in the duplication and dissipation of engineering effort. The lack of a globally harmonised spectral allocation policy increases engineering risk. This requires vendors to adopt more aggressive ROI policies. As a result, engineering costs increase.

21.19   Standards Policy, Spectral Policy and RF Economies of Scale

Cellular and mobile broadband handsets benefit from substantial economies of scale. Cellular base stations and related network components have much lower scale economies. There are (several orders of magnitude) fewer base stations in the world than handsets. This makes it harder to recover nonrecurring technology and engineering costs. The result is that base stations and network components are necessarily expensive.

As network density has increased over time, the number of base stations has increased, providing the basis for more effective cost amortisation. Successive generations of radio-access technologies options compete with each other on the promise of offering efficiency gains and a cost/performance advantage. For example, GSM was justified on the basis of capacity and coverage benefits when compared to ETACS. CDMA and UMTS and HSPA were justified on a similar basis and more recently, LTE has been promoted on the basis of data rate and efficiency gains.

In practice, progress in terms of radio and network efficiency tends to be more incremental than market statement or sentiment would imply. Step function gains in efficiency just don't happen in practice. Performance gain is achieved as the result of technology maturation based on engineered optimisation combined with market volume.

Similarly, technology can deliver useful gains in terms of reduced component count and component cost, but again these gains are incremental and need to be coupled with engineering effort and market volume to be significant.

So, component-cost reductions are a product of technology, engineering effort and market volume. Improvements in access efficiency are a product of technology, engineering effort and market volume. The requirement for market volume implies that economies of scale are dependent on the ability of vendors to ship common products to multiple markets.

21.20   The Impact of IPR on RF Component and Subsystem Costs

Similarly, there is a need to have equitable intellectual property agreements in place that provide an economic return on company specific organisationally specific technology-specific research and development investment.

This is a troubled area made more troublesome by regional differences in IPR law and accepted good practice. The recent success by CSIRO[6] in Australia in pursuing OFDM patent rights is relevant. Overall, it would seem to make economic sense to develop effective patent pooling arrangements as part of an integrated international standards making activity, though such a suggestion might be considered over sanguine.

21.21   The Cost of 'Design Dissipation'

Resolving intellectual property rights through an adversarial and fragmented international IPR regime is just another example of how potential gains from technology and engineering innovation may fail to deliver their full cost efficiency and economic benefits due to poorly executed governmental and intergovernmental policy.

To summarise:

Technologies when first introduced have a potential 'efficiency gain'. The potential gain is only realised after considerable engineering effort is invested to ensure the technology actually works as originally intended.

In cellular and mobile broadband networks having to support multiple technologies deployed into multiple bands that are different from country to country results in an unnecessary dissipation of design effort and engineering resource. A lack of international clarity on IPR issues creates additional aggravation.

This implies a need for an Emerging Market Network Programme which includes internationally mandated spectral allocation, internationally mandated technology standards and an internationally mandated IPR regime based on well-established and demonstrably successful patent pooling principles. Such a programme would need to reflect the shift in influence from the US to Asia and the BRICS markets (**B**razil, **R**ussia, **I**ndia, **C**hina, **S**outh Africa). Such a requirement is largely at odds with present 'light-touch' regulatory policy and the continuing dominance of the US and Europe in standards making remains problematic.

The adoption of technology neutrality as a policy is an abrogation of international governmental responsibility – a policy to not have a policy is not a policy.

Regulators have a duty to ensure efficiencies of scale can be applied in the supply chain and to avoid the negative impact that design dissipation has on product efficiency, product

---

[6] http://www.csiro.au/news/ps2gw.html.

pricing and product availability. This is particularly true if we are serious about addressing cost-sensitive markets and, incidentally, serious about wireless competing with, or working effectively with, continuously improving fibre and copper access technologies.

Bridging the 'digital divide' in emerging nations requires cooperation not exhortation.

This in turn implies a return to prescriptive globally harmonised spectral allocation combined with prescriptive and closely integrated mandated technology standards and a globally harmonised approach to IPR management.

On the income side the focus has shifted dramatically over the past five years from voice and text value to content value to application value. There was a period around about 2008 when content was thought of as king. Vendors of billing products introduced content management and billing platforms and operators waited for the money to roll in. Content, however, has significant associated costs and the value is often resident with parties other than those that bear the costs either of origination and/or storage and delivery cost.

21.22   The Hidden Costs of Content – Storage Cost

For example, the assumption is that storage costs are decreasing over time, a function of the halving of memory costs on a 12-month cycle. This would only be true if content was expanding at a lower rate than memory bandwidth, and this is presently not the case.

Content bandwidth inflation is being caused by the transition to high-definition TV, a fourfold bandwidth expansion. This is compounded by the move to higher-fidelity audio, a composite of enhanced MP3 and five- or seven-channel surround sound.

Still-image content expansion is being driven by ever higher-resolution image capture platforms, 44-megapixel cameras being an extreme but relevant example. This bandwidth expansion hits every stage of the content production chain from original capture through post production through to storage and distribution.

21.23   The Hidden Costs of User-Generated Content – Sorting Cost

This includes user-generated content. For example, the BBC has a programme called Autumn Watch, a study of how weather affects people and plants.

A request for viewers to send in their gardening photos generates several hundred e mails with attachments, a significant percentage of which contain uncompressed files. Someone has the Herculean task of sorting through these pictures, deciding which ones to keep, which ones to use and how to describe them in the data base. User-generated content is therefore not free and indeed has a significant cost that is increasing over time.

21.24   The Hidden Cost of Content – Trigger Moments

Broadcast content may also have trigger moments, voting for example in a talent show. Instead of an avalanche of photos the problem is now an avalanche of SMS messages or phone calls that have to arrive and be dealt with by a specified time in a specified way.

An organisation now exists to monitor and manage and regulate the operation of phone-in promotions. Several producers lost their jobs for failing to manage this process and participation revenues in the UK reduced significantly as a result.

21.25   The Hidden Cost of Content – Delivery Cost

Trigger moments can create loading issues that can only be fully addressed by over provision-ing store and forward and onward delivery bandwidth. Overprovisioning store and forward and delivery bandwidth both over the air and through the network implies substantial un-derutilisation for most of the time. Additionally, content has different delivery requirements ranging from best effort (lowest delivery cost, highest buffer cost), streaming (audio and video downloading), interactive (gaming) and conversational (highest delivery cost, no buffer cost).

Supporting all four types of content simultaneously adds substantial load to the network. The administrative effort counted in software clock cycles is significant and outweighs any benefits theoretically available from multiplexing best-effort data into the mix.

21.26   The Particular Costs of Delivering Broadcast Content Over Cellular Networks

The DTV alliance analysed delivery costs and delivery revenues in a White Paper. The purpose of the White Paper was to illustrate the costs of delivering broadcast content either as unicast (eye wateringly expensive) or multicast (less but still too expensive).

Vendors and operators might argue that broadcast over cellular standards will reduce these costs but until these standards are implemented the costs are real and actual and the comparisons in the White Paper are valid. The thesis of the White Paper is that there is a pain threshold, which is the point at which delivery costs exceed delivery revenues excluding spectral cost amortisation.

The pain threshold point was calculated to be a 6-minute low-resolution video.

The bandwidth delivery cost of delivering a 6-minute high-resolution TV programme would be \$2.76 or \$13.80 for a 30 minute programme. To achieve margins equivalent to voice the operator would need to charge \$13.80 for the 6-minute video and \$33.60 for the 30-minute programme.

This implies that the opportunity cost has to be factored in of any possible impact the audio and/or video service might have on existing voice traffic. A 3-G BTS was taken as an example with a busy hour capacity of 2.5 Mbps times 3600 seconds, equivalent to 9 gigabits of delivery bandwidth. If 5% of the subscribers watched two-minute clips at 128 kbps then they will consume 4.8 Gigabits in an hour, half the capacity of the cell site. Double the data rate or number of users or length of the clip and the voice traffic completely disappears.

Now, you might argue with some of these numbers but the essentially valid point is that rich content is expensive to deliver. This is particularly true in mobile broadband networks when a user is close to the edge of the cell.

21.27   Summary – Cost and Value Transforms

Essentially, we are saying that content has hidden costs and that these costs are increasing rather than decreasing over time. Costs may be apparently reducing but may reappear in other areas. For example, the quality of service mechanisms needed in IP networks to handle rich media substantially increase the overall cost of delivery.

In parallel, value is decreasing including content with presently highly contested acquisition costs, premier league football for example. We should not assume that so-called premium content maintains that premium over time. As humans we have a finite absorption bandwidth and we must be close to the safe absorption limit, particularly as far as football is concerned.

There is an argument that the value is still there but is realised in different ways. O2 Telefonica sponsor events at the Millenium Dome. This includes concerts by artists such as Prince, the Rolling Stones and Led Zeppelin. Some artists, for example Prince, give their music away for free or at a deep discount but the live concert grosses millions. The majority of this value goes to the artists (typically 110% of the ticket costs) and venue owner, the American Anschutz Entertainment Group. These are examples of cost and value transforms.

Costs do not necessarily disappear but may reappear in other areas. Value may be realised by third parties who may be the unintended though possibly deserving beneficiaries of the original investment process.

And transforms are probably a good place to end this narrative, particularly as it is more or less where we started. When asked to make some closing remarks at a wireless conference held coincidentally at St Johns, my old college in Cambridge it struck me that the college was having its 500th anniversary.

The presentations at the college reflected many of the themes that we have explored in this book but also introduced some broader perspectives. Sometimes the obvious can be profound. For example, the fact that most families living in the slums of India have no water or sanitation but own a TV, that the most used function in a mobile phone is often the FM receiver, that India prospers despite its lack infrastructure while China prospers because of it,[7] that SMS is used in India to send greetings, that countries are different and cultures are different and that means markets are different.

From where we were sitting at that particular time in that particular place we could see St Johns, Trinity College and Caius College with Clare College and Kings College in the distance, nestling along the River Cam.

Over the centuries those colleges had hosted Isaac Newton 1643 to 1727, who did that stuff with the apple, but more importantly developed the science of differential and integral calculus, Charles Babbage 1791 to 1871 the man who built the world's first mechanical computer unless you count the abacus several thousand years before, Augustus Morgan 1806 to 1871 a founding father of algebraic and algorithmic innovation, Srinivasa Ramanujan 1887 to 1920 the Indian mystic and number theory man, Alan Turing 1917 to 1954 the master of encryption, Paul Dirac 1927 to 1969 a pioneer of quantum mechanics and nanotechnology and Stephen Hawking . . . . . . cosmology and string theory.

These seven mathematicians of course only represent a small fraction of the legacy of intellectual innovation upon which telecommunications value is ultimately built, a future that will be built on a combination of materials innovation, manufacturing innovation and mathematical and algorithmic innovation.

As we started this book with a science fiction author we may as well end it with one, William Gibson. Gibson invented the term cyberspace in a short story called Burning Chrome that was then included in his debut novel Neuromancer in 1984.

In a 1993 interview he is said to have said – 'The future is already here – it's just not very evenly distributed'.

Telecommunications has the power to unlock and redistribute the intellectual and innovation power of the world's population – a truly transformative transform.

---

[7] R Swaminathan Reliance Communications India and Kanwar Chadha CMO CSR.

Index

Making Telecoms Work: From Technical Innovation to Commercial Success, First Edition. Geoff Varrall.
© 2012 John Wiley & Sons, Ltd. Published 2012 by John Wiley & Sons, Ltd.